Harmonic Grammar and Harmonic Serialism

Advances in Optimality Theory
Editors: Armin Mester, University of California, Santa Cruz, and Vieri Samek-Lodovici, University College London

Optimality Theory is an exciting new approach to linguistic analysis that originated in phonology but was soon taken up in syntax, morphology, and other fields of linguistics. Optimality Theory presents a clear vision of the universal properties underlying the vast surface typological variety in the world's languages. Cross-linguistic differences once relegated to idiosyncratic language-specific rules can now be understood as the result of different priority rankings among universal, but violable constraints on grammar.

Advances in Optimality Theory is designed to stimulate and promote research in this provocative new framework. It provides a central outlet for the best new work by both established and younger scholars in this rapidly moving field. The series includes studies with a broad typological focus, studies dedicated to the detailed analysis of individual languages, and studies on the nature of Optimality Theory itself. The series publishes theoretical work in the form of monographs and coherent edited collections as well as pedagogical texts and reference texts that promote the dissemination of Optimality Theory.

Published:

Optimality Theory, Phonological Acquisition and Disorders
Edited by Daniel A. Dinnsen and Judith A. Gierut

Modeling Ungrammaticality in Optimality Theory
Edited by Curt Rice and Sylvia Blaho

Phonological Argumentation
Essays in Evidence and Motivation
Edited by Steve Parker

Hidden Generalizations
Phonological Opacity in Optimality Theory
John J. McCarthy

Conflicts in Interpretation
Petra Hendriks, Helen de Hoop, Irene Krämer, Henriëtte de Swart and Joost Zwarts

The Phonology of Contrast
Anna Łubowicz

Prosody Matters
Essays in Honor of Elisabeth Selkirk
Edited by Toni Borowsky, Shigeto Kawahara, Takahito Shinya and Mariko Sugahara

Blocking and Complementarity in Phonological Theory
Eric Baković

Linguistic Derivations and Filtering
Minimalism and Optimality Theory
Edited by Hans Broekhuis and Ralf Vogel

Understanding Allomorphy
Perspectives from Optimality Theory
Edited by Eulàlia Bonet, Maria-Rosa Lloret and Joan Mascaró

Layering and Directionality
Metrical Stress in Optimality Theory
Brett Hyde

Harmonic Grammar and Harmonic Serialism

Edited by
John J. McCarthy and Joe Pater

SHEFFIELD UK BRISTOL CT

Published by Equinox Publishing Ltd.
UK: Office 415, The Workstation, 15 Paternoster Row, Sheffield, South Yorkshire S1 2BX
USA: ISD, 70 Enterprise Drive, Bristol, CT 06010

www.equinoxpub.com

First published 2016

British Library Cataloguing-in-Publication Data

A catalogue record for this book is available from the British Library.
ISBN-13 978 1 84553 149 2 (hardback)

Library of Congress Cataloging-in-Publication Data
McCarthy, John J., 1953- | Pater, Joe.
Harmonic grammar and harmonic serialism / edited by John J. McCarthy and Joe Pater.
Sheffield, UK; Bristol, CT: Equinox Publishing Ltd, [2016] |
Advances in Optimality Theory | Includes bibliographical references and index.
LCCN 2015030854 | ISBN 9781845531492 (hb)
LCSH: Optimality theory (Linguistics) | Grammar, Comparative and general – Syntax.
LCC P158.42 .H37 2016 | DDC 415.01/8–dc23
LC record available at http://lccn.loc.gov/2015030854

Typeset by S.J.I. Services, New Delhi
Printed and bound by Lightning Source Inc. (La Vergne, TN), Lightning Source UK Ltd. (Milton Keynes), Lightning Source AU Pty. (Scoresby, Victoria).

Contents

Preface

As even the title of Prince and Smolensky (1993/2004) makes clear, Optimality Theory (OT) is a theory of constraint interaction in generative grammar. Representational constraints had of course long existed as a means of formally expressing restrictions on linguistic structures, but the notion that one constraint could compel the violation of another had never been given much attention before OT. Constraint interaction has deep consequences for the analysis of individual languages, for the study of language typology, and for the study of language learning. The constraints used to analyze a given language can often be formulated in a more general fashion when their failure to apply in some context can be attributed to constraint conflict. When general constraints are used to analyze individual languages, they can form the basis of substantive theories of Universal Grammar, which make testable predictions about the range of possible variation amongst languages. The greater generality of OT constraints as compared to inviolable ones has allowed for the development of typological theories with greater coverage and precision of prediction than those that came before. Finally, a view of language learning as changing the relative priorities of constraints, instead of turning them on or off, allows for the creation of new and arguably better theories of learnability, which aim to show that explicit algorithms can find correct grammars, and of acquisition, which aim to characterize the human learning path.

There are many ways that violable constraints can be implemented in a theory of grammar. Although the two theories that give this book its title are in some ways relatives of OT in that they incorporate much of its structure (e.g., candidate comparison by markedness and faithfulness constraints), they actually antedate the 1993 Prince and Smolensky manuscript. Harmonic Grammar (HG) is the name of the weighted constraint predecessor of OT first characterized in Legendre, Miyata, and Smolensky (1990), while Harmonic Serialism (HS) was used in Prince and Smolensky's early analysis of Berber syllabification that became the first substantive chapter of their book.

Despite these connections, HS and HG differ in what they change about Prince and Smolensky's standard OT model. HS questions the choice of parallel over serial evaluation, while HG questions the assumption that constraints are ranked rather than weighted. Because HS and HG revise different aspects of the standard OT architecture, it is also possible to entertain a kind of HS/HG hybrid, called Serial HG, for which see Pater's and Elfner's chapters in this volume, as well as other works cited there, and also Staubs and Pater's chapter on learning. A further variation on the standard architecture, OT with candidate chains (McCarthy 2007), is the topic of Wolf's chapter.

To facilitate readers' entry into this realm of alternatives to standard OT, each of us has written introductory chapters on HG (Pater) and HS (McCarthy). These introductions are followed in Part II by chapters working out various consequences of the revised architectures for phonological analysis and typology. Part II begins with two chapters on prosody (McCarthy, Pater, and Pruitt; Elsman). These chapters principally address arguments that have been made for standard OT's parallel derivations over serial theories like HS and Serial HG. They are followed by four chapters on language typology (Staubs; Jesney; Kimper; Moore-Cantwell). These chapters share a concern with the 'too many repairs' problem (Blumenfeld 2006, Lombardi 1995/2001, Pater 1999, Steriade 2001/2008, Wilson 2001, and others): the observation that free permutation of otherwise plausible markedness and faithfulness constraints will sometimes predict typologically implausible patterns. Part II concludes with two chapters about phonological opacity (Elfner; Torres-Tamarit) and one about the phonology-morphology interface (Wolf). Part III presents work about learning of serial grammars (Staubs and Pater) and weighted-constraint grammars (Boersma and Pater).

We acknowledge the support of the National Science Foundation through grants BCS-0813829 and BCS-1424077 to the University of Massachusetts Amherst.

References

Legendre, Géraldine, Yoshiro Miyata and Paul Smolensky (1990) Harmonic Grammar – A formal multi-level connectionist theory of linguistic well-formedness: Theoretical foundations. *Proceedings of the Twelfth Annual Conference of the Cognitive Science Society*, 388–395. Mahwah, NJ: Lawrence Erlbaum Associates.

McCarthy, John J. (2007) *Hidden Generalizations: Phonological Opacity in Optimality Theory*. London: Equinox Publishing.

Prince, Alan and Paul Smolensky (1993/2004) *Optimality Theory: Constraint Interaction in Generative Grammar*. Malden, MA, and Oxford, UK: Blackwell.

Chapter 1
Universal grammar with weighted constraints[1]

Joe Pater, University of Massachusetts, Amherst

1.1 Introduction

The aim of this paper is to show that weighted constraints hold much promise as the basis of models of Universal Grammar (UG), in the sense of Optimality Theory (OT; Prince and Smolensky 1993/2004). In OT, a universal constraint set is used to analyze individual languages, and to generate predictions about the range of possible languages, that is, about language typology. A ranking of the universal constraint set determines the form of a single language, and every ranking corresponds to some possible language. The view that this framework could benefit from the adoption of weighted rather than ranked constraints is controversial. Prince and Smolensky (1993/2004: 232) claim that this would lead to typologically implausible patterns of constraint interaction – versions of this claim also appear in Prince and Smolensky (1997), Legendre, Sorace and Smolensky (2006), Smolensky (2006a), and elsewhere in the

1 This paper reconfigures and slightly revises portions of Pater (2009a) and Pater (2009b), and adds some new material (especially secs. 1.3 through 1.6). Special thanks to Karen Jesney and John McCarthy for extended discussion of most of this work and for helpful comments on the manuscript, and to Paul de Lacy and Chris Potts for discussion of the Section 1.3 research, as well as my other collaborators on Potts *et al.* (2009) and Staubs *et al.* (2010) – Rajesh Bhatt, Michael Becker, Patrick Pratt and Robert Staubs – both projects pushed this work forward considerably. Along with all others thanked in the acknowledgments of the above papers, I particularly appreciate the feedback of participants in our grant group meetings, and in a 2011 LSA summer institute course. This research was supported by NSF Grant 0813829 to the University of Massachusetts Amherst.

OT literature. The cited arguments are assessed in Sections 1.2 and 1.4 below. Weighted constraints were first applied to generative grammar in a theory of gradient syntactic well-formedness named Harmonic Grammar (HG; Legendre, Miyata, and Smolensky 1990; see also Goldsmith 1990, Goldsmith 1993a, and some of the papers collected in Goldsmith 1993b for early applications to phonology). To recognize this innovation, I adopt HG as the name for a version of OT that uses weighted constraints, but as the above citations should make clear, prior work by the original proponents of HG does not take the position that weighted constraints form a plausible basis for OT-style theories of UG.

Part of the promise of HG for UG derives from inherent restrictions on cumulative interaction in this framework. In Section 1.2, building on observations of Prince (2003), I discuss the consequences of what I call the asymmetric trade-off requirement on gang effects. Only certain patterns of constraint violation can produce gang effects, and hence contain the potential for OT-HG differences. To further demonstrate this inherent restrictiveness, I contrast HG with Smolensky's (2006b) Optimality Theory with Local Constraint Conjunction (OT-LC), which does not impose similar restrictions on cumulative constraint interaction, and thus generates a set of implausible typological predictions that are not shared by HG. I further demonstrate the restrictiveness of HG by pointing out that meeting the asymmetric trade-off requirement is not a guarantee of an OT-HG difference: some gang effects are vacuous, in that they yield no differences in the predictions of the two frameworks.

A second part of the promise of weighted constraint theories of Universal Grammar derives from the ability of HG to generate attested patterns that fall out of the reach of OT using the same set of constraints. That is, HG permits new theories of Con, the universal constraint set. In Section 1.3, I discuss the compatibility of scalar constraints with HG (building on Flemming 2001), and their incompatibility with OT (building on Prince and Smolensky 1993/2004, McCarthy 2003). Because a fleshed-out HG theory of any domain will likely be operating with a different constraint set than an OT one, the assessment of the relative success of OT and HG as general frameworks for typological study becomes more complicated, and more interesting.

A further complication in making OT-HG comparisons arises when we take into account the limits on patterns of constraint interaction that follow when Prince and Smolensky's (1993/2004) parallel theory of candidate generation and evaluation is replaced by a serial one (see McCarthy this volume for an introduction to Harmonic Serialism and an overview of results). In Section 1.4, I show that some clearly undesirable predictions of HG are eliminated when only a single application of an operation is allowed in constructing a candidate. The upshot of this discussion is that some pathological predictions of parallel HG are plausibly due to the globality of parallel evaluation, rather than to the power of weighted constraint interaction.

Typological research using weighted constraints may well have been hindered in the past by the relative difficulty of finding by hand a correct set of constraint weights for a given set of linguistic data, and especially of determining what all of the languages are that a given set of constraints can generate. This paper introduces the basic techniques for doing both of these analytic tasks, and discusses how they can be aided by using computational techniques, in particular the implementation of Potts *et al*'s (2010) Linear Programming methods in OT-Help (Staubs *et al.* 2010). A great deal of the prior and ongoing computational work on HG focuses on its use in learning with probabilistic versions of the framework (Maximum Entropy Grammar, Goldwater and Johnson 2003, Noisy HG, Boersma and Pater this volume). Because the goal of this paper is to examine the differences between the typological predictions of standard OT and a minimally different version with weighted constraints, I adopt a categorical version of the theory here. Section 1.5 briefly discusses the translation of the present model and the results obtained with it to probabilistic variants of HG.

1.2 Asymmetric trade-offs

1.2.1 Background

We start with the simple HG tableau in (1), which has an input with a pair of voiced obstruents, and as output candidates the result of changing the voicing of either one, of both, or of neither. The optimum is the candidate with the highest *Harmony*, which in HG is the weighted sum of constraint violations. Constraint violations are indicated with negative integers, and the constraint weights are given immediately beneath the constraint names: 3 for *CODA-VOICE, and 2 for IDENT-VOICE. The Harmony score for each candidate is given in the rightmost column. In this tableau, the candidate with final devoicing, [bat], receives a score of −1 on IDENT-VOICE, which penalizes each input consonant whose voice specification is changed in the output (McCarthy and Prince 1999).

Since IDENT-VOICE has a weight of 2, [bat] has Harmony of –2. This candidate has the highest score, and is thus optimal. In particular, the faithful [bad], which violates a constraint against voiced codas, *CODA-VOICE, gets a lower Harmony score (–3) because that constraint's weight is higher.

(1) *Final devoicing in HG*

	/bad/	*CODA-VOICE 3	IDENT-VOICE 2	
a.	bad	–1		–3
b.	pad	–1	–1	–5
c. →	bat		–1	–2
d.	pat		–2	–4

The weights in the above tableau are partially arbitrary; there is an infinite set of weights that could be used to make [bat] optimal – e.g weights (30, 20) and (0.111, 0.110) instead of (3, 2) would produce the same optima. The non-arbitrary aspect is that for [bat] to be optimal, the weight of *CODA-VOICE must be greater than that of IDENT-VOICE. With the reverse relationship, [bad] becomes optimal.

More generally, given a desired optimum, or 'Winner', a set of failed candidates, or 'Losers', and their associated vectors of scores – which can also be positive rewards – on a set of constraints, we can produce a set of linear inequalities, or *weighting conditions*, that must obtain if the Winner is to be made correctly optimal (like Prince's 2002 OT *ranking conditions*). We can obtain a useful representation for examining weighting conditions by subtracting the (unweighted) scores of the Winner from those of a Loser, thus producing an HG comparative vector (see again Prince 2002 in OT; in HG see Goldwater and Johnson 2003, Potts *et al.* 2010, Bane and Riggle 2012, and Boersma and Pater this volume). The comparative vectors from the candidates in (1), with [bat] as the Winner, are shown in (2).

(2) Comparative vectors for (1)

	W ~ L	*CODA-VOICE	IDENT-VOICE
a.	[bat] ~ [bad]	+1	–1
b.	[bat] ~ [pat]		+1
c.	[bat] ~ [pad]	+1	

The weighting conditions derive from the requirement that for the Winner to be correctly optimal, the weighted sum of the scores in each row must be greater than zero. For example, for row (2a.), we have the requirement that 1 times the weight of *CODA-VOICE, plus –1 times the weight of IDENT-VOICE, must be above zero, which is satisfied by the weights from (1) – *i.e.* $1 \times 3 + -1 \times 2 = 1$. When all of the non-zero scores are +1 and –1, as in this row and all others in (2), we can also simply say that the sum of the weights of the constraints preferring the Winner must be greater than the sum of the weights preferring the Loser.

Our comparative vectors have not introduced any new weighting conditions beyond the one we noted before: as indicated by vector (2a.), the weight of *CODA-VOICE must be greater than that of IDENT-VOICE. So long as weights are limited to positive values, it is impossible to make either [pat] or [pad] beat [bat]; any positive weights applied to the vectors in (2b.) and (2c.) will lead to weighted sums above zero. These are instances of simple harmonic bounding, in that [pat] and [pad] are both harmonically bounded by [bat] because they each have a proper superset of its constraint violations. As Prince (2003) points out, positively weighted constraints preserve simple harmonic bounding relations from OT.[2] Zero weights could make a simply harmonically bounded candidate tie for optimality, but could not make it the sole optimum. Negative weights can make a harmonically bounded candidate solely optimal in its tableaux, and thus need to be banned in a version of HG that aims to function anything like OT.

With more candidates and more tableaux, the set of weighting conditions can become more complex, making it difficult to find a correct set of weights by hand. Weighting conditions are linear inequalities, and a system of linear inequalities can be solved by Linear Programming's simplex algorithm. Potts *et al.* (2010) show how to translate HG learning

2 In cases of 'collective bounding' (Samek-Lodovici and Prince 2005), where a candidate cannot be made optimal because it loses under some weightings or rankings to one candidate, and under some other weightings/rankings to another, whether HG and OT yield the same results depends on the particular violation profiles in question. Anticipating the discussion below, under a symmetric trade-off between constraint violations, OT and HG produce identical results. For example, neither can make [dak.bad] or [dag.bat] optimal for /dag.bad/ in the example in (10). One exception is that Maximum Entropy Grammar can give such collectively bounded candidates non-negligible probability – see Section 1.5 for further discussion. When the trade-off is asymmetric, OT and HG can diverge – see Section 1.4 for examples of candidates that are collectively bounded in OT that are optimal in HG.

problems into systems solvable by the simplex, essentially by making the same transition from tableaux to weighting conditions that we have made here. Not only can the simplex be used to find a set of weights that meets a set of weighting conditions of arbitrary complexity, but it can also detect when no correct weighting exists because the weighting conditions are inconsistent. As such, it does for HG what Recursive Constraint Demotion (RCD, Tesar and Smolensky 2000) does for OT. The inconsistency detection property of RCD makes it particularly useful in calculating the set of languages produced by a set of constraints, that is, the sets of candidates across tableaux that can be made jointly optimal. It was first applied in this way in the typology calculator implemented in OT-Soft (Hayes, Tesar and Zuraw 2003). Staubs *et al.*'s (2010) OT-Help uses Potts *et al.*'s simplex application, as well as RCD, to calculate and compare OT and HG typologies.

It is important to note that both RCD and the Linear Programming application in Potts *et al.* (2009) only work for cases in which there is a single optimum per candidate set. Because they rely on these algorithms to calculate typologies, OT-Soft and OT-Help will not find languages with tied optima. A limit to single optima (for non-identical candidates) is in fact a useful idealization for a comparison between the standard version of OT and a minimally different version with weighted constraints, since the standard version of OT generally produces only single optima. Throughout this paper I discuss only languages with one optimum per tableau, but return to this issue of ties in the discussion of probabilistic versions of HG in Section 1.5.

1.2.2 Loanword devoicing in Japanese as cumulative constraint interaction

To see how weighted constraints can produce results that diverge from ranked ones, we can consider a slightly more complicated example in the phonology of obstruent voicing. In Japanese, only a single voiced obstruent is usually permitted in a word (see Ito and Mester, 1986, 2003). This restriction is termed Lyman's Law (Lyman, 1894). In loanwords, however, multiple voiced obstruents are permitted (Kawahara, 2006, 2011; Nishimura, 2003, 2006; all data are from Kawahara 2006):

(3) Violations of Lyman's Law in loanwords

[bagi:] 'buggy'	[bogi:] 'bogey'	[bobu] 'Bob'
[bagu] 'bug'	[dagu] 'Doug'	[giga] 'giga'

Japanese also has a restriction against obstruent voicing in geminates. But again, in loanwords, voiced geminate obstruents are permitted:

(4) Voiced/voiceless obstruent geminate near-minimal pairs in Japanese loanwords

[web:u] 'web'	[wip:u] 'whipped (cream)'
[sunob:u] 'snob'	[sutop:u] 'stop'
[hab:uru] 'Hubble'	[kap:uru] 'couple'
[kid:o] 'kid'	[kit:o] 'kit'
[red:o] 'red'	[autoret:o] 'outlet'
[hed:o] 'head'	[met:o] 'helmet'

Devoicing occurs just when a loanword contains both a voiced geminate and another voiced obstruent. The geminate is optionally, but categorically, devoiced:

(5) Optional devoicing of a geminate in Lyman's Law environment

[gud:o] ~ [gut:o] 'good'	[dog:u] ~ [dok:u] 'dog'
[bed:o] ~ [bet:o] 'bed'	[deibid:o] ~ [deibit:o] 'David'
[dored:o] ~ [doret:o] 'dredlocks'	[bag:u] ~ [bak:u] 'bag'
[bad:o] ~ [bat:o] 'bad'	[bud:a] ~ [but:a] 'Buddha'
[dorag:u] ~ [dorak:u] 'drug'	[big:u] ~ [bik:u] 'big'

According to Nishimura (2003, 2006) and Kawahara (2006, 2011), such devoicing is judged unacceptable (or much less acceptable; Kawahara 2011) in the word types illustrated in both (3) and (4).

In HG, this devoicing pattern can be analyzed as being due to two independently motivated constraints. This analysis draws on Nishimura's (2003, 2006) account using Smolensky's (2006b) OT with Local Conjunction; the possibility of an HG reanalysis was suggested by Shigeto Kawahara (p.c.). The first constraint, *VCE-GEM expresses a cross-linguistically common ban against voiced obstruent geminates, which can be held responsible for their absence in native Japanese words. The other constraint is OCP-VOICE, which Ito and Mester (1986) propose to account for the Lyman's Law restriction in native Japanese words. This constraint penalizes every sequence of voiced obstruents, even ones separated by any number of segments (other than voiced obstruents). For example, [deibit:o] and [dored:o] each have one violation of OCP-VOICE ([d...d:]), and [deibid:o] has two ([d...b], [b...d:]), while [deipit:o] and [doret:o] have none. The weight of IDENT-VOICE is greater than that of each of OCP-VOICE and *VOICE-OBS, so that a word can have either a pair of voiced obstruents, as in (6), or a voiced geminate as in (7).

(6) Multiple voiced obstruents permitted

	/bobu/	IDENT-VOICE 3	OCP-VOICE 2	*VCE-GEM 2	
a. →	bobu		−1		−2
b.	bopu	−1			−3
c.	pobu	−1			−3

(7) Voiced geminates permitted

	/web:u/	IDENT-VOICE 3	OCP-VOICE 2	*VCE-GEM 2	
a. →	web:u			−1	−2
b.	wep:u	−1			−3

In the tableau in (8), we have a word with both a voiced geminate and another voiced obstruent. As the summed weight of OCP-VOICE and *VCE-GEM is greater than that of IDENT-VOICE, the geminate devoices (the probabilistic versions of HG discussed in Section 1.5 could generate the observed optionality; see Pater 2009a for an explicit analysis).

(8) Devoicing of geminate in the context of another voiced obstruent

	/dog:u/	IDENT-VOICE 3	OCP-VOICE 2	*VCE-GEM 2	
a.	dog:u		−1	−1	−4
b. →	dok:u	−1			−3
c.	tog:u	−1		−1	−5
d.	tok:u	−2			−3

In this *gang effect*, we have two constraints ganging up to overcome a third one with higher weight. Due to strict domination, OT cannot express this sort of constraint interaction. In this case, if IDENT-VOICE were ranked above each of *VCE-GEM and OCP-VOICE, as required for the first two tableaux, the optimal output for /dog:u/ would be [dog:u], rather than [dok:u].

The OT-HG difference can be clearly seen in the comparative vectors in (9). For HG, the requirements that the weight of each of OCP-VOICE and *VOICE-GEM be lower than that of IDENT-VOICE (9a. and 9b.) do not contradict the requirement that their summed weight be greater (9c.). In OT, on the other hand, (9c.) requires that either OCP-VOICE or *VCE-GEM dominate IDENT-VOICE, which would contradict one of (9a.) or (9b.).

(9) Comparative vectors for Japanese loanword devoicing

	W ~ L	OCP-VOICE	*VOICE-GEM	IDENT-VOICE
a.	[bobu] ~ [bopu]	−1		+1
b.	[web:u] ~ [wep:u]		−1	+1
c.	[dok:u] ~ [dog:u]	+1	+1	−1

Many other linguistic patterns have been analyzed in terms of weighted constraint interaction (in phonology, see e.g., Guy, 1997 as a precedent, and more recently Farris-Trimble 2008, Potts *et al.* 2010, Jesney and Tessier 2011, and Jesney 2012/this volume). This example provides particularly striking evidence for HG because OCP-VOICE and *VCE-GEM are independently motivated in the phonology of Japanese. As noted above, the OCP-VOICE constraint was posited by Ito and Mester (1986) to account for the fact that the native vocabulary is restricted by Lyman's Law, and for the morphological alternations that show its effects. In generative linguistics outside of OT and HG, an active constraint is a true statement about the domain in which it applies – in other words, an active constraint is inviolable. Given the examples of loanwords with pairs of voiced obstruents like [bobu], OCP-VOICE would have to be considered inactive, at least for loanwords (see Ito and Mester, 1995). The prediction, then, is that OCP-VOICE should not have any effect on the devoicing of geminates. Similarly, if OCP-VOICE is outranked by IDENT-VOICE for the OT account of [bobu]-type words, then as we have seen, it cannot participate in the devoicing of geminates. Nishimura's (2003, 2006) discovery of the cumulative effect of OCP-VOICE and *VCE-GEM in Japanese loanwords thus falsifies these predictions of frameworks with inviolable or ranked constraints, and counts in favor of HG's weighted ones.

The predictions of these frameworks depend, of course, on the contents of the constraint sets. With a different constraint set, one could analyze the Japanese case with either inviolable constraints or ranked ones. Nishimura (2003, 2006) posits a conjoined version of OCP-VOICE and *VCE-GEM, which is violated if and only if a word contains violations of both constraints. In the following section, we will turn to a discussion of the problematic typological predictions made by OT with constraint conjunction. Less obviously problematic is Kawahara's (2006) introduction of a singleton-specific IDENT-VOICE, which fails to apply to geminates, and can thus rank above OCP-VOICE to protect singletons in the Lyman's Law environment, while still allowing devoicing of geminates in that context. The point, though, is that the cumulative interaction was *predicted* only by HG.

1.2.3 The necessity of asymmetric trade-offs

Japanese loanword devoicing involves an *asymmetric trade-off*: the choice between [dok:u] and [dog:u] in (8) trades a single IDENT-VOICE violation against violations of OCP-VOICE and *VCE-GEM. Because asymmetric trade-offs are required to produce gang effects, in their absence HG produces the same typology as OT.

We will now look at this key restrictiveness result in some detail. We can call a constraint that distinguishes an optimum from another candidate – that is, on which the optimum and some other candidate have different violation scores – a distinguishing constraint. A gang effect can be generally defined simply as an HG tableau in which the distinguishing constraint with highest weight[3] does not prefer the optimal candidate – e.g. IDENT-VOICE does not prefer the optimal [dok:u] in (8). If we convert the weight values to OT ranks, the resulting OT hierarchy will choose a different optimum than the HG weighting did – e.g. OT chooses [dog:u], HG chooses [dok:u]. This is a 'gang' effect because the lower-valued constraint violations are ganging up to overcome the higher weighted one.

The asymmetric trade-off requirement falls out directly from this definition. Let us assume that the HG optimum in this scenario has just one more violation mark on the highest weighted distinguishing constraint than the OT optimum – (8) again serves as an example, since HG optimal [dok:u] violates just IDENT-VOICE, which the OT optimum [dog:u] satisfies. For the HG optimum to in fact be optimal, it must then be the case that the OT optimum incurs at least two constraint violations that are not shared by the HG optimum: at least two unshared violations – OCP-VOICE and *VCE-GEM in our example – are required for their summed weight to be greater than that of the highest weighted distinguishing constraint (e.g. IDENT-VOICE). More generally, in a gang

3 I assume a version of OT in which every constraint ranking is a total order, and there is thus only one highest ranked distinguishing constraint. For simplicity, I set up the discussion with the parallel assumption that there is a single highest weighted constraint, but this is not crucial to any of the results.

effect, if the difference between the OT and the HG optima on the highest weighted distinguishing constraint is n, then there must be at least $n+1$ constraint violations incurred only by the OT optimum. Therefore, a set of violation profiles that can produce a gang effect must have a pair of candidates in which n violations of one constraint trade against at least $n + 1$ violations of some other constraint(s), that is, they must contain an asymmetric trade-off.

The candidate set in (10) demonstrates a situation in which the asymmetric trade-off requirement is not met. For every violation of *CODA-VOICE that is taken away, one violation of IDENT-VOICE is added, and *vice versa.* Keeping aside the possibility of tied optima in both theories, no set of weights will produce an OT-HG difference: if *CODA-VOICE has a higher weight than IDENT-VOICE, all underlying voiced consonants in surface coda position undergo devoicing, and with the reverse weighting, they will all surface with voicing intact. This is just the same as the effect of ranking the constraints with respect to one another.

(10) Symmetric trade-off

	/dagbad/	*CODA-VOICE	IDENT-VOICE
a.	dag.bad	−2	
b.	dak.bad	−1	−1
c.	dag.bat	−1	−1
d.	dak.bat		−2

The equivalence between HG and OT when constraint violations trade one-to-one was first noted by Prince (2003), who points out that in this situation, the result of the 'greater than' relation between positive weights in HG is the same as the 'dominates' relation in OT – any set of weights that meets these inequalities will behave just like the ranked constraints (i.e. 'Anything Goes').

Prince (2003) is concerned only with the issue of translating between ranking and weighting, that is, of finding a set of weights that will behave as an OT hierarchy. Here we see that there is also a typological consequence of this observation. This restrictiveness result is far from trivial, and is in some ways non-intuitive. This is highlighted by the fact that at least one of the differences between ranking and weighting that Prince and Smolensky (1997) use to motivate OT does not in fact exist, due to the asymmetric trade-off requirement.

In making the case for OT to an interdisciplinary audience, Prince and Smolensky (1997: 1604) draw the generalization that:

(11) In a variety of clear cases where there is a strength asymmetry between two conflicting constraints, no amount of success on the weaker constraint can compensate for failure on the stronger one.

They attribute this type of phenomenon to strict domination property of ranked constraints. As an example, they discuss the interaction of NoCoda and Parse. Parse is a faithfulness constraint that demands that input segments be parsed into output syllable structure; a consonant that is unparsed is unpronounced (≈deleted). Prince and Smolensky (1997: 1606) state that:

(12) Domination is clearly 'strict' in these examples: No matter how many consonant clusters appear in an input, and no matter how many consonants appear in any cluster, [the grammar with NoCoda ≫ Parse] ... will demand that they all be simplified by deletion (violating Parse as much as is required to eliminate the occasion for syllable codas), and [the grammar with Parse ≫ NoCoda] ... will demand that they all be syllabified (violating NoCoda as much as is necessary). No amount of failure on the violated constraints is rejected as excessive, as long as failure serves the cause of obtaining success on the dominating constraint.

In this passage, Prince and Smolensky offer as one illustration of strict domination the observation that the number of consonant clusters, that is, of potential codas, does not affect whether deletion occurs or not – see Section 1.4 on the second case, the number of consonants in a single cluster. The table in (13) shows that the trade-offs across candidates' violation profiles have the same symmetry as coda devoicing in (9). Unparsed segments are placed between angled brackets; as they are unparsed, they do not violate NoCoda.

(13) Symmetric trade-off

	/dagbadga/	NoCoda	Parse
a.	dag.bad.ga	−2	
b.	da<g>.bad.ga	−1	−1
c.	dag.ba<d>.ga	−1	−1
d.	da<g>.ba<d>.ga		−2

Regardless of the number of potential codas, there are only two outcomes: if Parse has a greater weight than NoCoda, the language will permit codas (as in [dag.bad.ga]), and if the relationship is reversed, all potential codas will fail to be pronounced (as in [da<g>.ba<d>.ga]]). Thus, strict domination is irrelevant to the irrelevance of the number of potential codas; the lack of that number's effect on whether deletion applies is a prediction of an optimization system with these constraints, be they ranked or weighted.

I can also point to some anecdotal evidence that this HG restrictiveness result is non-intuitive. When I first started investigating the relationship between weighted and ranked constraints, many of the cases that I imagined would yield OT-HG differences turned out to involve symmetric trade-offs and thus failed to distinguish the frameworks. Such cases have also sometimes been offered as potential instances of OT-HG differences by participants in courses and audience members at talks in which I have presented this material. For me, and maybe others, the source of this misleading intuition was the failure to appreciate the role of optimization in affecting what HG can and cannot do. This version of HG, like OT, is choosing the best outcome in a candidate set, rather than simply imposing a numerical cut-off on degree of violation (though see sections 1.3 and 1.4 on situations in which it can do something like this). There is no lower bound on the Harmony of an output – any amount of ill-formedness will be tolerated if the cost of avoiding it is higher.

1.2.4 The relative restrictiveness of HG and OT with Local Conjunction

This section illustrates the force of the asymmetric trade-off requirement in HG through a comparison with Smolensky's (2006b) OT with locally conjoined constraints (henceforth OT-LC), a framework that does not require asymmetric trade-offs in order to generate effects of cumulative violation. This comparison builds on that of Legendre, Sorace and Smolensky (2006), though does come to the opposite conclusion about the relative merits of OT-LC and HG as frameworks for Universal Grammar.

Smolensky (2006b: 43) defines local conjunction as in (14). This operation of conjunction yields a new constraint that is separately rankable in the constraint hierarchy.

(14) Local conjunction within a domain *D*
*A&$_D$*B is violated if and only if a violation of *A and a (distinct) violation of *B both occur withing a single domain of type *D*.

The original motivation for OT-LC, and the one that Smolensky (2006b) focuses on, is the reduction of complex markedness constraints to more basic primitives. Following Ito and Mester (2003: 26), we can take as an example *CODAVOICE. This can be expressed as the conjunction of two independently needed constraints, NOCODA and *VOICEOBS ('assign a violation mark to a voiced obstruent'). When these are conjoined in the domain of a segment, the resulting constraint penalizes voiced obstruent codas, but neither voiceless codas nor voiced obstruents. If the conjoined constraint is ranked above IDENT-VOICE, which is in turn ranked above *VOICEOBS, the result is coda devoicing (the position of NOCODA is arbitrary):

(15) Local conjunction analysis of final devoicing

	/bad/	NOCODA&$_{SEG}$ *VOICEOBS	IDENT-VOICE	*VOICEOBS	NOCODA
a.	bad	−1		−2	−1
b.	pad	−1	−1	−1	−1
c. →	bat		−1	−1	−1
d.	pat		−2		−1

Because a single violation of IDENT-VOICE cannot be used to avoid violations of both NOCODA and *VOICEOBS, with just the unconjoined contraints in (15), HG cannot generate this pattern. This is shown clearly by the comparative vectors for [bat] as a winner in (16). The shared NOCODA violations cancel out, and the remaining violation pattern is a one-to-one trade between *VOICEOBS and IDENT-VOICE. No set of positive weights will pick [bat] as optimal.

(16) Comparative vectors for unconjoined constraints in (15)

	W ~ L	*VOICEOBS	IDENT-VOICE	NOCODA
a.	[bat] ~ [bad]	+1	−1	
b.	[bat] ~ [pat]	−1	+1	
c.	[bat] ~ [pad]			

Legendre, Sorace and Smolensky (2006) point to similar cases of HG not being able to reduce a constraint to the activity of more basic ones. One example they give is that HG cannot ban [x] with general constraints against velars and fricatives, when the inventory contains both velars and non-velar fricatives, and in which constraint violations trading off against *VELAR and *FRICATIVE are non-overlapping (e.g.

IDENT-PLACE and IDENT-CONTINUANT). They argue that OT-LC's success in this regard counts in its favor.

OT-LC's success on cases like these does not clearly pick it over HG and standard OT. Whether complex markedness constraints like *CODAVOICE and *[x] should be reduced to more basic constraints is a matter of some general controversy. The idea that *CODAVOICE is the sum of the effects of a constraint against codas and one against voiced obstruents harks back in some ways to theories of prosodic licensing (e.g. Ito 1986, Goldsmith 1990, Lombardi 1991, Steriade 1995), which see contextual markedness as the inability of marked prosodic contexts to license marked segments (see Ito and Mester 2003: 28–29 on this connection). Much work in OT has questioned this approach, analyzing contextual markedness as the effect of rather specific, substantively motivated constraints (e.g. Pater 1999, Steriade 1999). The empirical motivation for these alternatives is that contextual markedness displays asymmetries that are not captured by prosodic licensing: the set of marked contexts is not the same for every marked segment, and markedness relationships between segments can be reversed across contexts (see Barnes 2006 for an extensive recent critique of prosodic licensing theory). Like prosodic licensing, this reductionist application of OT-LC also fails to express these asymmetries. Whether building a phonological theory with very specific phonetically grounded universal constraints is the right response to such asymmetries is of course also a matter of controversy (see Blevins 2004 and Hayes, Kirchner and Steriade 2004 for two poles of the debate). However, the greater reductionism possible in an OT-LC account of contextual markedness does not seem to be a knockdown argument for it over OT and HG, especially since that reductionism is only obtained in a somewhat abstract sense, given the presence in the grammar of the conjoined constraint, and the absence of a learning mechanism that yields conjunction.

Legendre, Sorace and Smolensky (2006) point out that in comparison with HG's linear model (the harmony function is a linear equation), OT-LC is a *superlinear* theory of constraint interaction. In linguistic theory, greater power is of course a double-edged sword. The superlinearity of OT-LC also allows it to generate unattested linguistic patterns not generated by HG. One example, discussed first by Ito and Mester (1998), can be produced by conjoining NOCODA with IDENT-VOICE in the domain of the segment (thanks to Matt Wolf for bringing this to my attention).

(17) Local conjunction analysis of initial devoicing

	/bad/	NoCoda&$_{\text{SEG}}$ Ident-Voice	*VoiceObs	Ident-Voice	NoCoda
a.	bad		−2		−1
b. →	pad		−1	−1	−1
c.	bat	−1	−1	−1	−1
d.	pat	−1		−2	−1

The ranking *VoicedObs ≫ Ident-Voice leads to devoicing, which is blocked by the dominant conjoined constraint when it occurs in coda position, where it violates both NoCoda and Ident-Voice. This pattern is referred to as a markedness reversal, since the feature is now protected in only the marked environment (see Lubowicz 2005 for further cases). NoCoda and Ident-Voice cannot interact in this way in HG because the NoCoda violation is shared by all of the candidates, and is hence irrelevant to the outcome.

The fact that [pad] can be the optimal output for /bad/ in OT-LC but not in HG points to an important difference between the frameworks. As discussed in Section 1.2.1, simple harmonic bounding relations from OT are preserved in a version of HG with positive weights. This example shows that they are not maintained under constraint conjunction. Critiques of OT-LC have generally focused on its excessive power (see e.g. McCarthy 1999, 2003, Padgett 2002), and have brought out two main issues: that it does not require cumulative interactions to be *local*, or *co-relevant*. We will now see that locality and co-relevance restrictions emerge from the structure of HG, due to the asymmetric trade-off requirement.

An example of the locality problem for OT-LC arises when the domain of the conjoined constraint used for coda devoicing is widened from the segment to the word. The tableaux in (18) and (19) illustrate the result: a consonant in any position will devoice if there is a coda anywhere in the word.

(18) Onset consonant devoices when the word contains a coda

	/balatak/	NoCoda&$_{\text{Wd}}$ *VoiceObs	Ident-Voice	*VoiceObs	NoCoda
a.	balatak	−1		−1	−1
b. →	palatak		−1		−1

(19) Onset consonant is voiced when the word has no coda

	/balata/	NoCoda&$_{Wd}$ *VoiceObs	Ident-Voice	*VoiceObs	NoCoda
a. →	balata			−1	−1
b.	palata		−1		−1

HG cannot replicate this result, again because the violations of NoCoda and *VoiceObs cannot both be removed with a change in voicing.

The co-relevance problem refers to the ability of OT-LC to express cumulative interaction between constraints that are in fact independent of one another. We can construct an example by conjoining a constraint demanding place agreement (Agree-Place) with *Coda-Voice (which itself could be the conjunction in (15)). The tableau in (20) shows that when Ident-Voice is ranked beneath the conjoined constraint but above *Coda-Voice, consonants only devoice in the context of a following heterorganic consonant.

(20) Coda devoicing in the presence of place disagreement

	/wadmad/	*Coda-Voice&$_{Seg}$ Agree-Place	Ident-Voice	*Coda-Voice	Agree-Place
a.	wadmad	−1		−2	−1
b. →	watmad		−1	−1	−1
c.	watmat		−2		−1

Although coda devoicing and place agreement are common patterns cross-linguistically, they do not interact in this way. Once again, this pattern cannot be expressed in HG: Ident-Voice violations trade with *Coda-Voice violations, not with Agree-Place.

Baković (2000) and Lubowicz (2005) propose restrictions on OT-LC that are aimed at addressing the co-relevance problem. Tellingly, the restriction that Lubowicz (2005) imposes on OT-LC cumulative interactions emerges from the nature of HG cumulativity: for two markedness constraints to have a cumulative effect, they must be satisfied by violating a single other (faithfulness) constraint. The relative merits of OT-LC and HG remain to be fully assessed, and this comparison is complicated by the fact that it depends on what limitations one applies to conjunction. For example, OT-LC with self-conjoined constraints forming a 'Power Hierarchy' (Smolensky 2006b) can generate the same counting effects in stress placement that Legendre, Sorace and Smolensky (2006) present as fatal for HG (though see Section 1.4). Paul Smolensky (p.c.) suggests that one might entertain a version of OT-LC that operates without this sort of self-conjunction. It is important too to note in this context that it is not the case that the full version of OT-LC is in a superset relation with HG in terms of the languages it generates. As McCarthy (2002) points out, HG does not require constraints in a gang effect to share a domain, and it is possible to construct patterns that only HG generates (for example, on the basis of cumulative effects of Max and markedness constraints, which Moreton and Smolensky 2000 claim are impossible in OT-LC). However, the inherent locality and co-relevance properties of HG, which derive from the asymmetric trade-off requirement, restrict HG in such a way that the freedom to express cumulativity without a shared domain does not appear to lead to anything like the kind of overgeneration that arises from superlinear OT-LC.

1.2.5 *The insufficiency of asymmetric trade-offs*

In order to provide a reasonably accurate general depiction of the relationship between the patterns produced by weighted and ranked constraint interaction, it is important to emphasize that while an asymmetric trade-off is a necessary condition for an HG–OT difference, it is not a sufficient one. We get a simple example by adding *VoiceObs to the constraint set used for coda devoicing in (1).

(21) Asymmetric trade-off with voicing constraints

	/bad/	*Coda-Voice	*VoiceObs	Ident-Voice
a.	bad	−1	−2	
b.	bat		−1	−1
c.	pad	−1	−1	−1
d.	pat			−2

Here we have an asymmetric trade-off between the violations of *CodaVoice and *VoiceObs in [bad], and the violation of Ident-Voice in [bat]. The shared violation of *Voice cancels out, resulting in the same two-to-one trade that we saw in Japanese; the comparative vector in (22a.) looks just like the [dok:u] ~ [dog:u] vector in (9c.).

(22) Comparative vectors for coda devoicing

	W ~ L	*CODA-VOICE	*VOICEOBS	IDENT-VOICE
a.	[bat] ~ [bad]	+1	+1	−1
b.	[bat] ~ [pat]		−1	+1
c.	[bat] ~ [pad]	+1		

By adding *VOICEOBS to the constraint set, we have changed the weighting conditions (compare the vectors in (2), which have only the leftmost two constraints of (22)): IDENT-VOICE now must have a greater weight than *VOICEOBS (22b.), and the sum of *CODA-VOICE and *VOICEOBS must be greater than that of IDENT-VOICE.

We have not, however, introduced a difference in the typologies produced by OT and HG: both theories can make [bad], [bat] and [pat] optimal, but not [pad]. *CODA-VOICE and *VOICE can participate in a gang effect to make [bat] win, but that gang effect would be vacuous in that it does not lead to an OT-HG difference.

OT can deal with the comparative vectors in (22) because there are no data that force *CODA-VOICE to be dominated by IDENT-VOICE: *CODA-VOICE prefers only Winners. More generally, it is impossible to create a dataset in which the sum of *CODA-VOICE and *VOICE are required to be greater than IDENT-VOICE, but in which *CODA-VOICE alone must have a lower weight, because every instance of a *CODA-VOICE violation is also a *VOICE violation. Many constraints in the OT literature stand in the specific-to-general relationship exemplified by *VOICE and *CODA-VOICE, in which the specific constraint assigns violations to a proper subset of the forms violated by the general constraint (see e.g. de Lacy 2006). The gang effect between any of these pairs of constraints will usually be vacuous, indistinguishable from the pattern produced by obedience to the specific constraint alone (though see Pater 2010). In most of these cases, then, asymmetric trade-offs will fail to yield HG–OT differences in typology (though see Jesney and Tessier 2011 on advantages of HG for modeling language learning when faithfulness constraints are in a specific-to-general relationship).

As another type of example of a vacuous gang effect, we turn to a case of constraint interaction that Prince and Smolensky (1993/2004) point to as illustrating the difference between ranking and weighting. It involves the pair of candidates in their tableau (183A), which forms part of their analysis of Lardil final vowel truncation (Hale 1973):

(23) Tableau 183A from Prince and Smolensky (1993/2004)

	/yiliyili/	Free-V	Align	Parse	NoCoda
a. →	yi.li.yil.<i>		−1	−1	−1
b.	yi.li.yi.li	−1			

Parse and NoCoda are violated in the first candidate because the final vowel is unparsed and the last consonant is syllabified as a coda. These violations serve to satisfy Free-V, which demands that the word-final vowel be unparsed. Satisfaction of Free-V also forces a violation of the constraint Align, which requires the edge of the word to coincide with a syllable boundary. Prince and Smolensky make two comments about this tableau. The first is on p. 144:

(24) The relative harmonies of .yi.li.yil.<i> (183 A.i) and .yi.li.yi.li. (183 A.ii) pointedly illustrate the strictness of strict domination. Fully parsed .yi.li.yi.li. is less harmonic than truncated .yi.li.yil.<i> even though it violates only one constraint, while the truncated form violates three of the four lower ranked constraints...

The second is on p. 148:

(25) *Strictness of strict domination.* In several examples the correct analysis violates many constraints, and its optimality rests crucially on the fact that competitors with a cleaner record overall happen to violate some single dominant constraint. Recall the discussion of /yiliyili/ in 7.3.2: a strong contender violating just one constraint is bested by an optimal parse violating three of the four less dominant constraints. This effect highlights the content of the central evaluative hypothesis, and sets the theory apart from others in which richer notions of 'weighting' and 'trade-off' are entertained.

It is in fact not clear how the Lardil example is meant to set ranking apart from weighting. First, we can obviously assign a set of weights to the constraints to pick the correct optimum: for any finite set of data, any set of OT optima can also be made optimal in HG (Prince and Smolensky 1993/2004). So long as the weight of Free-V is greater than the summed weights of Align, Parse, and NoCoda, [yi.li.yil.<i>] will emerge as optimal in an HG version of (23).

Less obviously, any gang effect between the three constraints violated by [yi.li.yil.<i>] would be vacuous. This is due to another kind of specific-to-general relationship that obtains between Free-V and both Align and Parse: Any candidate that satisfies Free-V necessarily violates Align and Parse, but not vice versa (Prince and Smolensky, 1993/2004: 7.2.1). Because Free-V satisfaction entails the violation of these constraints, a gang effect that involves ALIGN and/or PARSE with NOCODA in blocking deletion would be vacuous: The sum of the effects of ALIGN and/or PARSE with NOCODA in forcing the violation of FREE-V would be the same as the effect of NOCODA alone. These constraints fail to provide a one-to-one trade-off between NOCODA and FREE-V. In the absence of this one-to-one trade-off, there would be no occasion for the lower weight of NOCODA than FREE-V to show its effect. As this gang effect is vacuous, it does not produce a divergence between the typological predictions of HG and OT.

1.3 OT and HG with different scalar constraint sets

In the last section, I mentioned Prince and Smolensky's (1993/2004) observation that HG is in a superset relation with OT: any finite set of OT optima can also be made optimal in HG (p. 236), but some HG optima cannot be OT optima (p. 233).[4] One might conclude from this that OT is inherently more restrictive in terms of its typological predictions. However, this follows only if OT and HG share the same constraint set. As we have already seen in the Japanese loanword devoicing example, the greater power of weighting can allow a set of constraints to capture an attested language that OT would fail to generate. It is thus likely that for any fleshed out theory of some domain, the HG constraint set will be different from the one assumed in OT. When HG and OT are operating with different constraint sets, there is no necessary relation of relative restrictiveness.

A related inference that one might draw is the one stated by Coetzee and Pater (2005: 114):

(26) One of the most striking results of the typological research that has been conducted in Optimality Theory is that there seems to be very little counter-evidence for strict domination.

4 For an infinite dataset, ranking can produce patterns that elude weighting. For example, the ranking WEIGHT-TO-STRESS ≫ ALIGN-R discussed in Section 1.4 produces a pattern in which a heavy syllable will get stressed, no matter how far it is from the right edge of the word, and therefore no matter how many violations of ALIGN-R it incurs. This is impossible to reproduce in HG: regardless of how much higher the weight of WEIGHT-TO-STRESS is (up to infinity), there will be some distance from the right edge of the word that will make the cost of satisfying that constraint too high relative to ALIGN-R. For directly observable typology, however, it is correct to say that with a given set of constraints, HG will produce a superset of the languages that OT does.

That is, the large body of successful linguistic analysis with the more restrictive ranked constraints stands as an argument for the premise that ranking suffices (a similar observation is made by Prince and Smolensky 1993/2004: 94). The flaw in this argument is that OT can capture any pattern that HG can, if we expand the constraint set appropriately (see also the discussion of Japanese loanword devoicing above).

A comparison of the two frameworks then must necessarily involve careful analysis of attested languages with both weighted and ranked constraints, and a comparison of the resultant typological predictions. This sort of research is just in its infancy. Jesney (2012, this volume) shows that HG can generate more of the attested typology of positional restrictions with positional markedness constraints than OT can, raising the possibility of avoiding unwelcome typological predictions of positional faithfulness. Potts *et al.* (2010) provide a similar comparison of an HG analysis of Lango vowel harmony with an OT-LC one, again showing that the more restricted constraint set in HG generates a tighter typology than OT with a less restricted constraint set.

Scalar constraints are known to undergenerate in OT (Prince and Smolensky 1993/2004, McCarthy 2003). In Pater (2012), I show that HG allows them to capture the basic type of attested pattern that makes them incompatible with OT. Here I use that case to provide an explicit example of how OT and HG operating with different constraint sets produce partially overlapping typological predictions. I also discuss some of the issues in making scalar constraints accountable for existing data patterns. The investigation of scalar constraints in HG began with Flemming (2001), who uses HG to allow for real-valued scales, a distinct, but closely related application of weighted constraints – see more recently Cho (2011), McAllister Byun (2011) and Ryan (2011) on weighted scalar constraints.

We start with a review of the example from Pater (2012): the differences between the sets of languages generated by HG and OT with a constraint on the sonority of syllable nuclei. McCarthy (2003) suggests the following restatement of Prince and Smolensky's (1993/2004) H-Nuc, which I rename *C-Nuc in Pater (2012), to make it clear that the two constraints have different effects.[5]

5 Prince and Smolensky's H-Nuc directly orders candidates rather than simply assigning violation marks in the fashion of *C-Nuc and maybe all other OT constraints. This allows it to deal with aspects of the Berber syllabification data that *C-Nuc fails on in a parallel theory. See Pater (2012) for discussion, and for a serial HG analysis using *C-Nuc that has some advantages over the OT analyses of Berber in Prince and Smolensky (1993/2004).

(27) *C-Nuc
Assign a violation mark to a nucleus for each degree of sonority separating it from [a]

Because *C-Nuc assigns multiple violations to a single structure (that is, it is gradient, in McCarthy's 2003 sense), we can get asymmetric trade-offs between just it and another constraint. Assuming a simple sonority scale in which sonorant consonantal nuclei are one step away from [a] and obstruents are two, we get the *C-Nuc violations for [tN] and [tS] in (28), where capitalization indicates nuclear status. The comparisons in (28) are with candidates with epenthesis of the vowel [a], which avoids a *C-Nuc violation. These violate Dep, so the faithful candidates are preferred by a margin of +1.

(28) Comparative vectors for consonantal nuclei and epenthesis

	W ~ L	*C-Nuc	Dep
a.	[tN] ~ [tan]	−1	+1
b.	[tS] ~ [tas]	−2	+1

In OT, there are only two possibilities: either *C-Nuc dominates Dep, and both epenthetic candidates in (28) win, or Dep dominates *C-Nuc, and neither does. That is, OT cannot generate a cut-off in this scale. It is this problem that leads Prince and Smolensky (1993/2004: sec. 8.1) to replace the scalar H-Nuc with constraints in a fixed ranking, and McCarthy (2003) to propose a general prohibition against scalar constraints in OT.

HG does not suffer from this problem, as illustrated by the weighted comparative vectors in (29). The final column shows the weighted sum of violation differences, or margin of separation. Because these numbers are positive, the Winners are correctly preferred. With these weights, a penalty of −2 on *C-Nuc, as would be incurred by the sub-optimal *[tS], is sufficient to force epenthesis, but the penalty of −1, as assessed for [tN], is not. English is a close-by example of a language that imposes this sort of restriction, though a full account would need to deal with the fact that only syllabic [ɹ] is permitted in stressed syllables.

(29) Weighted comparative vectors for a language with only sonorant nuclei

	W ~ L	*C-Nuc 2	Dep 3	
a.	[tN] ~ [tan]	−1	+1	+1
b.	[tas] ~ [tS]	+2	−1	+1

While scalar constraints in HG do not have the basic problem that they do in OT, there are more subtle issues that remain to be dealt with. De Lacy (2004) discusses a pattern he calls 'conflation,' where some sonority differences fail to play a role in a given language. He provides as one example stress placement in Nganasan. In words with three light syllables, stress generally falls on the penultimate syllable. However, when the penult is of low sonority, containing one of the vowels [i y u ə ɨ], and the antepenult is of higher sonority, containing mid [e o] or low [a], stress falls on the higher sonority antepenult (30a.). Conflation is observed in that not all sonority differences upset the basic stress pattern in this way. Antepenult low vowels fail to attract stress away from penult mid ones (30b.), and antepenult high vowels fail to attract stress from penult central ones (30c.). De Lacy shows that in other languages, there are sonority sensitive stress patterns that distinguish higher sonority low vowels from mid ones, and high from central.

(30) a. [cétəmti] 'fourth'; [kóntuɟa] 'carry'; [négyʃa] 'tease'; [ániʔə] 'large'; [hʲásɨrə] 'fishing rod'

b. [bacébsa] 'breathing'; [kacéməʔ] 'examine'; [lʷamóbtuʔ] 'spill, splash'

c. [cintə́ɟi] 'stroke'; [kunsɨ́nɨ] 'inside'; [hytə́ðə] 'torso'

If scalar constraints assign violations according to universal linear scales, this sort of pattern cannot be accounted for. We will focus just on the low/mid/high distinctions, and assume the constraints in (31) and (32):

(31) STRESS-TO-SONORITY-LINEAR (STRESS-SON-LIN)
Assign a violation to the head of a foot for each degree of sonority separating it from [a] (low vowel = 0, mid vowel = −1, high vowel = −2)

(32) PENULT
Assign a violation for antepenultimate stress

The problem is that the difference between the penalties assigned to stressed high vowels (−2) *vs.* stressed mid ones (−1) is the same as the difference between those assigned to stressed mid vowels (−1) *vs.* low ones (0). This leads to the inconsistent weighting conditions illustrated in (33): antepenultimate stress in [négyʃa] requires STRESS-SON-LIN to outweigh PENULT, but penultimate stress in [kacéməʔ] requires the reverse.

(33) Inconsistent comparative vectors with a linear scalar constraint

	W ~ L	STRESS-SON-LIN	PENULT
a.	[négyʃa] ~ [negýʃa]	+1	−1
b.	[kacéməʔ] ~ [kácəməʔ]	−1	+1

One solution would be to allow language-specific conflation in the violation scales. For example, in Nganasan mid vowels would be conflated with low vowels in not violating STRESS-SON-LIN. Another solution is to adopt nonlinear violation scales. With the revised STRESS-SON-NONLIN in (34), the distance between mid and high is 2, while the distance between low and mid remains 1.

(34) STRESS-TO-SONORITY-NONLINEAR (STRESS-SON-NONLIN)
Assign a violation to the head of a foot for each degree of sonority separating it from [a] (low vowel = 0, mid vowel = −1, high vowel = −3)

This revision allows us to find a correct set of weights, as shown in (35).

(35) Weighted comparative vectors with a nonlinear scalar constraint

	W ~ L	STRESS-SON-NONLIN 2	PENULT 3	
a.	[négyʃa] ~ [negýʃa]	+2	−1	+1
b.	[kacéməʔ] ~ [kácəməʔ]	−1	+1	+1

De Lacy (2004) points out that some patterns of conflation are a problem for fixed rankings of constraints in OT, and argues that constraints in a stringency relation can get the full range of patterns. The constraints required in OT for the patterns we have been analyzing with the scalar Stress-Son constraints are given in (36).

(36) OT Stress-Sonority constraints
a. *STRESS-[I]
Assign a violation to a head of a foot that is a high vowel
b. *STRESS-[I,E]
Assign a violation to a head of a foot that is a high or mid vowel

We can thus compare OT with the set of constraints in (36) to HG with the constraint in (34), each interacting with the Penult constraint. The first row of the table in (37) shows three inputs, which illustrate the three possible distributions of penultimate and antepenultimate vowel sonority in which the antepenult has greater sonority than the penult.

The first two columns differ in whether the difference is between mid and high (/teniti/) or low and mid (/taneti/); the third column pits low against high (/taniti/). The rows beneath show the output stress patterns generated by HG and OT with different constraint sets. Asterisks indicate the instances of antepenultimate stress, where the demands of a stress-sonority constraint prevail over the penult positional preference. Checkmarks indicate which theories generate each pattern. Our present comparison is between OT and HG-1.

(37) Typological predictions of OT and HG with different constraint sets

	/teniti/	/taneti/	/taniti/	OT	HG-1	HG-2
a.	téniti*	táneti*	tániti*	✓	✓	✓
b.	teníti	taníti	taníti	✓	✓	✓
c.	téniti*	tanéti	táneti*	✓	✓	✓
d.	teníti	táneti*	tániti*	✓		✓
e.	teníti	taníti	tániti*		✓	✓

The pattern in (37a.) is one in which sonority fully determines the outcome, and the one in (37b.) is the converse case where the positional preference is always respected. The next row, (37c.) is the Nganasan conflation pattern, where low and mid vowels attract stress from high ones, but in which the low-mid difference is ignored.

The last two rows of (37) illustrate OT-HG differences. As the checkmarks indicate, OT is in more restrictive in one way, and HG in another. The difference illustrated in (37e.) arises from the fundamental difference between ranked and weighted constraints: weighted constraints can model a 'sufficient reward' threshold, in which a general preference is overridden only to gain a sufficient benefit on another dimension. Here that benefit is stressing the best type of vowel, low [a], instead of the worst one, high [i]. A gain from worst to intermediate – high [i] to mid [e] – or intermediate to best – mid [e] to low [a] – is insufficient to compensate for placing stress on the dispreferred antepenultimate position. De Lacy's (2004, 2006) typological survey appears to include no vowel quality-based stress pattern with a sufficient reward threshold. Further research is required to determine whether that gap is accidental, as predicted by HG, or is a reflection of a general restriction on constraint interaction, as predicted by OT. For sonority-stress interactions, it seems that the existing set of typological data is too sparse to make this determination.

The row in (37d.) shows that HG fails to generate a pattern in which conflation is between mid and high vowels, rather than between low and mid. This is an example of a prediction that derives from nonlinear scalar constraints. When there is a greater difference in the number of violations on one step of the scale than another (e.g. with STRESS-SON-NONLIN, low-mid = 1, mid-high = 2), if the smaller difference favors an optimum, so will the larger difference (all else being equal). Here, it is impossible to make [táneti] beat [tanéti] while at same time making [teníti] beat [téniti], as the inconsistent comparative vectors in (38) show.

(38) Inconsistent comparative vectors for mid-high conflation

	W ~ L	STRESS-SON-NONLIN	PENULT
a.	[táneti] ~ [tanéti]	+1	–1
b.	[teníti] ~ [téniti]	–2	+1

For this particular example, the difference counts in favor of OT: de Lacy (2006) points to Gujarati as an example of a language that conflates the sonority scale in this way in arbitrating stress placement. The overall resulting OT prediction, that there are no implicational relationships of this type in conflation, remains to be critically examined. *A priori*, it seems plausible that some adjacent points on scales, on sonority and other dimensions, are more easily conflated than others, and nonlinearities in the scales seem a like natural way to express this.

The column labeled 'HG-2' in (37) shows the result if we include two nonlinear STRESS-TO-SONORITY constraints, one that has a greater difference between mid and high vowels, and one that has a greater difference between low and mid. These are labeled STRESS-TO-SONORITY-13 and STRESS-TO-SONORITY-23 in (39) respectively. With these constraints, HG gets the full set of patterns.

(39) a. STRESS-TO-SONORITY-13
Assign a violation to the head of a foot for each degree of sonority separating it from [a] (low vowel = 0, mid vowel = –1, high vowel = –3)

b. STRESS-TO-SONORITY-23
Assign a violation to the head of a foot for each degree of sonority separating it from [a] (low vowel = 0, mid vowel = –2, high vowel = –3)

While this may appear to be be a brute force solution, mid vowels are independently known to vary in their featural affiliation with low and high vowels, hence their usual designation as [–high, –low], with features that allow them to be classed with either.

To further paint what is likely to be the general picture of partially overlapping typological predictions between OT and HG with different constraint sets, we can add competing constraints that apply to another scalar dimension: syllable weight. Languages typically make only a two-way distinction between heavy and light syllables, but some languages make a three-way distinction. For example, in the variety of Hindi described by Kelkar (1968) and analyzed by Hayes (1995) and Prince and Smolensky (1993/2004), syllables with long vowels or codas count as heavy, and syllables with both of these count as superheavy. For HG, the typology calculation used the scalar Weight-to-Stress constraint in (40), interacting with the two Stress-to-Sonority constraints in (39). The nonlinear scale in (40) makes conflation between heavy and light entail conflation between heavy and superheavy.

(40) Weight-to-Stress
Assign a violation to an unstressed heavy syllable (–2 if heavy, –3 if superheavy)

For OT, the weight-to-stress constraints in the typology calculation are the ones in (41), which interact with the stress-to-sonority constraints in (36).

(41) a. Heavy-to-Stress
Assign a violation to an unstressed heavy or superheavy syllable.
b. Superheavy-to-Stress
Assign a violation to an unstressed superheavy syllable.

The results are illustrated in (42). For every input, the syllable on the left is heavier than the syllable on the right, while the one on the right is higher in sonority. The first six languages are generated both by OT and HG with their respective constraint sets. The first language is one in which the sonority-sensitive constraints fully determine the outcome, and the last of the six is one in which the weight-sensitive constraints are fully obeyed. In between those two are the four languages that display intermediate outcomes that both theories can generate, while in the last 19 languages are nine mixed patterns that only OT can produce, and the 10 that only HG gets.

(42) OT and HG with different constraint sets

ti:nma:	ti:nma	ti:me	te:nma:	ti:nme	te:nma	te:ma	ti:ma	ti:nme:	OT	HG
ti:nmá:	ti:nmá	ti:mé	te:nmá:	ti:nmé	te:nmá	te:má	ti:má	ti:nmé:	✓	✓
ti:nmá:	ti:nmá	ti:mé	te:nmá:	ti:nmé	té:nma	té:ma	ti:má	ti:nmé:	✓	✓
ti:nmá:	ti:nmá	tí:me	te:nmá:	tí:nme	te:nmá	te:má	ti:má	ti:nmé:	✓	✓
ti:nmá:	tí:nma	tí:me	te:nmá:	tí:nme	té:nma	té:ma	tí:ma	tí:nme:	✓	✓
ti:nmá:	tí:nma	tí:me	té:nma:	tí:nme	té:nma	té:ma	tí:ma	ti:nmé:	✓	✓
tí:nma:	tí:nma	tí:me	té:nma:	tí:nme	té:nma	té:ma	tí:ma	tí:nme:	✓	✓
ti:nmá:	ti:nmá	ti:mé	te:nmá:	tí:nme	te:nmá	te:má	ti:má	tí:nme:	✓	
ti:nmá:	ti:nmá	ti:mé	té:nma:	ti:nmé	té:nma	te:má	ti:má	ti:nmé:	✓	
ti:nmá:	ti:nmá	ti:mé	té:nma:	ti:nmé	té:nma	té:ma	ti:má	ti:nmé:	✓	
ti:nmá:	ti:nmá	tí:me	te:nmá:	tí:nme	te:nmá	te:má	ti:má	tí:nme:	✓	
ti:nmá:	tí:nma	tí:me	te:nmá:	tí:nme	té:nma	té:ma	tí:ma	ti:nmé:	✓	
ti:nmá:	tí:nma	tí:me	té:nma:	tí:nme	té:nma	té:ma	tí:ma	ti:nmé:	✓	
tí:nma:	tí:nma	ti:mé	té:nma:	tí:nme	té:nma	te:má	ti:má	tí:nme:	✓	
tí:nma:	tí:nma	ti:mé	té:nma:	tí:nme	té:nma	té:ma	ti:má	tí:nme:	✓	
tí:nma:	tí:nma	tí:me	té:nma:	tí:nme	té:nma	te:má	ti:má	tí:nme:	✓	
ti:nmá:	tí:nma	tí:me	té:nma:	tí:nme	té:nma	té:ma	tí:ma	tí:nme:		✓
ti:nmá:	tí:nma	tí:me	te:nmá:	tí:nme	té:nma	té:ma	ti:má	tí:nme:		✓
ti:nmá:	tí:nma	tí:me	té:nma:	tí:nme	té:nma	té:ma	ti:má	ti:nmé:		✓
ti:nmá:	tí:nma	tí:me	te:nmá:	tí:nme	té:nma	té:ma	ti:má	ti:nmé:		✓
ti:nmá:	ti:nmá	tí:me	te:nmá:	tí:nme	té:nma	té:ma	ti:má	ti:nmé:		✓
ti:nmá:	ti:nmá	ti:mé	te:nmá:	tí:nme	té:nma	té:ma	ti:má	ti:nmé:		✓
ti:nmá:	ti:nmá	tí:me	te:nmá:	tí:nme	té:nma	te:má	ti:má	ti:nmé:		✓
ti:nmá:	ti:nmá	ti:mé	te:nmá:	tí:nme	té:nma	te:má	ti:má	ti:nmé:		✓
ti:nmá:	ti:nmá	ti:mé	te:nmá:	ti:nmé	té:nma	te:má	ti:má	ti:nmé:		✓
ti:nmá:	ti:nmá	ti:mé	te:nmá:	tí:nme	te:nmá	te:má	ti:má	ti:nmé:		✓

This typological study illustrates the general way in which OT and HG with differing constraint set can generate partially overlapping predictions, and some of the particular differences between OT with constraints in a stringency relation, and HG with explicitly scalar constraints. Unfortunately, currently available typological data do not seem to choose one of these theories of constraints and their interaction over the others, since both sonority sensitivity and scalar weight sensitivity are rare. The choice between HG and OT will likely thus have to be made on the basis of other data.

1.4 Unbounded trade-offs and locality

Legendre, Sorace and Smolensky (2006) provide an example of an unattested linguistic system produced by HG but not OT. The example involves what can be referred to as an unbounded trade-off: satisfaction of one constraint can require a potentially unbounded number of violations of another. The constraints at issue are ones that determine stress placement. The first requires that a particular kind of syllable – a heavy one – be stressed (WEIGHT-TO-STRESS; Prince, 1990). Languages vary in which syllables fall into the heavy category; for the abstract example below, heavy syllables are ones that have codas. The other constraint penalizes a stressed syllable according to how far away it is from a word edge: it assigns a violation mark for each syllable that intervenes. I adopt the name MAINSTRESSRIGHT from Legendre, Sorace and Smolensky (2006) (cf. EDGEMOST in Prince and Smolensky 1993/2004, and ALIGNMENT in McCarthy and Prince 1993).

The table in (43), adapted from Legendre, Sorace and Smolensky (2006), compactly illustrates the unbounded trade-off. The syllable [ban] stands in for any heavy syllable and the coda-less syllable [ta] stands in for any nonheavy one. The variable σ_n is a string of a number *n* of nonheavy syllables. As in Legendre, Sorace and Smolensky (2006), we only consider candidates with a single stress per word. Stress on the final syllable violates WEIGHT-TO-STRESS and satisfies MAINSTRESSRIGHT. Stress on the initial syllable satisfies WEIGHT-TO-STRESS and violates MAINSTRESSRIGHT once for every syllable separating it from the right edge of the word. Stress on any of the syllables in σ_n would be harmonically bounded by final stress, as it would add at least one violation of MAINSTRESSRIGHT without compensating improvement on WEIGHT-TO-STRESS. Therefore, we need only consider the two candidates in (43).

(43) An unbounded trade-off

	ban$\sigma_n$ta	WEIGHT-TO-STRESS	MAINSTRESSRIGHT
a.	ban.σ_n.tá	−1	
b.	bán.σ_n.ta		−1−n

As there is no theoretical upper bound on the size of words, there is theoretically no upper bound on the number of MAINSTRESSRIGHT violations that can be traded off against the single violation of WEIGHT-TO-STRESS.

Legendre, Sorace and Smolensky (2006) point out that in OT there are only two possible languages given by the two rankings. The number of syllables intervening between a nonfinal stressed heavy syllable and the edge of the word is irrelevant; either stress will fall on the rightmost syllable or on the heavy syllable.

(44) Two languages in OT
WEIGHT-TO-STRESS ≫ MAINSTRESSRIGHT → bán.σ_n.ta
MAINSTRESSRIGHT ≫ WEIGHT-TO-STRESS → ban.σ_n.tá

They also note that HG produces a theoretically infinite set of languages with these constraints. With appropriate weights, stress can be limited to a 'window' of any number of syllables at the right edge of the word (see also Prince 1993, 2007 on Goldsmith 1994). In HG, the number of intervening syllables is crucial, as shown by the fact that this number is included in the weighting conditions in (45). In these inequalities, the constraint names stand for their weights. Stress will fall on a nonfinal heavy syllable only if the weight of WEIGHT-TO-STRESS is greater than the weight of MAINSTRESSRIGHT times the number of syllables separating the heavy syllable from the right edge. For example, if WEIGHT-TO-STRESS = 3.5 and MAINSTRESSRIGHT = 1, then a heavy syllable will get stressed if it is followed by three light syllables, but not four. No known language has such a four-syllable window.

(45) An infinite typology in HG
WEIGHT-TO-STRESS > $(n+1)$*MAINSTRESSRIGHT → bán.σ_n.ta
$(n+1)$*MAINSTRESSRIGHT > WEIGHT-TO-STRESS → ban.σ_n.tá

Taken on its own, this case is not particularly persuasive as an argument against HG as a framework for typological study. One problem is that there are, in fact, attested three-syllable windows (e.g., in Macedonian: Comrie, 1976; and Pirahã: Everett and Everett, 1984), and these cannot be generated by OT rankings of the set of constraints for stress in Prince and Smolensky (1993/2004) and McCarthy and Prince (1993) – see Hyde (2007) and Kager (2012). One might take the ability of HG to account for them as a positive result and seek an explanation for the absence of the larger windows, perhaps in terms of the size of words that a learner would need to hear to acquire the pattern (cf. Hammond, 1991), or in terms of the relative difficulty of acquiring the weight ratios needed to represent the pattern (cf. Prince, 1993: 91;

Prince, 2007a: 41); see Staubs (2014) for further discussion and arguments for this approach.[6]

Another problem is that gradient Alignment constraints are controversial, even in OT. In counting the distance between two portions of the representation, these constraints assign violation scores in an unusual manner, both in comparison with other OT constraints, and with ones elsewhere in GL (Eisner 1998, Potts and Pullum 2002, Biró 2003, and McCarthy 2003). For the other constraints discussed in this paper, which are typical of OT, each locus of violation incurs a bounded number of violation marks (see McCarthy 2003 on the formalization of 'locus of violation'). For a stressed syllable evaluated by a MAINSTRESSRIGHT, the number of violations depends on the distance from the edge of the word, and is thus unbounded. Not only are the gradient Alignment constraints formally unusual, but they also produce undesired typological predictions, in OT as well as HG. Based on these considerations, McCarthy (2003) proposes a revised theory of OT constraints in which gradient Alignment constraints are banned. Removing gradient Alignment constraints from OT has the effect of also removing a large class of potential problem cases for a version of the theory with weighted constraints.

Unbounded trade-offs, and divergences between the typological predictions of OT and HG, can also emerge from the interaction of constraints that only assign a single violation per locus. In some situations, the satisfaction of an output constraint can require a number of faithfulness violations with no theoretical upper bound. As an example, we can consider the interaction of NOCODA with the faithfulness constraint LINEARITY (McCarthy and Prince 1999), which assigns a violation mark for every pairwise reordering of the segmental string.

6 In terms of the present model, a typological restriction on the size of windows can be obtained by imposing a maximum value on the weights and a minimum difference between the harmony scores of the optima and their competitors (see Boersma and Pater this volume and Potts *et al.* 2010 on the margin of separation of harmony; see relatedly Albright, Magri and Michaels 2008). For example, if we require the optimal candidate's harmony to exceed that of any of its competitors by at least 0.5, then a three-syllable window requires a minimum constraint weight of 2.5, while a four-syllable window requires 3.5. The requirements of a large margin of separation and a maximum value on weights are commonly imposed in learning algorithms in the machine learning and neural modeling literature, and seem quite plausible as biological limitations. For example, in terms of Boersma and Pater's (this volume) Noisy HG, a minimum margin of separation is needed to overcome a given amount of noise. Probabilistic versions of these limits would further increase the learning difficulty for larger stress windows mentioned in the text.

(46) LINEARITY
If segment x precedes segment y in the input, x precedes y in the output

In (47), word-internal syllable boundaries are again indicated with periods, and NOCODA violations again occur when a syllable ends in a consonant (clusters are assumed to be split between syllables). Here we see that satisfaction of NOCODA can require two reorderings of the segmental string, as (47d.).

(47) Asymmetric trade-off between NOCODA and LINEARITY

	/apekto/	NOCODA	LINEARITY
a.	[a.pek.to]	−1	
b.	[pa.ek.to]	−1	−1
c.	[ap.ke.to]	−1	−1
d.	[pa.ke.to]		−2

It is also possible to create strings in which only one violation of LINEARITY would be needed to satisfy NOCODA (e.g., /ekto/, [ke.to]), as well as ones in which any higher number is needed (e.g., /idapekto/ requires three violations, as in [di.pa.ke.to]). Appropriate weightings of the constraints can create systems in which NOCODA is satisfied at the cost of n violations of LINEARITY, but not $n + 1$ violations, where n is any nonnegative integer. OT only produces two systems: one in which NOCODA is satisfied at any cost in terms of LINEARITY violations, and one in which even a single LINEARITY violation is worse than a violation of NOCODA.

McCarthy (2007a) in fact provides this example as a case in which the standard version of OT produces the wrong result. Although languages do employ local pairwise reorderings of segments to satisfy output constraints like NOCODA (see Hume, 2001 for a survey), none use a double reordering of the type illustrated in the final candidate in (47), which would be optimal under the ranking NOCODA ≫ LINEARITY (as well as with weights respecting the condition NOCODA > 2*LINEARITY).

McCarthy (2007a) shows that the correct typology is obtained in the alternative version of OT that Prince and Smolensky (1993/2004) call Harmonic Serialism (HS) – see McCarthy (this volume) for an overview of other typological advantages of HS. In this theory, Gen is limited to a single application of an operation; here it can produce a single pairwise reordering, but not two at once. Multiple applications of an operation can occur serially, if each one results in an improvement in harmony. The tableau in (48) shows the first candidate set that would be produced in an HS derivation, which lacks the candidate with a double reordering. The faithful candidate (48a.) harmonically bounds the others, so it would always be picked as optimal regardless of the constraint ranking. In HS a derivation terminates when the faithful candidate is chosen; (48) is thus both the first and last step.

(48) No double reordering in Harmonic Serialism

	/apekto/	NoCoda	Linearity
a. →	[a.pek.to]	−1	
b.	[pa.ek.to]	−1	−1
c.	[ap.ke.to]	−1	−1

McCarthy's solution extends to a weighted constraint version of the theory, as [a.pek.to] is equally guaranteed to win in (48) with any set of positive weights. It also eliminates the difference between OT and HG typology mentioned beneath (47), since in a serial version of either OT or HG, only single reorderings (e.g. /ekpo/, [ke.po]) can be used to satisfy NoCoda.

We now return the example that Prince and Smolensky (1997) point to as distinguishing ranked from weighted constraints: the interaction of NoCoda and Parse. In Section 1.2.3, we saw that the number of potential codas did not in fact lead to a difference between HG and OT: in either theory, all are deleted, or all are retained. A difference can emerge, though, from the second part of their example: the number of potential consonants in a single coda. If we adopt a version of NoCoda that assigns only single violation for the entire coda, we can create a system in HG in which NoCoda is satisfied at the cost of n violations of Parse, but not n+1.

As the pair of tableau in (49) and (50) shows, the result of the interaction of these constraints in parallel HG can be quite bizarre: a language that has codas with two or more consonants (e.g. → [(apt)] in (50)), but not one (e.g. *[(ap)] in (49)). Further, this cut-off can be made at any point: languages with codas with no fewer consonants than three, or four, or five, or any other number can be modeled in this theory.

(49) A single consonant fails to be parsed into a coda

	ap	NoCoda 1.5	Parse 1	
a. →	(a)p		−1	−1
b.	(ap)	−1		−1.5

(50) Two consonants are parsed into a coda

apt	NoCoda 1.5	Parse 1	
a. (a)pt		−2	−2
b. (ap)t	−1	−1	−2.5
c. → (apt)	−1		−1.5

If consonants are added one at a time to a syllable with an adjunction operation, as in Pater (2012), then this pattern is impossible to recreate in a serial version of HG.[7] With our two constraints, the first application of adjunction will beat the fully faithful candidate if and only if NoCoda has a greater weight than Parse-Seg. The tableau in (51) illustrates the second step of the derivation for the UR /apt/ with the same weights as in (50). Here we already have the nucleus syllabified through the prior application of nuclear projection. Importantly, the candidate set includes only the faithful candidate and the single adjunction, and not the fully syllabified candidate (apt) that was optimal in the parallel HG tableau for /apt/ in (50). Since that candidate is missing from the candidate set, the optimum is now [(a)pt]; we no longer get the strange pattern in which a coda is formed only to syllabify some minimum number of segments.[8]

(51) Coda formation in serial HG

(a)pt	NoCoda 1.5	Parse 1	
a. → (a)pt		−2	−2
b. (ap)t	−1	−1	−2.5

This example illustrates a general consequence of the single change limitation on candidates imposed in a serial theory: that the set of

7 See McCarthy (2009, this volume) for arguments that syllabification must at least sometimes be done in parallel. One way of reconciling those arguments with the serial results of Elfner (2009) and Pater (2012) would be to have serial initial syllabification, and parallel resyllabification. This could be formalized as a 'free' resyllabification operation that applies whenever a segment is stranded by the application of some other operation. This approach is in fact taken by Torres-Tamarit (this volume).

8 Bane and Riggle (2012) point to other problematic patterns that arise in parallel HG with these sort of basic syllable structure constraints, and at least some of these are resolved by adopting a serial version of the theory (and some are also produced by standard OT with a larger constraint set).

possible constraint interactions, or trade-offs in the terminology of Section 1.2, is restricted relative to a parallel model. In particular, trade-offs involving multiple instances of violation of a given faithfulness constraint will never occur in a serial version of HG or OT, insofar as the operations that create candidates incur at most one violation of each faithfulness constraint (as in McCarthy 2007b). Thus, scenarios in which a markedness constraint is satisfied at the cost of *n* violations of a faithfulness constraint, but not *n*+1, which can be created in a parallel version of HG, are impossible in a serial version. The typological benefit accrued by serialism in the NoCoda/Parse example is likely typical of such cases. For example, a parallel version of HG with Agree constraints for assimilation can generate a pattern in which one of a sequence of two disagreeing segments will assimilate, but in which assimilation is blocked if some larger number of segments would have to be changed to achieve agreement across the whole string (Pater, Bhatt and Potts 2007, Bane and Riggle 2012). This sort of pattern does not arise in the serial HG approach to assimilation developed by Mullin (2011).

Both gradient Alignment and the parallel evaluation of multiple instances of faithfulness constraint violation can be characterized as producing unwanted global effects in the standard version of OT. We thus have unwanted globality introduced by particular types of constraint, and by particular assumptions about how candidates are generated and evaluated. In both cases, it seems likely that refinements to the theory that impose the desired locality restrictions in OT will result in the elimination of the most serious problems faced by weighted constraint versions of OT-style UG. The biggest challenge faced by HG appears to be in stress typology. Although as discussed above, the particular case of an HG/OT difference discussed by Legendre, Sorace and Smolensky (2006) does not incontrovertibly favor OT, the full predicted typology generated by parallel HG with the alignment constraints of McCarthy and Prince (1993) does not look to be a particularly promising for the success of this theory (see further Potts *et al.* 2010 and Bane and Riggle 2012). If the right solution to the other locality problems discussed in Section 1.4 is to adopt HS, then we run into the problem that gradient Alignment seems to be necessary in a standard serial theory (Pruitt 2012).[9]

9 Targeted constraints in HS (Wilson 2013) need not suffer from the indeterminacy problems for stress placement in HS that Pruitt (2012) notes for Lapse constraints and other alternatives to gradient Alignment.

1.5 Probabilistic HG

It appears to be the standard assumption in current phonological theory that a realistic model of phonology must be able to cope with variation in outcomes across instances of production (and/or perception and/or experimental judgment) – see Coetzee and Pater (2011) for a recent overview of the data that motivate this view, and of OT and HG models of grammar that have stochastic output. There are two versions of probabilistic HG currently being pursued: Maximum Entropy Grammar (MaxEnt; Goldwater and Johnson 2003, Wilson 2006, Jäger and Rosenbach 2006, Jäger 2007, Hayes, Zuraw, Siptár and Londe 2009, Staubs and Pater this volume) and Noisy HG (see Boersma and Pater this volume for a description and references). These theories have gained attention primarily because of the existence of their associated learning algorithms. In this section I discuss how the behavior of their grammatical components relates to the general properties of the categorical version of HG explored in this paper.

In the categorical version of HG, the candidate with the highest Harmony, or weighted sum of constraint violations, is chosen as the optimum. This choice does not change across instances of evaluation. In Noisy HG, at each instance of evaluation the constraint weights are sampled from normal distributions around mean values, and the candidate with highest Harmony is chosen as optimal – because of the sampled weights, the optimum can vary across instances of evaluation. In MaxEnt the probability of a candidate is proportional to the exponential of its Harmony, and each time the grammar is used a candidate is sampled from this distribution. In both of these probabilistic versions of HG, the probability of a candidate increases with its (mean) Harmony, so to make one candidate more probable than another, it must be given higher Harmony. Since the relative Harmony of candidates is always at issue, by understanding the patterns produced by categorical HG, we are also learning about the patterns that can be produced by the probabilistic theories.

If we limit a MaxEnt or Noisy HG grammar to positive weights, then the sets of single optima that the categorical version of HG can generate are the same sets of single candidates to which the stochastic theories can give highest probability within their candidate sets, putting aside tics. Thus, we can do fairly direct mapping from the typological predictions of categorical HG discussed above. All and only the languages generated by categorical HG are generated by the probabilistic theories, where a 'language' has a single optimal candidate in each tableau in the categorical case, and a single candidate with greatest probability in the probabilistic case.

MaxEnt gives non-zero probability to candidates that are harmonically bounded in categorical HG (Jäger and Rosenbach 2006, Jesney 2007), that is, to candidates that can never have higher Harmony than all of their competitors. For simply harmonically bounded forms – those that have a proper superset of the violation marks of another candidate – the resulting prediction is that they can only ever surface as a minority variant (see Goldrick and Daland 2009 for related discussion with respect to speech errors and a version of Noisy HG that permits negative weights). As usual, the prediction critically depends on the contents of the constraint set, which may make formulating a definitive test of this prediction of MaxEnt difficult. In Noisy HG, on the other hand, simply harmonically bounded forms always have zero probability, since they can never win in any single evaluation, no matter what the positive weights are.

Putting aside ties again for the moment, the candidates that are collectively bounded in HG also have zero probability in Noisy HG. The tableau in (52) demonstrates that MaxEnt can give collectively bounded candidates maximum probability within a candidate set. Candidates (52b) [dak.bad] and (52c) [dag.bat] are collectively bounded in HG in that there is no weighting that can make either of them singly optimal: with *Coda-Voice > Ident-Voice, (52a) [dak.bat] wins, and with the reverse relationship, (52d) [dag.bad] does. The harmony of each candidate continues to be indicated as the value at the end of each row, and the MaxEnt probability assigned to each candidate is shown to its left. Under an equal weighting of the constraints, candidates (52b) and (52c) tie with (52a) and (52d), which translates into equal MaxEnt probability. This is as much probability as they can get: their probability decreases with respect to (52a) or (52d) as the weights are shifted in favor of *Coda-Voice or Ident-Voice respectively – neither (52b) nor (52c) can ever gain higher probability than both (52a) and (52d).

(52) Collectively bounded candidates with maximum probability in MaxEnt

	/dagbad/	*Coda-Voice 2	Ident-Voice 2	
a. 0.25	dag.bad	−2		−4
b. 0.25	dak.bad	−1	−1	−4
c. 0.25	dag.bat	−1	−1	−4
d. 0.25	dak.bat		−2	−4

As Jäger and Rosenbach (2006) and Jesney (2007) point out, it may be to the advantage of MaxEnt that it can make candidates like (52b) and (52c) surface. For many processes, 'local optionality', as illustrated by devoicing of only one of two codas, is attested (see Kimper 2011 for a recent overview). However, as Jesney notes, the issue is complicated by the fact that there are alternative analyses of local optionality. In particular, Kimper (2011) develops a serial account in HS, and as discussed in Section 1.4, HS has independent benefits for HG typology (see Staubs and Pater this volume on learning in serial probabilistic HG).

As mentioned at the end of Section 1.2.1, I abstract from ties in optimality in all of the discussion of categorical HG. In categorical HG, tableau (52) would be an instance of such a tie, and this one would be impossible if the constraints were in a totally ordered ranking as in standard OT. If we were committed to a categorical theory, then we would need to say something about the greater tie-generating capabilities of HG than standard OT. When we move to a more realistic stochastic version of the theory, ties are unremarkable. In MaxEnt they are simply interpreted as equal probability, as in (52). In Noisy HG, they are vanishingly improbable because they require precise weight values (so 52b and 52c would have vanishingly little probability in Noisy HG), and when they occur, can be simply resolved by random choice, as in Boersma and Pater (this volume).

1.6 Constraint universality

Throughout this paper I have been discussing the standard OT view of UG, that it includes a universal set of constraints. Most learnability research in OT, starting with the work presented in Tesar and Smolensky (2000), takes this constraint set as given to learners. An alternative view is that UG consists of constraint schema, which are filled on the basis of learning data, and this alternative assumption is the basis of another active line of research with weighted constraints. An explicit demonstration of the viability of this sort of constraint induction is provided by Hayes and Wilson (2008), who work with a version of Maximum Entropy Grammar that determines a probability distribution over the space of possible words, and thus functions as a model of phonotactics (see also Moreton 2010, Pizzo 2013 and Pater 2014 on constraint induction with weighted constraints). Even though the premises and goals of this research are quite different from those of standard OT (see Pater and Moreton 2012 for extended discussion), a better understanding of the patterns that a given set of constraints can and cannot generate will no doubt be useful for the purpose of building of models of constraint induction. Such an understanding can be gained by further examining the predictions of particular sets of constraints, whose universality may be taken as a useful idealization. A choice between ranking and weighting also has to be made for induced constraints (see Adriaans and Kager 2010 on constraint induction with OT), though the arguments for one or the other become quite subtle if one is not aiming to derive typological generalizations directly from the constraint set.

1.7 Conclusions

This paper aims to reopen the discussion of whether weighted constraints are suitable as a framework for phonological analysis and typological study, a discussion which may have seemed to have ended when Prince and Smolensky (1993/2004) rejected weighted constraints in favor of ranked ones. I have tried to show that the choice between OT and HG is far more difficult to make than one might initially assume, and that HG has considerable untapped potential in this domain. The asymmetric trade-off requirement imposes inherent restrictions on the types of cumulative interaction that HG can express – restrictions that are not shared by the alternative formalization of cumulativity in OT with local constraint conjunction. In addition, the greater expressive power of weighted than ranked constraints allows for new hypotheses about the contents of the universal constraint set. The study of OT and HG operating with different constraint sets is still a new area of research, and the above comparison of OT and HG accounts of scalar phenomena indicates some of the open paths for further work in this area. Finally, a full tally of the relative benefits of weighed and ranked constraints must also take into account the simultaneous choice of parallel *vs.* serial candidate evaluation, since the locality restrictions imposed by serialism eliminate some of the differences between HG and OT.

References

Adriaans, Frans and René Kager (2010) Adding generalization to statistical learning: The induction of phonotactics from continuous speech. *Journal of Memory and Language* 62: 311–321.

Albright, Adam, Giorgio Magri and Jennifer Michaels (2008) Modeling doubly marked lags with a split additive model. In Harvey Chan, Heather Jacob, and Enkeleida Kapia (eds), *BUCLD 32: Proceedings of the 32nd annual Boston University conference on language development*, 36–47. Somerville, MA: Cascadilla Press.

Baković, Eric (2000) *Harmony, dominance, and control.* Doctoral dissertation, Rutgers University.

Bane, Max, and Jason Riggle (2012) The typological consequences of weighted constraints. In Nikki Adams, Adam Cooper, Fay Parrill and Thomas Wier (eds), *Proceedings from the Annual Meeting of the Chicago Linguistic Society* 45, 13–27. Chicago, IL: University of Chicago.

Barnes, Jonathan (2006) *Strength and weakness at the interface: positional neutralization in phonetics and phonology.* Berlin and New York: Mouton de Gruyter.

Biró, Tamas (2003) Quadratic alignment constraints and finite state optimality theory. In *Proceedings of the workshop on finite-state methods in natural language processing*, 119–126. Budapest, Hungary. Available on Rutgers Optimality Archive, ROA-600.

Blevins, Juliette (2004). *Evolutionary Phonology: the emergence of sound patterns.* Cambridge: Cambridge University Press.

Boersma, Paul (2009) Some correct error-driven versions of the Constraint Demotion algorithm. *Linguistic Inquiry* 40: 667–686.

Boersma, Paul and Joe Pater (this volume) Convergence properties of a gradual learning algorithm for Harmonic Grammar.

Cho, Hyesun (2011) The timing of phrase-initial tones in Seoul Korean: A weighted-constraint model. *Phonology* 28: 293–330.

Coetzee, Andries and Joe Pater (2005) Lexically specific constraints: Gradience, leanability, and perception. In *Proceedings of the 3rd Seoul International Conference on Phonology*, 85–119. Seoul: The Phonology-Morphology Circle of Korea.

Coetzee, Andries and Joe Pater (2011) The place of variation in phonological theory. In John Goldsmith, Jason Riggle, and Alan Yu (eds), *The Handbook of Phonological Theory (2nd edn)*, 401–431. Malden, MA: Blackwell.

Comrie, Bernard (1976) Irregular stress in Polish and Macedonian. *International Review of Slavic Linguistics* 1: 227–240.

Eisner, Jason (1998) FootForm decomposed: Using primitive constraints in OT. In Benjamin Bruening (ed.), *Proceedings of SCIL VIII, MIT Working Papers in Linguistics* 31, 115–143). Cambridge, MA: Department of Linguistics, MIT.

Elfner, Emily (2009) Syllabification and stress-epenthesis interactions in Harmonic Serialism. Unpublished paper, University of Massachusetts Amherst. Available on Rutgers Optimality Archive, ROA-1047.

Everett, Daniel and Keren Everett (1984) On the relevance of syllable onsets to stress placement. *Linguistic Inquiry* 15: 705–711.

Farris-Trimble, Ashley (2008) Cumulative faithfulness effects: Opaque or transparent? In Ashley W. Farris-Trimble and Daniel A. Dinnsen (eds), *IUWPL6: Phonological Opacity Effects in Optimality Theory*, 119–145. Bloomington, IN: IULC Publications.

Flemming, Edward (2001) Scalar and categorical phenomena in a unified model of phonetics and phonology. *Phonology* 18: 7–44.

Goldrick, Matthew and Robert Daland (2009) Linking speech errors and phonological grammars: Insights from Harmonic Grammar networks. *Phonology* 26: 147–185.

Goldsmith, John (1990) *Autosegmental and metrical phonology*. Oxford and Cambridge, MA: Blackwell.

Goldsmith, John (1993a) Harmonic phonology. In John Goldsmith (ed.) *The last phonological rule*, 21–60. Chicago: University of Chicago Press.

Goldsmith, J. (1993b) *The last phonological rule: Reflections on constraints and derivations.* Chicago: University of Chicago Press.

Goldsmith, John (1994) A dynamic computational theory of accent systems. In Jennifer Cole and Charles Kisseberth (eds), *Perspectives in phonology*, 1–28. Stanford, CA: CSLI.

Goldwater, Sharon and Mark Johnson (2003) Learning OT constraint rankings using a maximum entropy model. In Jennifer Spenader, Anders Eriksson and Östen Dahl (eds), *Proceedings of the Stockholm Workshop on 'Variation within Optimality Theory'*, 111–120. Stockholm: Stockholm University.

Guy, Gregory R. (1997) Violable is variable: Optimality Theory and linguistic variation. *Language Variation and Change* 9: 333–347.

Hale, Kenneth (1973) Deep-surface canonical disparities in relation to analysis and change: An Australian example. In Thomas Sebeok (ed.), *Current trends in linguistics, volume 9: Diachronic, areal and typological linguistics*, 401–458. The Hague, The Netherlands: Mouton.

Hammond, Michael (1991) Parameters of metrical theory and learnability. In Iggy Roca (ed.), *Logical issues in language acquisition*, 47–62. Dordrecht, The Netherlands: Foris.

Hayes, Bruce (1995) *Metrical stress theory: Principles and case studies*. Chicago: The University of Chicago Press.

Hayes, Bruce, Robert Kirchner and Donca Steriade (eds) (2004) *Phonetically based phonology*. Cambridge: Cambridge University Press.

Hayes, Bruce, Bruce Tesar and Kie Zuraw (2003) *OTSoft 2.1*. Software package, http://www.linguistics.ucla.edu/people/hayes/otsoft/

Hayes, Bruce and Colin Wilson (2008) A Maximum Entropy model of phonotactics and phonotactic learning. *Linguistic Inquiry* 39: 379–440.

Hayes, Bruce, Kie Zuraw, Péter Siptár and Zsuzsa Londe (2009) Natural and Unnatural Constraints in Hungarian Vowel Harmony. *Language* 85: 822–863.

Hume, Elizabeth (2001) Metathesis: Formal and functional considerations. In Elizabeth Hume, Norval Smith and Jeroen van de Weijer (eds.), *Surface syllable structure and segment sequencing* (1–25). Leiden, The Netherlands: HIL.

Hyde, Brett (2007) An alignment approach to stress windows. Colloquium presentation, Rutgers University. [Available at: http://www.artsci.wustl.edu/bhyde/windowshandout.pdf.]

Ito, Junko (1986) *Syllable theory in prosodic phonology*. Doctoral dissertation, University of Massachusetts, Amherst. Published 1988, New York: Garland.

Ito, Junko and Armin Mester (1986) The phonology of voicing in Japanese: Theoretical consequences for morphological accessibility. *Linguistic Inquiry* 17, 49–73.

Ito, Junko and Armin Mester (1995) Japanese phonology. In John Goldsmith (ed.), *Handbook of phonological theory*, 817–838. Cambridge, MA: Blackwell.

Ito, Junko and Armin Mester (2003) *Japanese morphophonemics: Markedness and word structure*. MIT Press Linguistic Inquiry Monograph Series 41. Cambridge, MA: MIT Press.

Jäger, Gerhard (2007) Maximum entropy models and Stochastic Optimality Theory. In Annie Zaenen, Jane Simpson, Tracy Holloway King, Jane Grimshaw, Joan Maling, and Chris Manning (eds), *Architectures, rules, and preferences: A Festschrift for Joan Bresnan*, 467–479. Stanford, CA: CSLI.

Jäger, Gerhard and Anne Rosenbach (2006) The winner takes it all – almost. Cumulativity in grammatical variation. *Linguistics* 44: 937–971.

Jesney, Karen (2012) Licensing in multiple contexts: an argument for Harmonic Grammar. In Nikki Adams, Adam Cooper, Fay Parrill, and Thomas Wier (eds), *Proceedings from the Annual Meeting of the Chicago Linguistic Society* 45, 287–301. Chicago, IL: University of Chicago.

Jesney, Karen (this volume) Positional constraints in Optimality Theory and Harmonic Grammar.

Jesney, Karen and Anne-Michelle Tessier (2011) Biases in Harmonic Grammar: The road to restrictive learning. *Natural Language and Linguistic Theory* 29: 251–290.

Kager, René (2012) Stress in Windows: Language Typology and Factorial Typology. *Lingua* 122: 1454–1493.

Kawahara, Shigeto (2011) Japanese loanword devoicing revisited: A rating study. *Natural Language and Linguistic Theory* 29: 705–723.

Kelkar, Ashok R. (1968) *Studies in Hindi-Urdu. Vol. 1: Introduction and word phonology.* Poona: Deccan College.

Kimper, Wendell (2011) Locality and globality in phonological variation. *Natural Language and Linguistic Theory* 29: 423–465.

de Lacy, Paul (2004) Markedness conflation in Optimality Theory. *Phonology* 21: 145–199.

de Lacy, Paul (2006) *Markedness*. Cambridge: Cambridge University Press.

Legendre, Géraldine, Yoshiro Miyata, and Paul Smolensky (1990) Can connectionism contribute to syntax?: Harmonic grammar, with an application. Report CU-CS-485–90. Computer Science Department, University of Colorado at Boulder. [http://www.colorado.edu/ics/sites/default/files/attached-files/90-12.pdf.]

Legendre, Géraldine, Antonella Sorace, and Paul Smolensky (2006) The Optimality Theory-Harmonic Grammar connection. In Paul Smolensky and Géraldine Legendre (eds), *The harmonic mind: From neural computation to optimality theoretic grammar, vol. 2: Linguistic and philosophical implications*, 339–402. Cambridge, MA: MIT Press.

Lombardi, Linda (1991) *Laryngeal features and laryngeal neutralization*. Doctoral dissertation, University of Massachusetts, Amherst.

Lubowicz, Anna (2002) Derived environment effects in Optimality Theory. *Lingua* 112: 243–280.

Lubowicz, Anna (2005) Locality of conjunction. In John Alderete, Chung-hye Han, and Alexei Kochetov, *Proceedings of the 24th West Coast Conference on Formal Linguistics* 24, 254–262. Somerville MA: Cascadilla Proceedings Project.

Martin, Andrew (2007) *The evolving lexicon*. Doctoral dissertation, UCLA.

McAllister Byun, Tara (2011) A gestural account of a child-specific neutralisation in strong position. *Phonology* 28: 371–412.

McCarthy, John J. (1999). Sympathy and phonological opacity. *Phonology* 16: 331–399.

McCarthy, John J. (2003) OT constraints are categorical. *Phonology* 20: 75–138.

McCarthy, John J. (2007a) Restraint of analysis. In Sylvia Blaho, Patrick Bye and Martin Kramer (eds), *Freedom of analysis*, (203–231). Berlin: Mouton de Gruyter.

McCarthy, John J. (2007b) *Hidden generalizations: Phonological opacity in Optimality Theory.* London: Equinox.

McCarthy, John J. (2009) Studying Gen. *Journal of the Phonetic Society of Japan* 13: 3–12.

McCarthy, John J. (this volume). The theory and practice of Harmonic Serialism.

McCarthy, John J. and Alan Prince (1993) Generalized Alignment. In Geert Booij and Jaap van Marle (eds), *Yearbook of Morphology*, 79–153) Dordrecht, The Netherlands: Kluwer. Excerpts appear in John Goldsmith (ed.) (1999), *Essential readings in phonology*, 102–136. Oxford: Blackwell.

McCarthy, John J. and Alan Prince (1999) Faithfulness and identity in Prosodic Morphology. In René Kager, Harry van der Hulst and Wim Zonneveld (eds), *The prosody–morphology interface*, 218–309. Cambridge, England: Cambridge University Press.

Moreton, Elliott (2010) Constraint induction and simplicity bias in phonological learning. Paper presented at the Workshop on Grammar Induction, Cornell University [http://www.unc.edu/~moreton/Papers/MoretonCornellWkshp2010.pdf]

Moreton, Elliott and Paul Smolensky (2002) Typological consequences of local constraint conjunction. In Line Mikkelson and Chris Potts (eds). *Proceedings of the 21st West Coast Conference on Formal Linguistics*, 306–319, Somerville MA: Cascadilla Proceedings Project.

Mullin, Kevin (2011) Strength in Harmony Systems: Trigger and Directional Asymmetries. Ms., University of Massachusetts Amherst. Available on the Rutgers Optimality Archive, ROA-1134.

Nishimura, Kohei (2003) Lyman's Law in loanwords. Master's thesis, Nagoya University.

Nishimura, Kohei (2006) Lyman's Law in loanwords. *Phonological Studies* [Onin Kenkyuu], 9: 83–90.

Padgett, Jaye (2002) Constraint conjunction versus grounded constraint subhierarchies in Optimality Theory. Ms, University of California, Santa Cruz. Available on the Rutgers Optimality Archive, ROA-530.

Pater, Joe (1999) Austronesian nasal substitution and other NC effects. In Kager, René, Harry van der Hulst and Wim Zonneveld (eds) (1999). *The prosody-morphology interface*, 310–343. Cambridge: Cambridge University Press.

Pater, Joe (2009a) Weighted Constraints in Generative Linguistics. *Cognitive Science* 33: 999–1035.

Pater, Joe (2009b) Review of 'The harmonic mind: From neural computation to optimality theoretic grammar'. *Phonology* 26: 217–226.

Pater, Joe (2010) Non-vacuous specific with general gang effects. Ms, University of Massachusetts, Amherst. Available on the Rutgers Optimality Archive, ROA-1072.

Pater, Joe (2012) Serial harmonic grammar and Berber syllabification. In Toni Borowsky, Shigeto Kawahara, Takahito Shinya and Mariko Sugahara (eds), *Prosody matters: Essays in honor of Elisabeth O. Selkirk*, 43–72. London: Equinox Press.

Pater, Joe (2014) Canadian raising with language-specific weighted constraints. *Language* 90: 230–240.

Pater, Joe, Rajesh Bhatt and Christopher Potts (2007) Linguistic optimization. Ms, University of Massachusetts Amherst. Available on the Rutgers Optimality Archive, ROA-924.

Pater, Joe and Elliott Moreton (2012) Structurally biased phonology: Complexity in learning and typology. *EFL Journal*: 1–44. [Available at http://people.umass.edu/pater/pater-moreton-2012-offprint.pdf]

Pizzo, Presley (2013) Learning Phonological Alternations with Online Constraint Induction. Paper presented at the Old-World Conference on Phonology 10, Istanbul [http://people.umass.edu/ppizzo/presentation12-27.pdf]

Potts, Christopher, Joe Pater, Karen Jesney, Rajesh Bhatt, and Michael Becker (2010) Harmonic Grammar with Linear Programming: From linear systems to linguistic typology. *Phonology* 27: 77–117.

Potts, Christopher and Geoffrey K. Pullum (2002) Model theory and the content of OT constraints. *Phonology* 19: 361–393.

Prince, Alan (1993) *In defense of the number i: Anatomy of a linear dynamical model of linguistic generalizations*. RuCCS Technical Report 1. New Brunswick, NJ: Rutgers University.

Prince, Alan (2002) Entailed ranking arguments. Ms, Rutgers University. Available on the Rutgers Optimality Archive, ROA-500.

Prince, Alan (2003) Anything goes. In Takeru Honma, Masao Okazaki, Toshiyuki Tabata and Shin-ichi Tanaka (eds), *A New Century of Phonology and Phonological Theory. A Festschrift for Professor Shosuke Haraguchi on the Occasion of His Sixtieth Birthday*, 66–90. Tokyo: Kaitakusha. Available on the Rutgers Optimality Archive, ROA-536.

Prince, Alan (2007) In pursuit of theory. In Paul de Lacy (ed.), *Cambridge handbook of* phonology, 33–60. Cambridge, England: Cambridge University Press.

Prince, Alan and Paul Smolensky (1997) Optimality: from neural networks to Universal Grammar. *Science* 275: 1604–1610.

Prince, Alan and Paul Smolensky (2004). *Optimality Theory: constraint interaction in generative grammar*. Malden, MA and Oxford: Blackwell. [Revision of 1993 technical report, Rutgers University Center for Cognitive Science]. Available on the Rutgers Optimality Archive, ROA-537.

Pruitt, Kathryn (2012) *Stress in Harmonic Serialism*. Doctoral dissertation, University of Massachusetts Amherst.

Ryan, Kevin (2011) Gradient syllable weight and weight universals in quantitative metrics. *Phonology* 28: 413–454.

Samek-Lodovici, Vieri and Alan Prince (2005) Fundamental properties of harmonic bounding. Unpublished paper, University of London and Rutgers University, London and New Brunswick, NJ. Available on the Rutgers Optimality Archive, ROA-785.

Smolensky, Paul (2006a) Harmony in linguistic cognition. *Cognitive Science* 30: 779–801.

Smolensky, Paul (2006b) Optimality in phonology II: Harmonic completeness, local constraint conjunction, and feature domain markedness. In Paul Smolensky and Géraldine Legendre (eds), *The harmonic mind: From neural computation to optimality theoretic grammar, vol. 2: Linguistic and philosophical implications*, 27–160. Cambridge, MA: MIT Press.

Smolensky, Paul and Géraldine Legendre (2006). *The harmonic mind: From neural computation to optimality theoretic grammar*. Cambridge, MA: MIT Press.

Staubs, Robert (2014) *Computational modeling of learning biases in stress typology.* Doctoral dissertation, University of Massachusetts Amherst. Available at http://scholarworks.umass.edu/dissertations_2/230/.

Staubs, Robert (this volume). Serial Restrictions on Feature/Stress Interactions.

Staubs, Robert and Joe Pater (this volume). Learning serial constraint-based grammars.

Staubs, Robert, Michael Becker, Christopher Potts, Patrick Pratt, John J. McCarthy and Joe Pater (2010) OT-Help 2.0. Software package. Amherst, MA: University of Massachusetts Amherst.

Steriade, Donca (1995) Underspecification and markedness. In John Goldsmith (ed.) *The handbook of phonological theory*, 114–174. Cambridge, MA and Oxford: Blackwell.

Steriade, Donca (1999) Phonetics in phonology: the case of laryngeal neutralization. *UCLA Working Papers in Linguistics: Papers in Phonology* 3. 25–145.

Tesar, Bruce and Paul Smolensky (2000) *Learnability in optimality theory*. Cambridge, MA: MIT Press.

Torres-Tamarit, Francesc (this volume) Compensatory and opaque vowel lengthening in Harmonic Serialism.

Wilson, Colin (2006) Learning phonology with substantive bias: An experimental and computational study of velar palatalization. *Cognitive Science* 30: 945–982.

Wilson, Colin (2013) A Targeted Spreading Imperative for Nasal Place Assimilation. In Yelena Fainleib, Nicholas LaCara, and Yangsook Park (eds), *Proceedings of the 41st Annual Meeting of the North East Linguistics Society*, 261–272. Amherst MA: GLSA Publications.

Chapter 2

The theory and practice of Harmonic Serialism[1]

John J. McCarthy

2.1 Introduction

Harmonic Serialism (HS) is a variant of Optimality Theory. The difference between HS and standard OT is that HS is a *serial* model and standard OT is a *parallel* model. In this chapter, I will explain what that means, how they differ, and how HS works.

This chapter begins (Section 2.2) with a simple example and an overview of HS's basic architecture, *modus operandi*, and provenance. Section 2.3 then provides a more extended illustration of HS, using the well-known syllabically conditioned alternations of Yawelmani Yokuts. Section 2.4 develops some results about the nature of HS's derivations, and Section 2.5 briefly treats the principal consequences that HS has for the formulation of constraints. Section 2.6 explains a key element of the process of analysis in HS: the trade-off between GEN and CON as the locus of linguistic explanation. Comparison of HS and parallel OT (P-OT) on empirical grounds is the focus of Section 2.7, and finally Section 2.8 addresses the question of whether HS's derivations make it a more successful theory of opacity than P-OT.

1 This research was supported by grant BCS-0813829 from the National Science Foundation to the University of Massachusetts Amherst. I am indebted to all of the participants in our weekly grant group for their advice about this chapter, with particular thanks going to Kathryn Pruitt and Joe Pater for their comments on the manuscript.

2.2 An overview of Harmonic Serialism

In P-OT, GEN can make many changes at once when it produces a candidate. This means that competing candidates can differ in many ways from each other and from the underlying representation. For example, Yawelmani Yokuts has an underlying→surface mapping in which /taxa:-kˀa/ 'bring!' becomes *ta.xakˀ* by apocope and closed syllable shortening (Newman 1944). (The period/full stop marks a syllable boundary.) In a P-OT analysis of Yawelmani, the candidates from underlying /taxa:-kˀa/ include some that differ from it by a single change (such as *ta.xa:kˀ*, which differs only by virtue of a single deletion) and some that differ by two or more changes (such as *ta.xakˀ*, which differs by both deletion and shortening, or *taxa*, which differs by two deletions and a shortening process). In P-OT, GEN applies one or more operations together, in parallel, when it generates candidates. In consequence, P-OT's candidate sets are quite diverse and involve competition between candidates that may not resemble each other very much.

In HS, GEN is limited to making just one change at a time. The candidate set from underlying /taxa:-kˀa/ consists of all of the ways of making no more than a single change in this form: *ta.xa:kˀ*, *ta.xa.kˀa*, and so on, but not *ta.xakˀ* or *taxa*, which require two or more changes. In consequence, HS's candidate sets are less diverse than P-OT's, and they involve competition between candidates that are not very different from one another. This property of HS's GEN is called *gradualness*.

HS would be unworkable as a theory of language if all it did was to impose this limitation on GEN. That is because underlying and surface representations can in fact differ from one another in several ways. For instance, the surface form from underlying /taxa:-kˀa/ is *ta.xakˀ*, which shows the effects of both vowel deletion and vowel shortening. As we just saw, *ta.xakˀ* is not one of the candidates that HS's GEN produces from /taxa:-kˀa/. So how does the grammar get to it?

The answer is that HS has a loop. In HS, the optimal candidate chosen by EVAL becomes a new input to GEN, which forms a candidate set that goes to EVAL, and so on. The loop continues until EVAL picks an optimum that is identical with the most recent input to GEN. So the HS derivation for /taxa:-kˀa/ goes something like this:

(1) Derivation for /taxa:-kˀa/ in HS

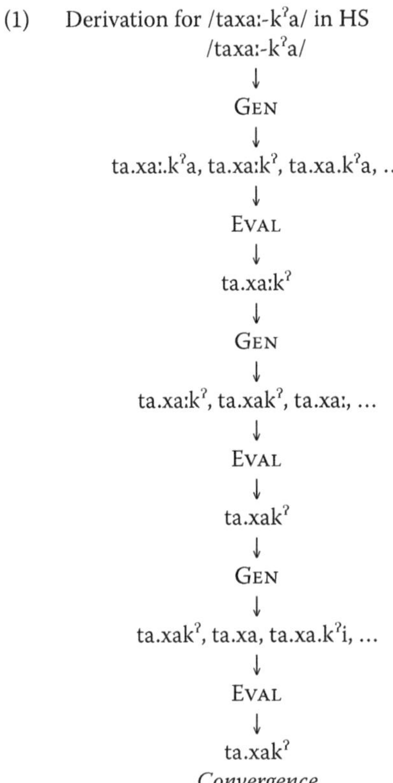

Each pass through Gen and Eval is called a *step*. The candidate set at each step includes the unchanged input to that step and all of the ways of making a single change in that input. (Syllabification, including resyllabification, gets special treatment. More about this in Section 2.6.) The derivation in (1) is over when a step begins and ends with the same form, *converging* on *ta.xakˀ*. This is the final output of the grammar – the surface form for underlying /taxa:-kˀa/. The whole model is summarized in the flowchart in (2).

(2) HS flowchart[2]

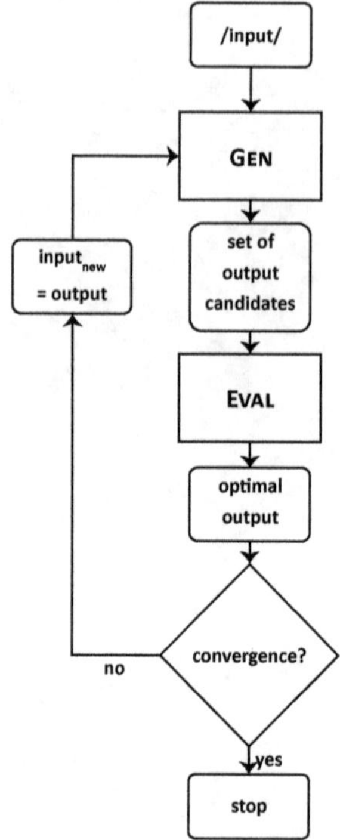

I will first illustrate HS with an example that is simpler than Yawelmani, but we will return in Section 2.3 to look in detail at how the derivation in (1) comes about.

In Classical Arabic, word-initial consonant clusters are prohibited. When they occur in underlying representations, glottal stop and a high vowel are preposed: /fʕal/ → *ʔif.ʕal* 'do!'. Two of the markedness constraints involved are *COMPLEX-ONSET, which is violated by faithful *fʕal*, and ONSET, which is violated by *if.ʕal*, with epenthetic *i* but not epenthetic *ʔ*.[3] In a P-OT analysis, both of these constraints are ranked at the top of the hierarchy, dominating DEP:[4]

2 Thanks to Kevin Mullin for creating the diagram in (2).
3 Other markedness constraints rule out other parses of faithful *fʕal*, such as those where *f* is a syllable nucleus or an appendix to the word.
4 Tableaux are in comparative format (Prince 2002). The winning candidate appears to the right of the arrow, and losers are in the rows below it. Integers stand for the number of violation marks incurred by a candidate, replacing the familiar strings of asterisks. In loser rows, the effects of the constraints are indicated by W and L, W if the constraint favors the winner and L if it favors the loser.

(3) P-OT analysis of Arabic /fʕal/ → *ʔif.ʕal*

	/fʕal/	*Comp-Onset	Ons	Max	Contig	Dep
a. →	ʔif.ʕal					2
b.	fʕal	1 W				L
c.	if.ʕal		1 W			1 L
d.	ʕal			1 W		L
e.	fi.ʕal				1 W	1 L

Tableau (3) includes two other constraints that are also active. Because Max dominates Dep, epenthesis is favored over deletion. And because Contiguity (Kenstowicz 1994, McCarthy and Prince 1995, 1999) dominates Dep, initial epenthesis of *i* is preferred to medial epenthesis, even though initial epenthesis of *i* ends up requiring epenthesis of *ʔ* as well, because of Onset.

Now we will look at the HS analysis, followed by comparison of the two. HS's Gen cannot epenthesize two segments at once, so at the beginning of the derivation the ultimate winner *ʔif.ʕal* is not in the candidate set. Instead, the candidate set is limited to forms that differ from /fʕal/ by at most one change: faithful *fʕal* and unfaithful *if.ʕal, ʕal, fi.ʕal,* etc. We want *if.ʕal* to win at this step of the derivation, because it is the only candidate that will get us eventually to *ʔif.ʕal.* And for *if.ʕal* to win, *Complex-Onset, Max, and Contiguity have to dominate Onset:

(4) HS analysis of /fʕal/ → *ʔif.ʕal* — Step 1

	/fʕal/	*Comp-Onset	Max	Contig	Ons	Dep
a. →	if.ʕal				1	1
b.	fʕal	1 W			L	L
c.	fal		1 W		L	L
d.	fi.ʕal			1 W	L	1

At the next step of the derivation, the input to Gen is *if.ʕal,* and the candidate set includes *if.ʕal* and all of the ways of effecting a single change in it: *fʕal, i.fi.ʕal, ʔif.ʕal,* etc. The same grammar is applied to this new candidate set and chooses *ʔif.ʕal* as the optimum because Onset dominates Dep:

(5) HS analysis of /fʕal/ → *ʔif.ʕal* — Step 2

	if.ʕal	*COMP-ONSET	MAX	CONTIG	ONS	DEP
a. →	ʔif.ʕal					1
b.	if.ʕal				1 W	L
c.	fʕal	1 W	1 W			L

Observe that the faithfulness violations are determined relative to the input to the current step of the derivation, not to the underlying representation. Hence, *ʔif.ʕal* receives only one violation mark from DEP in (5), and *if.ʕal* gets none at all. This is consistent with HS's basic assumptions and turns out to solve a problem with positional faithfulness, as shown by Jesney (2011). (See Section 2.7.1 for a brief summary of Jesney's argument.)

Although *we* know that we have reached the desired surface form, *the grammar* does not yet know that. So it submits the most recent optimum, *ʔif.ʕal,* as input for another pass through GEN and EVAL. Once again, the candidate set consists of *ʔif.ʕal* and all of the other forms that are one change away from it. None of the changed forms wins, and *ʔif.ʕal* once again emerges as victorious:

(6) HS analysis of /fʕal/ → *ʔif.ʕal* — Step 3

	ʔif.ʕal	*COMP-ONSET	MAX	CONTIG	ONS	DEP
a. →	ʔif.ʕal					
b.	ʔi.fi.ʕal					1 W
c.	ʔif.ʕa		1 W			

At this point, the derivation has converged on the final output *ʔif.ʕal.* The GEN/EVAL loop terminates.

It is instructive to compare the P-OT and HS analyses of these facts. The P-OT analysis in (3) does not assert any ranking between *COMPLEX-ONSET, MAX, and CONTIGUITY on the one hand and ONSET on the other. All four of these constraints are unviolated in all surface forms, and surface-unviolated constraints do not conflict in P-OT. Thus, they are unrankable.

In the HS analysis, however, *COMPLEX-ONSET, MAX, and CONTIGUITY must dominate ONSET. This ranking is necessary because, although ONSET is not violated by the faithful candidate at the beginning of the derivation or the optimal candidate at the end of the derivation, it is violated by the intermediate winner *if.ʕal.* This is by no means an unusual situation in HS. In fact, it is exactly what we should expect to see in a derivational theory with ranked, violable constraints. Constraint conflicts can emerge in the course of the derivation, and those conflicts are resolved in the usual OT fashion: the conflicting constraints are ranked, and the lower-ranking constraint is violated to spare a violation of the higher-ranking one. See McCarthy, Pater, and Pruitt (this volume) for further discussion.

I will conclude this brief introduction to HS with some remarks on its provenance. From the outset, it is important to understand that HS is really just a version of OT, not a full-blown alternative to it. Indeed, HS was first proposed in OT's *locus classicus,* Prince and Smolensky (1993/2004), but it was not pursued there and was in fact rejected in favor of P-OT. The case for HS was reopened in McCarthy (2000, 2002: 159–163, 2007b), where some general consequences of this theory are identified and discussed.

HS is distinct from, though related to, OT with candidate chains (OT-CC), in which an HS-like system is used to construct derivations that then compete against one another (McCarthy 2007a, Wolf 2008, this volume). Other efforts to implement OT with derivations should also be mentioned (Black 1993, Chen 1999, Kiparsky 2000, Norton 2003, Rubach 2000), as well as non-OT theories of derivational constraint satisfaction, such as Harmonic Phonology (Goldsmith 1990, 1993) and the Theory of Constraints and Repair Strategies (Paradis 1988a, b).

It is often asked how HS differs from rule-based phonology (RBP) in the tradition of Chomsky and Halle (1968). It would be better to ask how they are alike, because there are many differences and few similarities. In fact, there are just two shared properties: both theories posit derivations with intermediate representations; and both place limits on how much can change from one step of a derivation to the next.[5] The differences consist of all of the other ways that OT differs from RBP: in RBP, a grammar is a list of language-particular rules; in OT, a grammar is a ranking of universal constraints; in RBP, rules change one representation into another; in OT, constraints compare candidates; and so on. All of the arguments that support OT over RBP (e.g., McCarthy 2002: 66–138), other than arguments that crucially depend on parallelism, apply with equal force to HS. For more about this comparison, see McCarthy, Pater, and Pruitt (this volume).

5 Limits on how much a single rule can do were a later development in RBP. Examples include Prince's (1983) *Move x* and Archangeli and Pulleyblank's (1994) *Insert path.*

2.3 Doing Harmonic Serialism

Doing analysis in OT is hard because the required constraints are not always known in advance, so the analyst's task involves more than just finding a ranking. Indeed, one of the principal reasons for doing analysis in OT is to investigate the properties of the constraint component CON, because typologically informed analysis is the only way of studying CON.

Doing HS is harder than doing P-OT because the properties of GEN are so much more important in HS, and those properties are not always known in advance either. The goal of typologically informed analysis in HS is to investigate GEN, CON, and the relationship between the two. That sort of investigation is undertaken in Section 2.6 and in many of the other contributions to this volume. But first, the basic analytic techniques must be mastered. Chapter 2 of McCarthy (2008a) uses syllabically conditioned alternations in Yawelmani Yokuts (Newman 1944) to explain how to do analysis in P-OT. This section uses the same example, to facilitate comparison.

The syllabic phonology of Yawelmani is mainly guided by two surface-unviolated constraints against syllables bigger than CV: or CVC. One of these constraints, *SUPERHEAVY, is violated by both CV:C or CVCC syllables. The other, *COMPLEX-CODA, is violated by CVCC syllables only. (On why both constraints are needed, see Section 3.7 of McCarthy, Pater, and Pruitt (this volume).) One effect of *SUPERHEAVY is that long vowels are shortened in closed syllables:

(7) Closed syllable shortening (Kenstowicz and Kisseberth 1979: 83)

/la:n-hin/	lan.hin	'hear (nonfuture)'	cf. *la:.nal* 'id. (dubitative)'
/ṣa:p-hin/	ṣap.hin	'burn (nonfuture)'	cf. *ṣa:.pal* 'id. (dubitative)'

Another effect attributable to *SUPERHEAVY or *COMPLEX-CODA is that medial triconsonantal clusters are broken up by inserting *i* before the second consonant.

(8) Vowel epenthesis

/ʔilk-hin/	ʔi.lik.hin	'sing (nonfuture)'	cf. *ʔil.kal* 'id. (dubitative)'
/lihm-hin/	li.him.hin	'run (nonfuture)'	cf. *lih.mal* 'id. (dubitative)'

These examples are a good place to begin our HS analysis because the derivations are so shallow: the underlying and surface forms in each derivation differ by the effect of just one operation. The operations involved are mora deletion in (7) and vowel epenthesis in (8). Because the derivations are so shallow, the P-OT and HS tableaux are nearly identical. (On why it is possible to both syllabify and shorten/epenthesize/delete at Step 1, see Section 2.6.)

(9) Step 1 of /laːn-hin/ → *lan.hin*[6]

	/laːn-hin/	*COMP-CODA	*SUPER	PARSE-SEG	MAX	DEP	MAX-μ
a. →	lan.hin						1
b.	laːn.hin		1 W				L
c.	laː.ni.hin					1 W	L
d.	laː.hin				1 W		L
e.	laː.*n*.hin			1 W			L

(10) Step 1 of /ʔilk-hin/ → *ʔi.lik.hin*

	/ʔilk-hin/	*COMP-CODA	*SUPER	PARSE-SEG	MAX	DEP	MAX-μ
a. →	ʔi.lik.hin					1	
b.	ʔilk.hin	1 W	1 W			L	
c.	ʔil.hin				1 W	L	
d.	ʔil.*k*.hin			1 W		L	

In tableau (9), the faithful syllabification *laːn.hin* (9b) has a violation of *SUPERHEAVY. That violation could be avoided in various ways: by vowel shortening (9a), which violates MAX-μ, by vowel epenthesis (9c), which violates DEP, by consonant deletion (9d), which violates MAX, or by leaving *n* unsyllabified (9e), which violates PARSE-SEGMENT. Because the faithful candidate and the other alternatives violate constraints ranked higher than MAX-μ, (9a) is the winner. (Other candidates, not shown, might parse the *n* as a nucleus or as the onset of a degenerate syllable. They too violate constraints ranked higher than MAX-μ.)

In tableau (10), the situation is much the same, except that there is no long vowel to shorten. Because all of the other constraints dominate DEP, the winning candidate is (10a), with vowel epenthesis. (An alternative locus of vowel epenthesis, **ʔil.ki.hin*, is ruled out by ALIGN-RIGHT(stem, σ) (McCarthy and Prince 1993).)

At step 2, each of these derivations continues by submitting its output for another pass through GEN and EVAL. Readers can verify for

6 Unsyllabified consonants are italicized in the tableaux.

themselves that the now-faithful candidates *lan.hin* and *ʔi.lik.hin* are more harmonic than any other candidates that can be reached in a single step from them, so the derivations converge on these final outputs.

Yawelmani also has an apocope process. It is blocked when it would create a final consonant cluster:

(11) Final vowel deletion blocked (Kenstowicz and Kisseberth 1979: 98)

/xat-kʔa/	xat.kʔa, *xatkʔ	'eat!'
/xat-mi/	xat.mi, *xatm	'having eaten'

The constraint responsible for apocope – call it *V# – must dominate MAX to be active. But it must itself be dominated by *SUPERHEAVY or *COMPLEX-CODA and PARSE-SEGMENT, to account for the absence of apocope in (11). This ranking is shown in (12):

(12) Step 1 of /xat-kʔa/ → *xat.kʔa*

	/xat-kʔa/	*COMP-CODA	*SUPER	PARSE-SEG	*V#	MAX	DEP	MAX-μ
a. →	xat.kʔa				1			
b.	xatkʔ	1 W	1 W		L	1 W		
c.	xat.*kʔ*			1 W	L	1 W		

The candidate set in (12) was obtained under the same assumptions about GEN that produced the candidate sets in (9) and (10): it can (re)syllabify in parallel with unfaithful operations like deletion.

The candidate set in (12) is perhaps more noteworthy for what it does not contain as what it does. Missing are two candidates that might seriously challenge the winner: *xat,* with double deletion, and *xatikʔ*, with deletion and epenthesis. Neither of these forms is in the candidate set at step 1 because each is two steps away from the input. A P-OT analysis must take additional precautions to defeat these candidates, because unlike the intended winner they satisfy all four high-ranking markedness constraints, violating only low-ranking MAX and DEP.

Apocope may feed vowel shortening:

(13) Final vowel deletion with shortening (Kenstowicz and Kisseberth 1979: 98)

/taxa:-mi/	ta.xam	'having brought'
/taxa:-kʔa/	ta.xakʔ	'bring!'

The derivations of these forms involve two processes, segment deletion and mora deletion, and thus the derivations are longer than we have seen so far in Yawelmani.

For the HS analyst, the key questions in situations like this are:

(a) Does it matter which process happens first?
(b) If it does matter, what is the order?
(c) What constraint ranking is necessary to force that order?

The answers for Yawelmani are:

(a) Yes, it matters, because apocope creates the condition (a long vowel in a closed syllable) that necessitates shortening.
(b) For the reason just given, apocope must precede shortening.
(c) The ranking has to permit apocope even though it creates a configuration that requires subsequent shortening.

This ranking is a refinement of the ranking seen in the previous tableaux. *V# must dominate *Superheavy, the markedness constraint that CV:C syllables violate. Tableaux (14) and (15) show the first two steps of the apocope-shortening derivation under this ranking:

(14) Step 1 of /taxa:-mi/ → *ta.xam*

	/taxa:-mi/	*Comp-Coda	Parse-Seg	*V#	*Super	Max	Dep	Max-μ
a. →	ta.xa:m				1	1		
b.	ta.xa:.mi			1 W	L	L		
c.	ta.xa.mi			1 W	L	L		1 W

(15) Step 2 of /taxa:-mi/ → *ta.xam*

	ta.xa:m	*Comp-Coda	Parse-Seg	*V#	*Super	Max	Dep	Max-μ
a. →	ta.xam							1
b.	ta.xa:m				1 W			L

The derivation then converges at step 3.

The situation in tableau (14) recalls a point made in the discussion of Arabic in Section 2.2. It is not uncommon in HS to have derivational steps where a violation of a surface-true markedness constraint is introduced and later eliminated. For the analyst, it is important always to keep this possibility in mind, because without it many multi-process phonological derivations become unanalyzable. At step 1, Yawelmani swaps a violation of *V# for a violation of lower-ranking *SUPERHEAVY. That violation is eliminated at step 2, when the vowel shortens. The temporary violation of *SUPERHEAVY has to be tolerated because the ultimate winner, *ta.xam*, is not among the candidates that GEN produces at step 1.

2.4 Derivations in Harmonic Serialism

Derivations in HS are guaranteed to be finite and to show steady harmonic improvement until convergence. These two properties of HS derivations are related.

In HS, as in OT generally, harmony is the property that EVAL selects for. Given a constraint hierarchy H and two forms *A* and *B*, *A* is more harmonic than *B* if and only if the highest ranking constraint in H that distinguishes between *A* and *B* favors *A* over *B*. In other words, *A* is more harmonic than *B* relative to some constraint hierarchy if *A* satisfies that hierarchy better than *B* does.

Derivations in HS have to show steady harmonic improvement until convergence. The winner at step *n* has to be more harmonic than the winner at step *n*–1 (unless they are identical) because the step *n*–1 winner is a candidate at step *n*. That is because the step *n*–1 winner is the faithful candidate at step *n*, and by a standard assumption about GEN the faithful form is always a candidate.

Harmonic improvement tableaux like (16) are useful tools for studying derivations in HS. A harmonic improvement tableau shows the faithful candidate at the beginning and the winners at each successive step until convergence. By comparing each line in a harmonic improvement tableau with the one immediately above it, we can see why making the specified change is better than doing nothing at all. For example, at step 1 in (16) a violation of *COMPLEX-ONSET is eliminated at the expense of introducing violations of lower-ranking ONSET and DEP. At step 2, the ONSET violation is removed at the expense of introducing another DEP violation. At step 3, there is no further harmonic improvement, so the derivation has converged and terminates.

(16) Harmonic improvement tableau for /fʕal/ → *ʔif.ʕal*

		*Comp-Onset	Max	Contig	Ons	Dep	
Faithful	fʕal	1					
Step 1	if.ʕal				1	1	
Step 2	ʔif.ʕal					1	
Step 3	ʔif.ʕal						*convergence*

The harmonic improvement tableau is not a substitute for regular tableaux, because it does not show that the change made at each step is the best change. But the harmonic improvement tableau is useful for showing that harmony does indeed improve at each step of the HS derivation until convergence.

Because HS derivations must show steady harmonic improvement, they are guaranteed to converge in a finite number of steps. Under the standard OT assumption that all constraints either evaluate outputs (markedness) or require input-output identity (faithfulness), harmony cannot improve without limit (Moreton 2000, 2003). This result presupposes, of course, that the harmony criterion – that is, the ranking – does not change from one step to the next. If ranking can vary between steps, then eventual convergence is highly probable but not assured (Kimper 2011, Staubs and Pater this volume).

To see this more intuitively, it may be helpful to think about what it would take to produce a HS derivation that never converges. One imaginable way of getting a non-convergent derivation is to add structure without limit: /ta/ → *ta.ə* → *ta.ə.ə* → .... Another is to have a loop in the derivation, where a change is made and then immediately undone: /ta/ → *te* → *ta* → .... Neither is in fact possible in HS under standard assumptions about constraints and the constraint hierarchy.

For structure to be added without limit, there must be some markedness constraint that always favors a form with more structure over a form with less. Although there are certainly markedness constraints that favor having more structure under some circumstances – Onset is an example – there is not and cannot be any markedness constraint that favors having more structure under *all* circumstances, regardless of the length of the input. Anti-faithfulness constraints have the capacity to force the addition of structure to any input (cf. Alderete 1998, 2001), but they lie well outside the standard assumptions about what constraints can do. More to the point, it is clear that anti-faithfulness constraints are incompatible with HS precisely because they undermine the convergence guarantee. Positive constraints also have this capacity, under certain assumptions – see Kimper (this volume) for discussion.

For a HS derivation to have a loop like /ta/ → *te* → *ta* → ..., it must be the case that *ta* is more harmonic than *te*, and *te* is more harmonic than *ta*, according to the same constraint hierarchy. In a standard OT constraint hierarchy consisting only of markedness and faithfulness constraints, this is not possible. For the mapping /ta/ → *te* to occur, *te* must be less marked than the more faithful *ta*. And for the mapping *te* → *ta* to occur, *ta* must be less marked than the more faithful *te*. There is an obvious contradiction here, so the loop is an impossibility in HS. (Again, anti-faithfulness constraints would subvert this result.)

When a process is equally applicable in more than one location, HS derivations may exhibit a kind of arbitrariness that might at first seem disconcerting. Because HS's Gen is gradual, the process can only apply once in each step, though eventually it will apply everywhere. For example, Cairene Arabic has a process that shortens long vowels in unstressed syllables:

(17) Unstressed syllable shortening in Cairene

/itnaːʔiʃ-na/	ʔitna̲ʔíʃ-na	'we discussed'
/ʃaːf-it-ak/	ʃa̲fítak	'she saw you (m. sg.)'
/ma-ʃaːf-uː-niː-ʃ/	maʃa̲fu̲níːʃ	'they didn't see me'

As the last example shows, when there are several unstressed long vowels, all of them shortened.

I will ignore the part of the grammar that assigns stress and focus exclusively on the shortening process. Vowel shortening is a consequence of satisfying a constraint called Weight-to-Stress (abbreviated WSP), which is violated by unstressed heavy syllables (Prince 1990). WSP dominates Max-μ in Cairene:

(18) WSP dominates Max-μ (P-OT analysis)

	/ma-ʃaːf-uː-niː-ʃ/	WSP	Max-μ
a. →	maʃafuníːʃ		2
b.	maʃaːfuːníːʃ	2 W	L
c.	maʃaːfuníːʃ	1 W	1 L
d.	maʃafuːníːʃ	1 W	1 L

Now suppose we are analyzing the same data in HS. We will enter the derivation at step 2, after stress has been assigned. There are three relevant candidates from the input *maʃa:fu:ní:ʃ*: the unchanged input itself, *maʃafu:ní:ʃ*, and *maʃa:funí:ʃ*. The actual surface form is not a candidate, because it is two changes away from the current input. The candidates that have shortened a vowel are tied as the intermediate winners:

(19) Step 2 of HS analysis

	maʃa:fu:ní:ʃ	WSP	MAX-μ
a. →	maʃafu:ní:ʃ	1	1
b. →	maʃa:funí:ʃ	1	1
c.	maʃa:fu:ní:ʃ	2 W	L

Choose either one of the intermediate winners and input it to GEN. The winning candidate is one where both vowels have shortened, as desired:

(20) Step 3 with *maʃafu:ní:ʃ* as intermediate winner

	maʃafu:ní:ʃ	WSP	MAX-μ
a. →	maʃafuní:ʃ		1
b.	maʃafu:ní:ʃ	1 W	L

(21) Step 3 with *maʃa:funí:ʃ* as intermediate winner

	maʃa:funí:ʃ	WSP	MAX-μ
a. →	maʃafuní:ʃ		1
b.	maʃa:funí:ʃ	1 W	L

Intermediate *maʃafu:ní:ʃ* and *maʃa:funí:ʃ* are said to be in a *convergent tie*: they tie at an intermediate step of the derivation, and the ultimate result is the same either way. For further information about ties in HS, see Mullin *et al.* (2010: 7–10).

This example can also be used to illustrate one final point about HS derivations: the intermediate winner may be a form that is harmonically bounded in the P-OT analysis. Observe from (18) that *maʃafu:ní:ʃ* and *maʃa:funí:ʃ* are both harmonically bounded in P-OT under this constraint set. Yet both are intermediate winners in HS. This situation is by no means unexpected. The intermediate steps of HS derivations are local optima: they are the most harmonic member of HS's limited candidate set. HS has a different sense of harmonic bounding: *maʃafu:ní:ʃ* and *maʃa:funí:ʃ* are harmonically bounded because the derivation will not converge on them under either ranking of the two constraints. For further discussion of local and global optima, see Section 2.7.1.

2.5 Constraints in Harmonic Serialism

HS permits a rather different conception of faithfulness constraints than is standard in P-OT. Because the input and the output candidates can differ in unlimited ways in P-OT, Correspondence Theory was introduced to keep track of these differences and to assign faithfulness violations correctly (McCarthy and Prince 1995, 1999). In Correspondence Theory, the candidate set for input *in* consists, at a minimum, of all the strings over some alphabet and all of the ways of coindexing each of those strings with *in*. Thus, $d_1a_2g_3$ is a member of the candidate set for /$k_1æ_2t_3$/, as are $d_1a_3g_2$, d_1a_2g, and so on. Correspondence Theory's faithfulness constraints are defined in terms of these indices: the *d* of d_1a_2g is coindexed with /k/, so it violates IDENT(place) and IDENT(voice); the *g* of d_1a_2g has no index, so it violates DEP, and so on.

Although HS retains Correspondence Theory's *names* for the faithfulness constraints, the *definitions* of these constraints are quite different. For example, under a particular conception of HS's GEN, one of the candidates from input /kæt/ is *kæ*, which was produced from the input by applying a deletion operation. It is the application of this deletion operation that incurs the violation of MAX. This is how faithfulness is implemented in the OT-Help program (Staubs *et al.* 2010): each faithfulness constraint is associated with some operation(s) in GEN. Applying an operation incurs violations of its associated faithfulness constraint(s). Because faithfulness constraints look back only as far as the input to the most recent application of GEN, and not to the underlying representation,[7] the reasons why Correspondence Theory is required in P-OT do not carry over to HS.

All well-established faithfulness constraints are amenable to being redefined operationally. Violations of DEP and MAX are keyed to segment insertion and deletion operations. Applying insertion or deletion medially violates CONTIGUITY; applying these options peripherally

7 In McCarthy (2008b, c), I assume that faithfulness constraints in HS refer to the underlying representation, but I also observe that the same results could be obtained with faithfulness constraints that refer to the input to the current step. The latter view has become standard in subsequent work in HS.

violates ANCHOR. An operation that transposes adjacent segments is responsible for LINEARITY violations (McCarthy 2007b). The operations on distinctive features may involve changing feature values, thereby violating IDENT, or deleting and inserting featural elements, thereby violating DEP(feature) and MAX(feature) (see McCarthy 2008b for discussion).[8] This operational theory of faithfulness has consequences for the formulation of positional faithfulness constraints (see Section 2.7). Together with gradualness, it also limits the usefulness of local conjunction of faithfulness constraints (Kirchner 1996, Moreton and Smolensky 2002). See Section 2.8 for further discussion.

HS also has some consequences for markedness constraints; see Pruitt (2012) on *LAPSE and Pater (2012) on Prince and Smolensky's (1993/2004) scalar sonority constraint HNUC.

2.6 GEN in Harmonic Serialism[9]

Because of the gradualness requirement on GEN, HS's candidate sets are finite. Finiteness is assured as long as GEN contains no intrinsically iterative or recursive operations – that is, no operations that can add unbounded amounts of structure in a single step. This means that the effects of iteration or recursion have to be obtained over multiple steps of a derivation rather than all at once.

HS's GEN is gradual, but what exactly is meant by 'gradualness'? This is a central research question in HS. The claim is that GEN can make only one change at a time, so the candidate set from input *in* consists of *in* itself and all the forms differing from *in* by a single change. There are, however, important issues in specifying exactly what constitutes a single change. Fortunately, it is clear how we should go about finding an answer to this question, although the answer itself is not yet complete. The technique can be explained with an example.

In Cairene Arabic, syncope affects short high vowels in unstressed non-final open syllables in internal and external sandhi (see (22)). The consonant that preceded the deleted vowel is resyllabified as a coda:

8 Coalescence and breaking, which violate the correspondence constraints UNIFORMITY and INTEGRITY, have not yet been examined from an HS perspective.

9 Portions of this section come from McCarthy (2009).

(22) Cairene syncope (data from Watson 2002: 70–72)

/wiħiʃ-a/	ˈwiħ.ʃa	'bad (f. sg.)'
/xulusˤ-it/	ˈxul.sˤit	'she finished'
/tˤardi kibiːr/	ˈtˤar.dik.ˈbiːr	'my parcel is big'

Do syncope and resyllabification occur in separate steps? Is GEN able to resyllabify an onset at the same time that it deletes its nucleus? Which of the derivations in (23) is the right one?[10]

(23) Two possible derivations for /wiħiʃ-a/ → ˈ*wiħ.ʃa*

/wiħiʃ-a/ → wi.ħi.ʃa → ˈwi.ħi.ʃa → ˈwi.*ħ*.ʃa → ˈwiħ.ʃa

a. /wiħiʃ-a/ → wi.ħi.ʃa → ˈwi.ħi.ʃa → ˈwiħ.ʃa

Both derivations begin with initial syllabification followed by stress assignment. They then diverge, because each presupposes a different theory of GEN. The GEN that produces (23a) does syncope and resyllabification serially, in separate steps. The GEN that produces (23b) does syncope and resyllabification in parallel, in a single step.

I will argue that syncope and resyllabification must be simultaneous, as in (23b), by showing that the sequential derivation in (23a) is inconsistent with other facts of the language. The problem is that syncope is blocked after a consonant cluster: /ħagar kibiːr/ → ˈ*ħa.gar.ki.*ˈ*biːr* 'big stone'. The standard explanation for why syncope is blocked in this situation is that it would leave the *k* unsyllabified, because there is no way to adjoin it to the preceding or following syllable without creating a forbidden phrase-internal complex coda or complex onset: *ˈ*ħa.gark.*ˈ*biːr*, *ˈ*ħa.gar.*ˈ*kbiːr*. The sequential theory is unable to express this insight, however. In the sequential theory, the derivation reaches a point where the choice is between *ħa.gar.ki.*ˈ*biːr* with no syncope and **ħa.gar.k.*ˈ*biːr* with syncope. To block syncope as required, the latter candidate needs to be ruled out, and a constraint against unsyllabified consonants could do that. But this same constraint would also wrongly block syncope in (23a), where the derivation has to pass through a stage with an unsyllabified consonant.

10 The derivations in (23) do not exhaust the possibilities. For instance, (23a) could be elaborated further, with syllabification proceeding segment-by-segment or syllable-by-syllable, rather than all at once. Or (23b) could be further compressed, by allowing initial syllabification and assignment of stress to occur in a single step. Other contributions to this volume, particularly Moore-Cantwell's, explore other ideas about how syllabification is integrated into derivations.

This particular serial analysis has failed because HS has no ability to look ahead. The decision about whether or not to syncopate has to be made based on the conditions obtaining at that point in the derivation. Thus, in the sequential theory, there is no way of knowing whether it will eventually be possible to attach the unsyllabified consonant to a nearby syllable. The simultaneous theory does not have this problem because it does resyllabification in parallel with syncope. It can therefore use the markedness constraint against unsyllabified consonants to rule out ***ħa.gar.k.ˈbiːr*** without compromising the derivation of *wiħ.ʃa.*

This example illustrates a general method for deciding what GEN can and cannot do in a single step. GEN determines how much and what kind of information is available to EVAL at each step of the derivation. Since there is no look-ahead, all of the information necessary to determine whether the right candidate wins has to be available at the point where it is crucial for that candidate to win. In the case of Cairene, if a particular theory of GEN segregates syncope and syncope-triggered resyllabification into different steps, then information about the ultimate consequences of syncope for syllabification is unavailable at the point in the derivation where the syncope decision must be made. As a result, under this version of GEN it is not possible to capture a very familiar phonological generalization: syncope occurs unless it would leave an unsyllabifiable consonant (cf. Kisseberth 1970). As we saw, that generalization is attainable under a different theory of GEN in which syncope and resyllabification can be simultaneous. It is attainable because the consequences of syncope for syllabification are known at the derivational step where the syncope choice has to be made.

It is important to realize that arguments of this type depend on the details of the theory of CON as well as GEN. For example, if there were a constraint that specifically prohibited an unsyllabified consonant after a closed syllable, then the sequential theory would be viable, because now there would be a constraint that correctly allows *ˈwi.ħ.ʃa* and correctly blocks **ħa.gar.k.ˈbiːr.* On the trade-offs between GEN and CON and how to resolve them, see Section 2.7 of McCarthy, Pater, and Pruitt (this volume).

Although we have been focusing here on understanding a technique for studying GEN rather than drawing conclusions about GEN itself, we should not ignore the broader consequences of the argument presented above. Why is it possible to perform syncope and resyllabification in a single pass through GEN?

In McCarthy (2007a), I propose that GEN is limited to a single unfaithful operation at a time, but there is no limit on faithful operations. Syncope, epenthesis, feature change, and so on are unfaithful operations, so each of them requires a separate derivational step. But (re)syllabification is a faithful operation. It is therefore possible to combine syncope and (re)syllabification into a single derivational step.[11] This hypothesis about GEN can be tested by applying these techniques to other presumptively faithful operations, such as adjunction of an unstressed syllable to a foot, pruning of an empty node, or parsing a lexical word into a prosodic word.

2.7 Arguments for Harmonic Serialism[12]

Many of the arguments for HS over P-OT can be divided into two groups. Some arguments are based on the fact that HS, but not P-OT, has forms that are intermediate in a derivation. Others are based on differences between HS and P-OT in the typologies that they generate for a given constraint set. We will look at arguments of both types.

2.7.1 Evidence for HS from intermediate forms

The existence of intermediate derivational steps is one of the two characteristics of HS that distinguish it from P-OT. In this section, I briefly summarize two arguments for HS that are based on this difference. Several others are noted at the end of the section. These arguments share a common premise: certain generalizations cannot be expressed in underlying or surface representation, but those are the only two levels of representation that P-OT has. These generalizations are expressible in HS's intermediate representations, however.

The first argument comes from McCarthy (2008c). In many languages, some or all unstressed vowels delete. This simple generalization proves to be difficult to express in a P-OT analysis. The problem is that

11 The argument that resyllabification is a faithful operation is based on an observation about syllabification. Although languages differ in how they syllabify (e.g., [qa**b.**la] 'before' in Arabic vs. [ə.**bl**ɑjʤ] *oblige* in English), no known language has a contrast between monomorphemic [qab.la] and [qa.bla] (Blevins 1995: 221, Clements 1986: 318, Hayes 1989: 260, McCarthy 2003: 60–62). It is a basic tenet of OT that lack of contrast means lack of faithfulness.

12 This section originally appeared in McCarthy (2010b).

the generalization is inherently derivational: stress is assigned and then unstressed vowels are deleted. A P-OT grammar must optimize the effects of stress assignment and syncope simultaneously, and this turns out to be inadequate both descriptively and typologically.

In Macushi Carib (Hawkins 1950: 87), for example, words are parsed into iambic feet from left to right, and only then are unstressed vowels deleted:[13]

(24) Stress-syncope interaction in Macushi Carib

Underlying	Stress	Syncope	
piripi	(pirí)(pí)	(prí)(pí)	'spindle'
wanamari	(waná)(marí)	(wná)(mrí)	'mirror'
u-manari-rɨ	(umá)(narí)(rɨ́)	(má)(nrí)(rɨ́)	'my cassava grater'
u-wanamari-rɨ	(uwá)(namá)(rirɨ́)	(wá)(nmá)(rrɨ́)	'my mirror'

In HS, these are exactly the steps that the derivation follows. Because of gradualness, stress assignment and syncope cannot occur simultaneously. Stress assignment occurs first because syncope is *intrinsically ordered* after stress. Two processes are said to be intrinsically ordered if the applicability of one depends on the prior application of the other. In HS, this occurs when the markedness constraint implicated in the second process is not violated until the first process has applied. In the case of syncope and stress, the markedness constraint that is responsible for syncope, *V-PLACE$_{weak}$, is violated by a vowel in the weak syllable of a foot. (In other words, vowel place features are not licensed in this weak position.) Before foot structure has been assigned, all vowels vacuously satisfy this constraint. Therefore, *V-PLACE$_{weak}$ is not active until stress has been assigned, so stress is intrinsically ordered before syncope.

Let us assume that GEN includes operations that create a foot, remove a foot, or delete a vowel. This assumption about gradualness means that the candidate set at step 1, shown in tableau (25), includes candidates with foot parsing or syncope but not both. The imperative to parse into feet is provided by the constraint PARSE-SYLLABLE, which is violated by any unfooted syllable. (Prior or simultaneous syllabification may be assumed; see section 2.6.)

13 Not shown in (24) are lengthening of stressed vowels and main stress on the final foot.

(25) Step 1 of /wanamari/ → *(wná)(mrí)*

	/wanamari/	PARSE-SYLL	*V-PL$_{weak}$	MAX
a. →	(waná)mari	2	1	
b.	wanamari	4 W	L	
c.	wanmari	3 W	L	1 W

Tableau (25) shows how stress assignment is intrinsically ordered before syncope. Syncope prior to foot parsing, as in (25c), improves performance on PARSE-SYLLABLE, but less so than foot parsing in (25a). Foot parsing introduces a violation of *V-PLACE$_{weak}$, as in (25a), because *V-PLACE$_{weak}$ is vacuously satisfied by unfooted vowels. For that reason, PARSE-SYLLABLE must be ranked higher than *V-PLACE$_{weak}$.[14]

At step 2, satisfaction of PARSE-SYLLABLE is still the prime directive, so syncope is once again postponed:

(26) Step 2 of /wanamari/ → *(wná)(mrí)*

	/wanamari/	PARSE-SYLL	*V-PL$_{weak}$	MAX
a. →	(waná)(marí)		2	
b.	(waná)mari	2 W	1 L	
c.	(wná)mari	2 W	L	1 W
d.	(waná)mri	1 W	1 L	1 W

In a quadrisyllable like /wanamari/, full satisfaction of PARSE-SYLLABLE has been achieved by the end of step 2, so it is finally possible to attend to the requirements of *V-PLACE$_{weak}$, which is the next markedness constraint in the ranking. One of the unstressed, footed vowels deletes at step 3, with the other deleting at step 4 (not shown).

(27) Step 3 of /wanamari/ → *(wná)(mrí)*

	(waná)(marí)	PARSE-SYLL	*V-PL$_{weak}$	MAX
a. →	(wná)(marí)		1	1
b.	(waná)(marí)		2 W	L

Which vowel deletes first is unimportant, since ultimately both delete. As it happens, the constraint responsible for left-to-right foot parsing,

14 Other candidates, such as trochaic *(wána)mari* or right-to-left *wana(marí)*, are ruled out by constraints that are standard in the OT literature on stress systems. For textbook treatments, see Kager (1999: 142ff.) or McCarthy (2008a: 183ff.).

ALIGN-L(foot, word), also favors deleting from left to right – hence *(wná)(marí)* rather than *(waná)(mrí)* is shown as the winner in (27).

Finally, the derivation converges at step 5, with input and winner identical to one another. Alternatives to the intended winner, such as those in (28b) and (28c), reintroduce violations of the top-ranked markedness constraints or violate faithfulness constraints gratuitously:

(28) Step 5 of /wanamari/ → *(wná)(mrí)* — Convergence

	(wná)(mrí)	PARSE-SYLL	*V-PL$_{weak}$	MAX
a. →	(wná)(mrí)			
b.	wna(mrí)	1 W		
c.	(wnmrí)			1 W

This analysis shows that HS offers a viable approach to stress-syncope interactions. P-OT does not. The problem is that P-OT lacks the intermediate representation in which stress has been assigned prior to syncope. The P-OT analysis must therefore distinguish the intended winners from losing candidates that never even arise in the HS analysis, and this proves to be impossible. Here are some examples:

(29) Problematic losers in P-OT

Underlying	Intended winner	Problematic losers
piripi	(prí)(pí)	*(pí)(rpí)
wanamari	(wná)(mrí)	*(wá)(nmá)(rí)
u-manari-rɨ	(má)(nrí)(rɨ́)	*(ú)(mná)(rrɨ́)
u-wanamari-rɨ	(wá)(nmá)(rrɨ́)	*(ú)(wná)(mrí)(rɨ́)

The intended winners respect the generalization that syncope affects the odd-numbered non-final syllables – i.e., exactly the syllables that are left unstressed after the left-to-right iambic parse. The problematic losers follow the generalization that syncope affects the *even*-numbered non-final syllables – i.e., exactly the syllables that would be left unstressed by a left-to-right *trochaic* parse. The problem for P-OT is that no markedness constraint evaluating surface forms can systematically distinguish the two patterns of syncope. The reason for this failure is that P-OT has only two levels of representation to work with, underlying and surface, but capturing the generalization about which vowels are targeted for syncope requires an intermediate level, post-stress and pre-syncope.

Another argument for HS's intermediate levels of representation comes from Jesney (2011). Positional faithfulness constraints are like other faithfulness constraints except that their scope of action is limited to certain prominent positions, such as stressed syllables (Beckman 1998). For example, the positional faithfulness constraint IDENT$_{stress}$(nasal) is protective of nasalization contrasts in stressed syllables. When ranked above *V$_{nasal}$, which itself dominates the position-insensitive faithfulness constraint IDENT(nasal), the result is a language like Nancowry (Radhakrishnan 1981), where phonemic vowel nasalization is maintained in stressed syllables but neutralized in unstressed ones. In the following schematized example, stress is assumed to be trochaic, so TROCHEE is undominated:

(30) Attested positional faithfulness effect (P-OT)

	bãdõ	ID$_{str}$(nas)	PARSE-SYLL	TROCH	*V$_{nas}$	ID(nas)
a. →	(bã́do)				1	1
b.	(bádo)	1 W			L	2 W
c.	(bã́dõ)				2 W	L
d.	(badṍ)			1 W	1	1
e.	bado		2 W		L	2 W

Because of *V$_{nasal}$, nasalized vowels are neutralized to oral in unstressed syllables, as in (30a). But there is no neutralization in stressed syllables (cf. (30b)), because of IDENT$_{stress}$(nasal).

P-OT's problem, which was first recognized by Rolf Noyer (cited in Beckman (1998: fn. 37)), is that positional faithfulness constraints work as intended only when the position of greater faithfulness is held constant in those candidates where the positional faithfulness constraint is making a crucial comparison. That is certainly true in (30): the surface reflex of /ã/ is stressed in both (30a) and (30b). Candidates that are stressed differently or not at all, such as (30d) and (30e), are ruled out by other constraints, so they do not depend on IDENT$_{stress}$(nasal) to exclude them.

Now consider what happens when stress is allowed to differ among the viable candidates. In (31), TROCHEE is ranked below *V$_{nasal}$. The result is that stress is shifted from an underlying nasalized vowel onto an underlying oral one. This happens because the positional faithfulness constraint is crucially comparing two candidates, (31a) and (31b), that differ in stress:

(31) Unattested positional faithfulness effect (P-OT)

	/pãko/	ID_{str}(nas)	PARSE-SYLL	$*\text{V}_{nas}$	ID(nas)	TROCH
a. →	(pakó)				1	1
b.	(páko)	1 W			1	L
c.	(pã́ko)			1 W	L	L
d.	pako		2 W		1	L

When this same grammar is presented with any other combination of nasalized and oral vowels (i.e., /bãdõ/, /sato/, or /kafõ/), it defaults to trochaic stress. Thus, in this hypothetical language, stress is normally on the penult, but it is on the ultima when the penult vowel is underlying nasal and the final vowel is underlying oral, though both end up oral at the surface. No real language does anything remotely like this.

What is the source of this problem? Positional faithfulness constraints are sensitive to structure that is assigned by the grammar, such as stress. Since the surface form is the only grammar-derived level of representation in P-OT, P-OT's positional faithfulness constraints have to be defined like this: 'If a segment in the surface representation is in a stressed syllable, it must be faithful to its underlying correspondent'. When positional faithfulness constraints are defined in this way, the problem in (31) is unavoidable.

As Jesney shows, this otherwise intractable problem is solved if HS is adopted and if positional faithfulness constraints are defined to refer to the prosodic structure of the *input*: if a segment in the input to GEN is in a stressed syllable, it must be faithful in [nasal]. In HS, the input to GEN is not necessarily the underlying representation, so it can have structure that has been assigned by the grammar. Since the input is the same for all candidates being compared, problems like (31) cannot arise.

The HS derivation of /pã́ko/ proceeds as follows. At step 1, there is a choice between assigning stress or denasalizing *ã*. If $*\text{V}_{nasal}$ dominates PARSE-SYLLABLE, then denasalization takes precedence, and we have a language without nasalized vowels and without a positional faithfulness effect. If PARSE-SYLLABLE is ranked higher, as in tableau (32), then stress is assigned first. Stress (re)assignment and denasalization cannot cooccur, of course, because of gradualness.

(32) Step 1 from /pãko/

	/pãko/	ID_{str}(nas)	PARSE-SYLL	$*\text{V}_{nas}$	ID(nas)	TROCH
a. →	(pã́ko)			1		
b.	pako		2 W	L	1 W	
c.	(pãkó)			1		1 W

The derivation then converges at step 2, shown in (33). Input *(pã́ko)* has a stressed nasalized vowel. Since this vowel is stressed in the input to this pass through the GEN-EVAL loop, redefined IDENT_{stress}(nasal) protects it from denasalization:

(33) Step 2 from /pãko/

	(pã́ko)	ID_{str}(nas)	PARSE-SYLL	$*\text{V}_{nas}$	ID(nas)	TROCH
a. →	(pã́ko)			1		
b.	(páko)	1 W		L	1 W	
c.	(pãkó)			1		1 W

The failure of the final-stressed candidate *(pãkó)* in (32c) and (33c) is crucial to this argument for HS. If this candidate were to survive, it would change into *(pakó)* at the next step of the derivation, and HS would be making the same bad prediction as P-OT. In fact, it does not survive. The reasoning goes like this:

> The P-OT tableau (31) shows that *(pakó)* is the *global optimum* for underlying /pãko/ under this grammar. Another way of saying the same thing is that *(pakó)* is the *global minimum of potential for harmonic improvement.* The derivation (32)–(33) shows that this global optimum is inaccessible in HS because there is no harmonically improving path to it. The HS derivation gets stuck at a *local optimum, (pã́ko).* Equivalently, *(pã́ko)* is a *local minimum of potential for harmonic improvement.* This happens because HS's GEN cannot simultaneously assign final stress and denasalize the penult, and final stress is not harmonically improving unless the penult is simultaneously denasalized.

The overall lesson here is that HS has no look-ahead capability; a candidate that fails to improve harmony at step *n* cannot win simply because it would lead to greater harmonic improvement at step *n*+1. 'Getting stuck' at a local optimum sounds like a bad outcome, but it can actually be a good one. Positional faithfulness and many other typological results of HS depend on this property of the theory.

The examples of stress-syncope interaction and positional faithfulness show that HS's intermediate levels of representation are necessary to capture some basic generalizations about phonology. Staubs (this volume) shows how HS solves a problem with positional markedness that is similar to the positional faithfulness problem just discussed. Other arguments for HS's intermediate levels are based on discussions of opacity (Elfner this volume), the interaction of stress and syllabification (Elsman this volume), compensatory lengthening (Torres-Tamarit this volume), phonetically grounded constraints (McCarthy 2011), scalar constraints (Pater 2012), and local variation (Kimper 2011).

2.7.2 Evidence for HS from language typology

Language typology is a central concern of research in OT. Because the same constraints can be ranked differently in different languages, any proposed constraint system constitutes an implicit claim about the range of permissible variation among languages. The logic of language typology in HS is explained in McCarthy (2007b, 2009) and summarized here.

For identical constraint systems, P-OT and HS may predict different typologies. The source of the difference is HS's core properties, gradualness and harmonic improvement. Does a given constraint system Con yield a language in which underlying /A/ maps to surface *B*? In P-OT, the answer is yes if and only if there is some ranking of Con where *B* is more harmonic than *A* and every other candidate derived from /A/. In HS, this answer is sufficient only if *B* and /A/ differ by a single change. If it requires more than one change to get from /A/ to *B*, then there must also be a harmonically improving path of winning intermediate steps from /A/ to *B*. Sometimes, there is no such path. That is when P-OT and HS make different typological predictions.

This reasoning is important in HS's solution to some *too many repairs* (TMR) problems. A TMR problem is the observation that the actually attested ways of satisfying a markedness constraint are often more limited than we would expect from ranking permutation (Blumenfeld 2006, Lombardi 1995/2001, Pater 1999, Steriade 2001/2008, Wilson 2001, and others). For example, the markedness constraint Coda-Cond says that coda consonants do not license place of articulation (Goldsmith 1990: 123–128, Ito 1989). One way of satisfying this constraint is for a coda to share place with a following onset, since onset position does license place. This is the reason why place often assimilates in consonant clusters: in *pamta*, labial place is unlicensed in coda *m*, but in *pan͡ta* the *n*'s coronal place is licensed because it is shared with the onset (as indicated by the ligature). Unexplained is why place always assimilates from the onset to the coda and never the other way around: /pamta/ → *pan͡ta*, never *pam͡pa*. Tableau (34) illustrates the problem:

(34) /pamta/ → *pan͡ta*/*pam͡pa* in P-OT

	/pamta/	Coda-Cond	Id(place)
a. →	pan͡ta		1
b. →	pam͡pa		1
c.	pamta	1 W	L

P-OT predicts intra- or interlinguistic variation in direction of assimilation when violations of Coda-Cond are repaired, but the predicted variation is not usually observed.

HS offers an explanation for this asymmetry, once the process of place assimilation is properly understood in operational terms (McCarthy 2008b). Long before HS or even OT, it was claimed that place assimilation is a two-step process, with deletion of the unlicensed place feature prior to spreading of the licensed one (Cho 1990, Kiparsky 1993, Mascaró 1987, Poser 1982): /pamta/ → *paNta* → *pan͡ta*. (N denotes a placeless nasal.) If HS's Gen is restricted in this fashion, then the directional asymmetry in place assimilation follows automatically. At step 1, deletion of place from the coda consonant satisfies Coda-Cond, but deletion of place from the onset (yielding a placeless *ʔ*) does not:

(35) Step 1 of /pamta/ → *pan͡ta*

	/pamta/	Coda-Cond	Have-Pl	Id(place)
a. →	paNta		1	1
b.	pamta	1 W	L	L
c.	pamʔa	1 W	1	1

At step 2, placeless N becomes *n* by spreading place from the following *t*. This occurs to satisfy Have-Place, which N violates:

(36) Step 2 of /pamta/ → *pan͡ta*

	paNta	Coda-Cond	Have-Pl	Id(place)
a. →	pan͡ta			1
b.	paNta		1 W	L

The derivation then converges at step 3 (not shown).

This example illustrates the point about harmonic improvement that was made earlier in this section. In P-OT, *pan͡ta* and *pam͡pa* are both possible surface results from underlying /pamta/, since both satisfy CODA-COND and violate IDENT(place) equally. In HS, though, it is not enough for a surface form to be a P-OT winner; it must also be linked with the underlying form by a chain of harmonically improving intermediate forms. That is not the case with *pam͡pa*; under the stated assumption about GEN, it requires an intermediate form, *pamʔa*, that does not improve harmony relative to CODA-COND, as (35) shows. Because of gradualness and harmonic improvement, HS yields a more restrictive typology of place assimilation than P-OT does, all else being equal. This more restrictive typology better fits what we actually find in languages.

Another area where HS appears to have a typological advantage over P-OT is in accounting for locality effects. For example, Pruitt (2010) shows that metrical foot parsing exhibits locality effects that are hard to account for in P-OT but follow readily in HS from the assumption that GEN builds feet one at a time.

One such locality effect involves the interaction between foot parsing and vowel shortening. In quantity-sensitive languages, trochaic feet are usually limited to a pair of light syllables (ĹL) or a single heavy syllable (H́) (Hayes 1985, 1995, McCarthy and Prince 1986/1996, Prince 1990). (H́L) trochees are disfavored by a constraint dubbed RHHRM (for rhythmic harmony) by Prince and Smolensky (1993/2004: 70–71). In the following discussion, I will assume a language with left-to-right quantity-sensitive trochees.

The standard ranking for left-to-right foot parsing uses the constraint ALIGN-L(foot, word) (McCarthy and Prince 1993) to favor having all feet as far to the left as possible:

(37) Left-to-right parsing (P-OT)

	salamataka	FOOT-BIN	PARSE-SYLL	AL-L (ft, wd)	AL-R (ft, wd)
a. →	(sála)(máta)ka		1	2	4
b.	salamataka		5 W	L	L
c.	sa(láma)(táka)		1	4 W	2 L
d.	(sála)(máta)(ká)	1 W	L	6 W	4

If we include RHHRM at the top of the hierarchy and allow shortening of long vowels by ranking MAX-μ low, we get a language in which a long vowel in the first syllable is shortened only if it is followed by an odd number of light syllables. Compare (38) with (39):

(38) Shortening before odd L sequence (P-OT)

	pa:takasa	Foot-Bin	RhHrm	Parse-Syll	Al-L (ft, wd)	Max-μ
a. →	(páta)(kása)				2	1
b.	(pá:)(táka)sa			1 W	1 L	L
c.	(pá:ta)(kása)		1 W		2	L

(39) No shortening before even L sequence (P-OT)

	pa:takasafa	Foot-Bin	RhHrm	Parse-Syll	Al-L (ft, wd)	Max-μ
a. →	(pá:)(táka)(sáfa)				4	
b.	(páta)(kása)fa			1 W	2 L	1 W
c.	(pá:ta)(kása)fa		1 W	1 W	2 L	

No known language behaves in this highly non-local fashion, where avoiding an unfooted syllable at one end of the word — i.e., final *sa* in *(pá:)(táka)sa* (38b) — triggers vowel shortening at the other end of the word. Optimization of metrical structure can cause vowels to shorten, as it does in Latin (Allen 1973, Mester 1994, Prince and Smolensky 1993/2004, McCarthy, Pater, and Pruitt this volume) or Fijian (Dixon 1988, Hayes 1995, Schütz 1985), but these effects are always strictly local, involving a pair of adjacent syllables.

This example reflects a more general problem with P-OT, highlighted in McCarthy (2007b, 2008c, 2010a) and Pruitt (2010) and already encountered in section 2.7.1: it has excessive power to do global optimization. The reason why (37)–(39) is predicted to be a possible language in P-OT is that P-OT's Gen builds complete and final surface candidates in which the effects of vowel shortening and full metrical parsing are evaluated together. Thus, Eval gets to choose the best combination of shortening and parsing, no matter how distant the long vowel might be from the parsing problem.

In HS, if Gen is limited to building one foot at a time, then the language in (37)–(39) cannot be obtained with these constraints, as Pruitt (2010) demonstrates. We will first consider how iterative parsing works in HS with a word that contains no heavy syllables. At step 1 (tableau (40)), the best option is to build a disyllabic foot at the left edge of the word. Building no foot or a monosyllabic foot is disfavored by PARSE-SYLLABLE; the latter also violates FOOT-BINARITY. Building a foot non-initially violates ALIGN-L(foot, word):

(40) Step 1 of iterative parse

	/salamataka/	FOOT-BIN	PARSE-SYLL	AL-L (ft, wd)	AL-R (ft, wd)
a. →	(sála)mataka		3		3
b.	salamataka		5 W		L
c.	sa(láma)taka		3	1 W	2 L
d.	salama(táka)		3	3 W	L

At step 2 (tableau (41)), the best option is to build a disyllabic foot as far to the left as possible. PARSE-SYLLABLE requires construction of an additional foot, and ALIGN-L determines where it is built. After this, the derivation converges, as tableau (42) shows. The only remaining unfooted syllable is the last one, and FOOT-BINARITY ensures that nothing can be done about it.

(41) Step 2 of iterative parse

	(sála)mataka	FOOT-BIN	PARSE-SYLL	AL-L (ft, wd)	AL-R (ft, wd)
a. →	(sála)(máta)ka		1	2	4
b.	(sála)mataka		3 W	L	3 L
c.	(sála)ma(táka)		1	3 W	3 L

(42) Step 3 of iterative parse (convergence)

	(sála)(máta)ka	FOOT-BIN	PARSE-SYLL	AL-L (ft, wd)	AL-R (ft, wd)
a. →	(sála)(máta)ka		1	2	4
b.	(sála)(máta)(ká)	1 W	L	6 W	4

What happens when the underlying representation contains an initial long vowel? If we apply the same ranking as (38) and (39), but within the HS architecture, the result does not depend on whether the long vowel is followed by an odd number (43) or even number (44) of light syllables: the leftmost pair of light syllables is parsed into a foot, since this option best satisfies RHHRM and PARSE-SYLLABLE:

(43) Step 1 from /paːtakasa/ (odd sequence of Ls)

	/paːtakasa/	FOOT-BIN	RHHRM	PARSE-SYLL	AL-L (ft, wd)	MAX-μ
a. →	paː(táka)sa			2	1	
b.	paːtakasa			4 W	L	
c.	(páː)takasa			3 W	L	
d.	(páːta)kasa		1 W	2	L	
e.	patakasa			4 W	L	1 W

(44) Step 1 from /paːtakasafa/ (even sequence of Ls)

	/paːtakasafa/	FOOT-BIN	RHHRM	PARSE-SYLL	AL-L (ft, wd)	MAX-μ
a. →	paː(táka)safa			3	1	
b.	paːtakasafa			5 W	L	
c.	(páː)takasafa			4 W	L	
d.	(páːta)kasafa		1 W	3	L	
e.	patakasafa			5 W	L	1 W

These derivations continue, parsing pairs of light syllables from left to right, and then returning to parse the initial heavy syllable into a foot of its own. The derivations converge on *(páː)(táka)sa* and *(páː)(táka)(sáfa)*. With this ranking, there is no shortening, and there is certainly no dependency of shortening on whether an odd or even number of light syllables follow.

There will be shortening if the ranking of RHHRM and ALIGN-L(foot, word) is reversed, but still there is no dependency of shortening on the following syllables. At step 1, the first two syllables are parsed into a RHHRM-violating foot: *(páːta)kasa, (páːta)kasafa*. Foot parsing continues at step 2, yielding *(páːta)(kása)* and *(páːta)(kása)fa*. At this point, PARSE-SYLLABLE is as well satisfied as it can get, and RHHRM gets its chance to compel shortening of the long vowel (see McCarthy, Pater, and Pruitt this volume), but both the odd and even length words are affected.

This is another clear point of difference between P-OT and HS. Because P-OT optimizes globally, it allows long-distance dependencies like the one in (37)–(39). HS does not permit this dependency, at least with the standard constraints and the unremarkable version of GEN that we have been assuming. HS is more limited in this respect because decisions about foot parsing and shortening are made one at a time.

This example hints at an important connection between locality and serialism. Long-distance effects are often produced by iterative application of a process; the construction of metrical feet is one example, and successive cyclic *wh*-movement (Chomsky 1977) is another. In P-OT, process iteration is invisible to EVAL because it takes place entirely in GEN. EVAL sees only the final result. In HS, however, process iteration is visible to EVAL because the results of each iteration are presented in the candidate set. This enforces a kind of locality because each iteration must improve harmony or the process terminates (McCarthy 2007b, 2008c, 2010a, Pruitt 2010). The claim implicit in HS (and in successive cyclic *wh*-movement) is that visible iteration and its concomitant locality effects are a better theory of language typology than the global alternative.

Although the logic of typological research in P-OT and HS is clear, the practice can be difficult. When CON, GEN, or the input set are even moderately complex, accurate determination of typologies by hand becomes impossible. For this reason, computational tools for conducting typological research have been developed. It is instructive to compare the procedure for computing typologies in the two main systems in current use, OTSoft for P-OT (Hayes, Tesar and Zuraw 2003) and OT-Help for HS (Staubs *et al.* 2010).

In OTSoft, a language is a set of underlying forms and the surface forms that they map to. A typology of a constraint system is a set of languages and the rankings that produce each of them. For each underlying form, the user supplies a tableau with various surface candidates and the violation marks they receive from each constraint. OTSoft then uses Recursive Constraint Demotion (RCD) (Tesar and Smolensky 1998) to determine the possible optima for each underlying form. This is done by iterating through the surface candidates from each underlying form, setting each candidate as optimum and then using RCD to find a ranking. If RCD fails to find a ranking, then that candidate is not a possible optimum; if it succeeds, then that candidate is a possible optimum. The typology is computed by trying all combinations of possible optima, one for each underlying form. If RCD succeeds in finding a ranking, then this combination — this language – is in the typology. Otherwise it is not.

In HS, and consequently in OT-Help, a language is not just a combination of underlying forms and the surface forms that they map to; it also includes the derivations that link the underlying and surface forms. Because different rankings produce different derivations, and because two different derivations can sometimes produce the same underlying→surface mapping, derivations as well as mappings must be included in the typology. OT-Help must therefore iterate the procedure used in OTSoft, finding the possible optima at step 1, then finding the possible optima at step 2 by setting each of the step-1 optima as an input, and so on. A further point of difference is that OT-Help provides users with a language for defining operations in GEN and constraints in CON. This allows OT-Help to compute its own candidates (relying on the finiteness of the candidate set in HS) and to determine their constraint violations. Users can quickly and easily test hypotheses about GEN and CON, to address questions like those discussed above.

The study of language typology in HS is still at a relatively early stage. Besides the papers already mentioned, there has been work on typology in relation to autosegmental spreading (McCarthy 2007b, 2010a), apocope and metathesis (McCarthy 2007b), lexical structure (McCarthy and Pruitt 2012), and reduplication (McCarthy, Kimper and Mullin 2012). In this volume, the chapters by Moore-Cantwell, Staubs, and Kimper also address typological issues from a HS perspective.

2.8 Opacity in Harmonic Serialism

Certain types of phonological opacity are amenable to analysis in HS – see Elfner (this volume) and Torres-Tamarit (this volume), for example. HS is not a general theory of phonological opacity, however. I will illustrate this point with a pair of examples.

The derivation in (45) exemplifies the type of opacity known as *counterbleeding*.

(45) Counterbleeding order in Bedouin Arabic

Underlying	/ħaːkim-iːn/	
Palatalization (k→kʲ/__i)	ħaːkʲimiːn	
Syncope (i→ Ø/VC__CV)	ħaːkʲmiːn	
Surface	ħaːkʲmiːn	'ruling (masculine plural)'

This example is problematic in P-OT because deleting /i/ will satisfy two markedness constraints at once: the one that favors palatalizing *k* before *i*, and the one that favors deleting short high vowels in open syllables. Thus, the winning candidate should be *ħaːkmiːn*, with unpalatalized *k*. The actual winner, *ħaːkʲmiːn*, seems needlessly unfaithful because it has palatalized *k* even though the *i* that triggers palatalization is absent from the surface form.

HS fares no better on this example. If *ħa:kmi:n* is a candidate at step 1, then it will win for the same reason that it wins in P-OT. The derivation will then converge on it at step 2. At step 1, *ħa:kʲimi:n* loses to *ħa:kmi:n* because it violates the constraint against short high vowels in open syllables and both candidates do equally well on the constraint against plain *k* before *i*. And at step 2 *ħa:kʲmi:n* loses to *ħa:kmi:n* because it is unfaithful.

Counterfeeding opacity is also problematic in HS. Recall from (17) that Cairene Arabic shortens long vowels in unstressed syllables. Cairene also has a syncope process that deletes short high vowels in a VC___CV context, but short vowels derived from long ones do not delete:

(46) Cairene chain shift

a. Short high vowels delete in VC___CV

/fihim-u:/ fíhmu 'they understood' (cf. *fíhim* 'he understood')

b. But not if they are derived from long vowels

/ji-ʃi:l-u:-na/ jiʃilú:na 'they ask us' (cf. *jiʃí:l* 'he asks')

*jiʃlú:na

This combination of processes is called a *chain shift* because the output of one process is identical with the input to another: A → B and B → C in identical or overlapping contexts. This traditional name is somewhat misleading because the point is that the processes do not in fact chain together – underlying A does not map to surface C.

In P-OT, deletion of long vowels can be prevented by the constraint MAX(V:) (Gouskova 2003). This constraint is defined so that it checks a vowel's length in the input, not the output. Therefore, it protects underlying long vowels from deletion even when they have been shortened.

This move does not carry over to HS, however. The problem is that an underlying long vowel, once it has been shortened in the course of the derivation, is indistinguishable from an underlying short vowel. There is no obvious way of ruling out a derivation where a vowel first shortens (*jiʃi:lú:na* → *jiʃilú:na*) and then deletes (*jiʃilú:na* → **jiʃlú:na*). As far as MAX(V:) is concerned, the short *i* from /i:/ is indistinguishable from the short *i* in the middle syllable of /fihim-u/. Since /fihim-u/ → *fíhmu*, we expect /jiʃi:lu:na/ → **jiʃlú:na*.

One solution would be to take a different view of faithfulness constraints in HS. If MAX(V:) always looked back at the underlying representation instead of the input to the current derivational step, then it would see /ji-ʃi:l-u:-na/'s long /i:/ in the second syllable. The problem with this move is that is inconsistent with the improved theory of positional faithfulness in HS (see section 2.7.1).

Another approach, equally applicable in P-OT, is to reexamine purported chain shifts with an eye toward determining whether the B that is the output of the A → B mapping is truly identical with the B that is the input to the B → C mapping. If they are merely similar and not identical, a HS analysis may be possible. In Cairene, for example, shortened *i* is tense, like long *i:*, but *i* not derived from a long vowel is lax (Mitchell 1956: 10, 112). Also see Gouskova and Hall (2009) on possible phonetic differences between underlying and epenthetic vowels.

The most radical solution to the opacity problem is to adopt something like OT-CC (McCarthy 2007a, Wolf 2008, 2011). OT-CC is specifically a theory of opacity, based on evaluating derivations. It uses something like HS as its Gen, and it compares derivations using constraints on the order of operations. For further information, see Wolf (this volume).

2.9 Conclusion

As I noted earlier, OT is hard, and HS makes it even harder. In parallel OT, typology follows from hypotheses about the constraint set. In HS, typology follows from a combination of hypotheses about GEN and the constraint set. The results so far, in this volume and elsewhere, suggest that HS is worth this extra effort.

References

Alderete, John (1998) *Morphologically-Governed Accent in Optimality Theory*. Doctoral dissertation, University of Massachusetts Amherst, Amherst, MA. Available on Rutgers Optimality Archive, ROA-309. Published (2001) by Routledge.

Alderete, John (2001) Dominance effects as transderivational anti-faithfulness. *Phonology* 18: 201–253.

Allen, W. S. (1973) *Accent and Rhythm*. Cambridge: Cambridge University Press.

Archangeli, Diana and Douglas Pulleyblank (1994) *Grounded Phonology*. Cambridge, MA: MIT Press.

Beckman, Jill (1998) *Positional Faithfulness*. Doctoral dissertation, University of Massachusetts Amherst, Amherst, MA. Available on Rutgers Optimality Archive, ROA-234.

Black, H. Andrew (1993) *Constraint-Ranked Derivation: A Serial Approach to Optimization*. Doctoral dissertation, University of California, Santa Cruz. Available at http://www.sil.org/silepubs/abstract.asp?id=47751.

Blevins, Juliette (1995) The syllable in phonological theory. In John A. Goldsmith (ed.), *The Handbook of Phonological Theory*, 206–244. Cambridge, MA, and Oxford, UK: Blackwell.

Blumenfeld, Lev (2006) *Constraints on Phonological Interactions*. Doctoral dissertation, Stanford University, Stanford, CA. Available on Rutgers Optimality Archive, ROA-877.

Chen, Matthew (1999) Directionality constraints on derivation? In Ben Hermans and Marc van Oostendorp (eds), *The Derivational Residue in Phonological Optimality Theory*, 105–127. Amsterdam: John Benjamins. Available on Rutgers Optimality Archive, ROA-453.

Cho, Young-Mee Yu (1990) *Parameters of Consonantal Assimilation*. Doctoral dissertation, Stanford University.

Chomsky, Noam (1977) On *Wh*-movement. In Peter Culicover, Thomas Wasow and Adrian Akmajian (eds), *Formal Syntax*, 77–132. New York: Academic Press.

Chomsky, Noam and Morris Halle (1968) *The Sound Pattern of English*. New York: Harper & Row.

Clements, G. N. (1986) Syllabification and epenthesis in the Barra dialect of Gaelic. In Koen Bogers, Harry van der Hulst and Maarten Mous (eds), *The Phonological Representation of Suprasegmentals*, 317–336. Dordrecht: Foris.

Dixon, Robert M. W. (1988) *A Grammar of Boumaa Fijian*. Chicago: University of Chicago Press.

Goldsmith, John (1990) *Autosegmental and Metrical Phonology*. Oxford and Cambridge, MA: Blackwell.

Goldsmith, John (1993) Harmonic phonology. In John Goldsmith (ed.), *The Last Phonological Rule: Reflections on Constraints and Derivations*, 21–60. Chicago: University of Chicago Press.

Gouskova, Maria (2003) *Deriving Economy: Syncope in Optimality Theory*. Doctoral dissertation, University of Massachusetts Amherst, Amherst, MA. Available on Rutgers Optimality Archive, ROA-610.

Gouskova, Maria and Nancy Hall (2009) Acoustics of unstressable vowels in Lebanese Arabic. In Steve Parker (ed.), *Phonological Argumentation: Essays on Evidence and Motivation*. London: Equinox Publishing.

Hawkins, W. N. (1950) Patterns of vowel loss in Macushi (Carib). *International Journal of American Linguistics* 16: 87–90.

Hayes, Bruce (1985) Iambic and trochaic rhythm in stress rules. In M. Niepokuj, M. VanClay, V. Nikiforidou and D. Jeder (eds), *Proceedings of BLS 11: Parasession on Poetics, Metrics, and Prosody*, 429–446. Berkeley: Berkeley Linguistic Society.

Hayes, Bruce (1989) Compensatory lengthening in moraic phonology. *Linguistic Inquiry* 20: 253–306.

Hayes, Bruce (1995) *Metrical Stress Theory: Principles and Case Studies*. Chicago: The University of Chicago Press.

Hayes, Bruce, Bruce Tesar and Kie Zuraw. (2003) OTSoft 2.1. UCLA and Rutgers University, Los Angeles and New Brunswick,NJ. Software package available at http://www.linguistics.ucla.edu/people/hayes/otsoft/.

Ito, Junko (1989) A prosodic theory of epenthesis. *Natural Language & Linguistic Theory* 7: 217–259.

Jesney, Karen (2011) Positional faithfulness, non-locality, and the Harmonic Serialism solution. In Suzi Lima, Kevin Mullin and Brian Smith (eds), *Proceedings of the 39th Annual Meeting of the North East Linguistic Society*, 239–440. Amherst, MA: GLSA Publications. Available on Rutgers Optimality Archive, ROA-1018.

Kager, René (1999) *Optimality Theory*. Cambridge: Cambridge University Press.

Kenstowicz, Michael (1994) Syllabification in Chukchee: A constraints-based analysis. In Alice Davison, Nicole Maier, Glaucia Silva and Wan Su Yan (eds), *Proceedings of the Formal Linguistics Society of Mid-America 4*, 160–181. Iowa City: Department of Linguistics, University of Iowa. Available on Rutgers Optimality Archive, ROA-30.

Kenstowicz, Michael and Charles Kisseberth (1979) *Generative Phonology: Description and Theory*. New York: Academic Press.

Kimper, Wendell (2011) Locality and globality in phonological variation. *Natural Language & Linguistic Theory* 29: 423–465.

Kiparsky, Paul (1993) Blocking in non-derived environments. In Sharon Hargus and Ellen Kaisse (eds), *Studies in Lexical Phonology*, 277–313. San Diego: Academic Press.

Kiparsky, Paul (2000) Opacity and cyclicity. *The Linguistic Review* 17: 351–367.

Kirchner, Robert (1996) Synchronic chain shifts in Optimality Theory. *Linguistic Inquiry* 27: 341–350. Available on Rutgers Optimality Archive, ROA-66.

Kisseberth, Charles (1970) On the functional unity of phonological rules. *Linguistic Inquiry* 1: 291–306.

Lombardi, Linda (1995/2001) Why Place and Voice are different: Constraint-specific alternations in Optimality Theory. In Linda Lombardi (ed.), *Segmental Phonology in Optimality Theory: Constraints and Representations*, 13–45. Cambridge: Cambridge University Press. Originally circulated in 1995. Available on Rutgers Optimality Archive, ROA-105.

Mascaró, Joan. (1987) A reduction and spreading theory of voicing and other sound effects. Unpublished paper, Universitat Autònoma de Barcelona, Barcelona.

McCarthy, John J. (2000) Harmonic serialism and parallelism. In Masako Hirotani (ed.), *Proceedings of the North East Linguistics Society* 30, 501–524. Amherst, MA: GLSA Publications. Available on Rutgers Optimality Archive, ROA-357.

McCarthy, John J. (2002) *A Thematic Guide to Optimality Theory*. Cambridge: Cambridge University Press.

McCarthy, John J. (2003) Sympathy, cumulativity, and the Duke-of-York gambit. In Caroline Féry and Ruben van de Vijver (eds), *The Syllable in Optimality Theory*, 23–76. Cambridge: Cambridge University Press.

McCarthy, John J. (2007a) *Hidden Generalizations: Phonological Opacity in Optimality Theory*. London: Equinox Publishing.

McCarthy, John J. (2007b) Restraint of analysis. In Sylvia Blaho, Patrik Bye and Martin Krämer (eds), *Freedom of Analysis*, 203–231. Berlin and New York: Mouton de Gruyter. Also appears in Eric Bakovic, Junko Ito, and John J. McCarthy, eds (2006) *Wondering at the Natural Fecundity of Things: Essays in Honor of Alan Prince*. Linguistics Research Center, UC Santa Cruz. Available at http://repositories.cdlib.org/lrc/prince/10.

McCarthy, John J. (2008a) *Doing Optimality Theory*. Malden, MA, and Oxford, UK: Blackwell.

McCarthy, John J. (2008b) The gradual path to cluster simplification. *Phonology* 25: 271–319. doi:10.1017/S0952675708001486.

McCarthy, John J. (2008c) The serial interaction of stress and syncope. *Natural Language & Linguistic Theory* 26: 499–546. DOI: 10.1007/s11049-008-9051-3.

McCarthy, John J. (2009) Studying Gen. *Journal of the Phonetic Society of Japan* 13: 3–12.

McCarthy, John J. (2010a) Autosegmental spreading in Optimality Theory. In John Goldsmith, Elizabeth Hume and Leo Wetzels (eds), *Tones and Features*, 195–222. Berlin and New York: Mouton de Gruyter. Available at: http://works.bepress.com/john_j_mccarthy/100.

McCarthy, John J. (2010b) An introduction to Harmonic Serialism. *Language and Linguistics Compass* 4: 1001–1018.

McCarthy, John J. (2011) Perceptually grounded faithfulness in Harmonic Serialism. *Linguistic Inquiry* 42. Available at http://works.bepress.com/john_j_mccarthy/104/.

McCarthy, John J., Wendell Kimper and Kevin Mullin (2012) Reduplication in Harmonic Serialism. *Morphology* 22: 173–232.

McCarthy, John J. and Alan Prince (1986/1996) Prosodic Morphology 1986. New Brunswick, NJ: Rutgers University Center for Cognitive Science. Available at http://ruccs.rutgers.edu/pub/papers/pm86all.pdf.

McCarthy, John J. and Alan Prince (1993) Generalized Alignment. In Geert Booij and Jaap van Marle (eds), *Yearbook of Morphology*, 79–153. Dordrecht: Kluwer. Available on Rutgers Optimality Archive, ROA-7.

McCarthy, John J. and Alan Prince (1995) Faithfulness and reduplicative identity. In Jill Beckman, Laura Walsh Dickey and Suzanne Urbanczyk (eds), *University of Massachusetts Occasional Papers in Linguistics 18*, 249–384. Amherst, MA: GLSA Publications. Available on Rutgers Optimality Archive, ROA-103.

McCarthy, John J. and Alan Prince (1999) Faithfulness and identity in Prosodic Morphology. In René Kager, Harry van der Hulst and Wim Zonneveld (eds), *The Prosody-Morphology Interface*, 218–309. Cambridge: Cambridge University Press.

McCarthy, John J. and Kathryn Pruitt (2012) Sources of phonological structure. In Ralf Vogel and Hans Broekhuis (eds), *Linguistic Derivations and Filtering: Minimalism and Optimality Theory*. London: Equinox Publishing.

Mester, Armin (1994) The quantitative trochee in Latin. *Natural Language & Linguistic Theory* 12: 1–61.

Mitchell, T. F. (1956) *An Introduction to Egyptian Colloquial Arabic*. Oxford: Oxford University Press.

Moreton, Elliott. (2000) Faithfulness and potential. Unpublished paper, University of Massachusetts Amherst, Amherst, MA.

Moreton, Elliott (2003) Non-computable functions in Optimality Theory. In John J. McCarthy (ed.), *Optimality Theory in Phonology: A Reader*, 141–163. Malden, MA, and Oxford, UK: Blackwell. Available on Rutgers Optimality Archive, ROA-364.

Moreton, Elliott and Paul Smolensky (2002) Typological consequences of local constraint conjunction. In Line Mikkelsen and Christopher Potts (eds), *Proceedings of the 21st West Coast Conference on Formal Linguistics*, 306–319. Cambridge, MA: Cascadilla Press. Available on Rutgers Optimality Archive, ROA-525.

Mullin, Kevin, Brian W. Smith, Joe Pater and John J. McCarthy. (2010) OT-Help 2.0 User Guide. Unpublished paper, University of Massachusetts Amherst, Amherst, MA. http://people.umass.edu/kmullin/OTHelp2man.pdf.

Newman, Stanley (1944) *Yokuts Language of California*. New York: Viking Fund.

Norton, Russell J. (2003) *Derivational Phonology and Optimality Phonology: Formal Comparison and Synthesis*. Doctoral dissertation, University of Essex, Colchester, Essex. Available on Rutgers Optimality Archive, ROA-613.

Paradis, Carole (1988a) On constraints and repair strategies. *The Linguistic Review* 6: 71–97.

Paradis, Carole (1988b) Towards a theory of constraint violations. *McGill Working Papers in Linguistics* 5: 1–43.

Pater, Joe (1999) Austronesian nasal substitution and other NC effects. In René Kager, Harry van der Hulst and Wim Zonneveld (eds), *The Prosody-Morphology Interface*, 310–343. Cambridge: Cambridge University Press. Available on Rutgers Optimality Archive, ROA-160. Reprinted in *Optimality Theory in Phonology: A Reader*, (ed.) John J. McCarthy, Malden, MA and Oxford, Blackwell (2004).

Pater, Joe (2012) Serial Harmonic Grammar and Berber syllabification. In Toni Borowsky, Shigeto Kawahara, Takahito Shinya and Mariko Sugahara (eds), *Prosody Matters: Essays in Honor of Lisa Selkirk*. London: Equinox Publishing.

Poser, William J. (1982) Phonological representations and action-at-a-distance. In Harry van der Hulst and Norval Smith (eds), *The Structure of Phonological Representations*, 121–158. Dordrecht: Foris.

Prince, Alan (1983) Relating to the grid. *Linguistic Inquiry* 14: 19–100.

Prince, Alan (1990) Quantitative consequences of rhythmic organization. In M. Ziolkowski, M. Noske and K. Deaton (eds), *Parasession on the Syllable in Phonetics and Phonology*, 355–398. Chicago: Chicago Linguistic Society.

Prince, Alan (2002) Arguing optimality. In Angela Carpenter, Andries Coetzee and Paul de Lacy (eds), *University of Massachusetts Occasional Papers in Linguistics 26: Papers in Optimality Theory II*, 269–304. Amherst, MA: GLSA. Available on Rutgers Optimality Archive, ROA-562.

Prince, Alan and Paul Smolensky (1993/2004) *Optimality Theory: Constraint Interaction in Generative Grammar*. Malden, MA, and Oxford, UK: Blackwell.

Pruitt, Kathryn (2010) Serialism and locality in constraint-based metrical parsing. *Phonology* 27: 481–526.

Pruitt, Kathryn (2012) *Stress in Harmonic Serialism*. Doctoral dissertation, University of Massachusetts, Amherst, Amherst, MA.

Radhakrishnan, R. (1981) *The Nancowry Word: Phonology, Affixal Morphology and Roots of a Nicobarese Language*. Edmonton: Linguistic Research.

Rubach, Jerzy (2000) Glide and glottal stop insertion in Slavic languages: A DOT analysis. *Linguistic Inquiry* 31: 271–317.

Schütz, Albert J. (1985) *The Fijian Language*. Honolulu: University of Hawaii Press.

Staubs, Robert, Michael Becker, Christopher Potts, Patrick Pratt, John J. McCarthy and Joe Pater (2010) OT-Help 2.0 [computer program]. Computer program. Available at http://web.linguist.umass.edu/~OTHelp/.

Steriade, Donca (2001/2008) The phonology of perceptibility effects: The P-map and its consequences for constraint organization. In Kristin Hanson and Sharon

Inkelas (eds), *The Nature of the Word: Studies in Honor of Paul Kiparsky*, 151–180. Cambridge, MA: MIT Press. Originally circulated in 2001.

Tesar, Bruce and Paul Smolensky (1998) Learnability in Optimality Theory. *Linguistic Inquiry* 29: 229–268. Available on Rutgers Optimality Archive, ROA-155. Reprinted in *Optimality Theory in Phonology: A Reader*, (ed.) John J. McCarthy, Malden, MA and Oxford, Blackwell (2004).

Watson, Janet C. E. (2002) *The Phonology and Morphology of Arabic*. Oxford: Oxford University Press.

Wilson, Colin (2001) Consonant cluster neutralization and targeted constraints. *Phonology* 18: 147–197.

Wolf, Matthew (2008) *Optimal Interleaving: Serial Phonology-Morphology Interaction in a Constraint-Based Model*. Doctoral dissertation, University of Massachusetts Amherst, Amherst, MA. Available on Rutgers Optimality Archive, ROA-996.

Wolf, Matthew (2011) Limits on global rules in Optimality Theory with Candidate Chains. *Phonology* 28: 87–128.

Chapter 3

Cross-level interactions in Harmonic Serialism[1]

John J. McCarthy, Joe Pater, and Kathryn Pruitt
University of Massachusetts Amherst

3.1 Introduction

Phonological generalizations must often be stated in ways that refer to more than one level of the prosodic hierarchy. For example, pre-Classical Latin words of the form CVCV: shorten the final long vowel to make a better trochaic foot: /amo:/ → *(ˈamo)* 'I love'. This generalization refers to two levels of the prosodic hierarchy, foot and mora. We will refer to generalizations like this as *cross-level interactions* (CLIs).

In Kager (1999: 188–189), McCarthy (2002: 146–149), Mester (1994: 15), Pater (2000: 248–250), and Prince and Smolensky (1993/2004: 33–36, 148–149), CLIs are presented as crucial evidence for parallelism and against serialism in Optimality Theory (OT). The details of these arguments for parallelism will be discussed later, but for now it is sufficient to take note of what they have in common: they all involve CLIs that seem to require changing structure at two different levels of the prosodic hierarchy simultaneously. For instance, getting from /amo:/ to *(ˈamo)* in Latin seems to require building a foot and shortening a

1 This research was supported by grant BCS-0813829 from the National Science Foundation to the University of Massachusetts Amherst. We are indebted to all of the participants in our weekly grant group for their advice about this chapter, with particular thanks going to Matt Wolf for reviewing the manuscript.

vowel in parallel. As Kager (1999: 189) puts it, 'Parallelism establishes complete communication between all levels of the prosodic hierarchy. Consequently, changes at lower prosodic levels have effects at higher levels, and vice versa.' A serial analysis of the Latin facts has to shorten the vowel and build the foot in separate steps, but shortening before foot-building is unmotivated, and foot-building before shortening requires an intermediate stage *(ˈamoː)* with an ill-formed foot.

In this chapter, we review these arguments against serialism and show that they are not in fact arguments against at least one serial theory, Harmonic Serialism. Two ideas are key. One is *violation of the surface-true*: there are rankings in HS where a derivation introduces and later eliminates violations of a surface-unviolated constraint. The foot-structure constraint violated by Latin *(ˈamoː)* is an example. The other key idea is *full availability of structural operations*: in HS, structure is not built in a strictly bottom-up fashion, and conditions arising at later stages of the derivation can demand revision of structure inherited from earlier stages. Vowel shortening in the mapping *(ˈamoː)* → *(ˈamo)* is an example of such a revision.

To understand why CLIs have been seen as evidence for parallelism, it is necessary to understand why they are problematic in rule+constraint serial theories. In Sections 3.2 and 3.3, we explain the problem that rule+constraint theories have with CLIs and how this problem is solved in parallel OT. We then go on in Section 3.4 to argue that constraint violability, rather than parallelism, is essential to analyzing CLIs, leading to the conclusion that they are not problematic in HS. This is followed by discussion and further exemplification of violation of the surface-true in Section 3.5 and full availability of structural operations in Section 3.6. We conclude in Section 3.7 with a demonstration that language typology puts limits on what kinds of CLIs could and could not be analyzed in HS.

3.2 Cross-level interactions in rule+constraint theories

It is helpful to approach the analysis of CLIs in HS by way of a historical antecedent: the effort to incorporate constraints into rule-based phonology. As we will show, CLIs do present serious issues for a serial theory, but only if it operates with inviolable constraints.

One way that constraints interact with rules is by blocking their application. Blocking by constraints was introduced by Kisseberth (1970):

(1) Blocking by a derivational constraint in Kisseberth (1970: 305)
A rule applies just in case
(a) its structural description is satisfied by the input string, and
(b) the output string would not be in violation of the derivational constraint.

Similarly, Liberman and Prince (1977) propose the following condition on the interaction of well-formedness conditions and rules:

(2) Blocking by constraints in Liberman and Prince (1977: 290)
No rule may apply so as to produce an ill-formed representation

As an example of blocking, we can take the role of constraints in metrical theories of word stress that use both rules and constraints (Liberman and Prince (1977); see Hayes (1995), Kager (1995), and Kenstowicz (1994a) for further developments). In these theories, the parsing of syllables into feet is exhaustive, up to the limits imposed by the constraints that are active in the language. One CLI in this domain is the effect of the foot minimality constraint Foot-Binarity, which prohibits degenerate feet consisting of a single light syllable (Broselow 1982, Hayes 1995, McCarthy and Prince 1986/1996, Prince 1980, and much other work). Exhaustive parsing and the prohibition on degenerate feet come into conflict whenever odd-parity sequences of light syllables need to be parsed. Often, the leftover syllable remains unfooted, as in Hayes's (1995: 205–208) analysis of Hixkaryana (Derbyshire 1979). In this language, iambic feet are assigned from left to right, and final syllables are extrametrical (indicated by angled brackets). (The effects of a later rule of iambic lengthening are not shown in (3) but will be discussed shortly.)

(3) Hixkaryana foot structure (after Hayes 1995: 207)
(a) Even-parity sequence of light syllables before final
(neˈmo)(koˈto)<no> 'it fell'
(ˈtoh)(kuˈrʲe)(hoˈna)(haˈʃa)<ka> 'finally to Tohkurye'
(b) Odd-parity sequence of light syllables before final
(aˈʧo)wo<wo> 'wind'
*(aˈʧo)(ˈwo)<wo>
(ˈtoh)(kuˈrʲe)ho<na> 'to Tohkurye'
*(ˈtoh)(kuˈrʲe)(ˈho)<na>

Ungrammatical *(aˈʧo)(ˈwo)<wo>* contains the degenerate foot *(ˈwo)*. This is ruled out by Foot-Binarity, and the less complete foot parsing of *(aˈʧo)wo<wo>* is preferred instead.[2] This is an example of the *blocking* mode of interaction between constraints and rules: a rule that attempts to parse exhaustively is blocked when the result would contain a degenerate foot.

Besides blocking, there is a second mode of interaction between constraints and rules, *triggering*. Triggering was first recognized by Sommerstein (1974), who introduced the idea of rules that are 'positively motivated' by constraints:

(4) Triggering by constraints in Sommerstein (1974: 75)
A P-rule R is positively motivated with respect to a phonotactic constraint C just in case the input to R contains a matrix or matrices violating C AND the set of violations of C found in the output of R is null or is a proper subset of the set of such violations in the input to R.

According to Sommerstein's convention on rule application, rules apply only when they are positively motivated according to the definition in (4), that is, when they remove a violation of a constraint. In short, the constraint violation triggers the application of the rule.

In triggering mode, FOOT-BINARITY could be enforced by lengthening a short vowel. Assuming a moraic representation of length, the vowel lengthening rule can be stated as in (5):

(5) Vowel lengthening
Insert a mora

Rules of this generality are typical in rule+constraint theories like the ones we are discussing. Under the principle in (4), the vowel lengthening rule in (5) will only be applicable when some phonotactic constraint triggers it.

An example of triggering can be drawn from Hayes's (1995: 110–113, 208) analysis of what he calls the 'unstressable word syndrome' in Hixkaryana. Words consisting of just two light syllables, such as /kana/ 'fish', offer up only one light syllable for footing, because the final syllable is extrametrical. In this case, FOOT-BINARITY triggers an application of the rule of vowel lengthening:

2 **(aˈʧo)(ˈwo)<wo>* also violates *CLASH, but that cannot be the reason why it is ill-formed, because Hixkaryana is otherwise quite tolerant of clash: *(ˈnak)(ˈɲoh)(ˈyaʧ)(keˈna:)no* 'they were burning it'.

(6) Hixkaryana vowel lengthening triggered by Foot-Binarity
/kana/ → (ˈka)<na> → (ˈka:)<na>

This example immediately raises the fundamental problem of formalizing rule+constraint theories. In (6), the constraint violation was introduced not by the underlying representation, as in Sommerstein's examples, but rather by the prior application of the footing rule. If constraints are inviolable filters on the output of rules, the creation of the degenerate foot *(ˈka)* should have been blocked. In some sense, this temporary violation of Foot-Binarity could be said to be licensed by the presence of the lengthening 'repair', but in a theory with serially ordered rules and inviolable constraints, it is hard to see how that explanation could be formalized.

One approach to the problem of how repairs license apparent constraint violations is to relax serial ordering, so as to allow the repair to apply in parallel with the rule that would otherwise create a constraint violation. This is the solution proposed by Mester (1994) for the Latin problem mentioned in the introduction. In a process called 'iambic shortening', Plautine Latin avoids the degenerate foot of **(ˈa)<mo:>* by revoking the extrametricality of the final syllable and bringing it into the foot. The result would be an illicit light-heavy trochee **(ˈamo:)*, so Latin also shortens the final vowel, yielding *(ˈamo)*. In his analysis, Mester (1994: 15) conflates the building of the disyllabic foot and the shortening of the vowel into a single derivational step consisting of 'a structure-changing imposition of a foot ... on otherwise unfootable words. A structure-changing imposition of a foot is one which, simultaneous with constituent formation, invokes the designated repair strategy of the language (for Latin, REMOVE-μ ...) to achieve a well-formed result'. By having the two rules apply simultaneously, the derivation goes directly from unfooted /amo:/ to footed and shortened *(ˈamo)*, without intermediate stages that violate Foot-Binarity or the constraint against light-heavy trochees. Similar ideas are explored in Myers (1991) and Paradis (1988a, b).

While this locally solves the problem of temporary violation, as far as we know, this 'repair strategy' approach has never been fully formalized. Hixkaryana illustrates the challenges that any theory of constraints, rules, and repairs will have to face. On the one hand, Foot-Binarity triggers the lengthening repair in /kana/ → *(ˈka:)<na>*. But it also blocks footing of the penult in (3b) *(aˈʧo)wo<wo>*. Why isn't the lengthening repair available in this case, yielding **(aˈʧo)(ˈwo:)<wo>*? A clue is provided by Hayes's identification of /kana/ → *(ˈka:)<na>* as an example of the unstressable word syndrome: evidently the lengthening repair is available only when an entire word would otherwise go footless. (Mester's reference to 'otherwise unfootable words' says practically the same thing about Latin.) But this restriction on when the repair is available goes unexpressed in the formal theory.

To make matters even worse, the lengthening repair is available in all words – not just those that would otherwise be footless – when the trigger is not FOOT-BINARITY but rather the constraint SWP (for Stress to Weight Principle),[3] which prefers light-heavy iambs to light-light ones. As a result of this process of iambic lengthening, as it is known, the actual surface forms of the examples in (3) have lengthening of stressed open syllables, as shown in (7):

(7) Result of iambic lengthening in Hixkaryana
 (a) Even-parity sequence of light syllables before final
  (neˈmo:)(koˈto:)<no>
  (ˈtoh)(kuˈrʲe:)(hoˈna:)(haˈʃa:)<ka>
 (b) Odd-parity sequence of light syllables before final
  (aˈʧo:)wo<wo>
  (ˈtoh)(kuˈrʲe:)ho<na>

Extant theories of rule/constraint interaction are unable to describe this state of affairs, where it appears that one constraint (SWP) is able to trigger a given repair (lengthening) but the other (FOOT-BINARITY) is not always able to do so (i.e., **(aˈʧo)(ˈwo:)<wo>*).

This is not an isolated example: many conspiracies involve complex interactions between triggering and blocking, which raise exactly the same questions. See, for example, the discussion of the formal challenges raised by Kisseberth's (1970) Yawelmani conspiracy in Kiparsky (1973), McCarthy (2002, 2008), and Sommerstein (1977: 197–198).

Besides relaxing serialism, the other approach to the problem of violation and repair is to relax constraint inviolability. Something like this is suggested by Kisseberth (1970) in reference to an analysis of Tunica. In Tunica, some rules are blocked by a constraint against adjacent stressed syllables, while two other rules create these sequences, only to have them eliminated by later rules, thus raising for the first time the problem we are focusing on here. As a first step toward a solution Kisseberth

3 For copious references to previous work on SWP, see Gouskova (2003: 90fn.).

offers a distinction between two kinds of rules, those that are relatively obligatory and those that are absolutely obligatory:[4]

(8) Obligatoriness in Kisseberth (1970: 305)
A rule that is absolutely obligatory applies if its structural description is satisfied; it does not have to meet any added restrictions. A relatively obligatory rule is a rule which applies only if its structural description is satisfied and its output would not violate a given derivational constraint.

This idea will not help with blocking and repair by the constraint FOOT-BINARITY in Hixkaryana, however. The problem is that the same rule, foot formation, acts as absolutely obligatory with respect to FOOT-BINARITY in /kana/ → *(ˈka)<na>* → *(ˈka:)<na>*, but as only relatively obligatory with respect to FOOT-BINARITY in the blocked derivation /atʃowowo/ → *(aˈtʃo)wo<wo>* $\rightarrow_{\text{blocked}}$ **(aˈtʃo)(ˈwo)<wo>* → **(aˈtʃo)(ˈwo:)<wo>*.

The literature reviewed in this section was largely successful in identifying the ways in which phonological processes can interact with well-formedness constraints. Formalization proved problematic, however — no rule+constraint theory was able to account for the range of blocking and triggering effects that are found cross-linguistically, as illustrated by Hixkaryana.[5]

4 Kisseberth (1970: 305) does not claim to have solved the conspiracy problem – he offers his proposal about the formalization of (non-)blocking as a 'first tentative step in the construction of a theory of phonology employing the notion of derivational constraints'.

5 Calabrese (2009) proposes a constraint+repair system that appears to be capable of accounting for both blocking and triggering effects involving markedness constraints that are surface-inviolable. Each active markedness constraint is associated with a ranked list of repairs, but markedness constraints themselves are not ranked. If the input to the derivation violates some markedness constraint M1, its first repair $R1_{M1}$ is triggered. If the result of applying $R1_{M1}$ violates M2, then its first repair $R1_{M2}$ is triggered, and so on. The derivation terminates when all markedness constraints have been satisfied. If the derivation reaches a point where a markedness constraint M_B is violated and there is no viable repair, then a kind of blocking occurs: the derivation crashes and a new derivation begins in which M1's second-ranked repair $R2_{M1}$ is triggered.

The biggest problem with this theory is its commitment to surface-inviolability of active markedness constraints. As a result, FOOT-BINARITY cannot block foot parsing in Hixkaryana *(aˈtʃo)wo<wo>*, though it can trigger lengthening in *(ˈka:)<na>*. The problem is that the standard foot-parsing markedness constraint, PARSE-SYLLABLE ('assign one violation mark for every unfooted syllable') is not surface-true in Hixkaryama, so it could not be active, according to this theory. That problem could be circumvented if PARSE-SYLLABLE were

3.3 Cross-level interactions in parallel OT

It is clear from the discussion in Section 3.2 that rule+constraint theories have real problems with Hixkaryana's CLI. The worst of these problems are not shared by the parallel OT (P-OT) analysis of Hixkaryana in Kager (1999: Section 4.3), however.

Parallelism allows GEN to produce candidates that show the effects of more than one operation simultaneously. Thus, GEN(/kana/) includes *(ˈka:)na,* which shows the simultaneous effects of both the footing and the lengthening operations. This allows it to beat **(ˈka)na,* with its violation of FOOT-BINARITY, **(kaˈna),* which violates NON-FINALITY, and footless *kana,* which violates the constraint requiring every prosodic word to contain a main-stressed head foot, HEADEDNESS(PrWd). These three constraints must dominate the faithfulness constraint that opposes lengthening, DEP-μ:

(9) /kana/ → *(ˈka:)na* in P-OT

	/kana/	FT-BIN	NON-FIN	HEAD (PrWd)	DEP-μ
a. →	(ˈka:)na				1
b.	(ˈka)na	1 W			L
c.	(kaˈna)		1 W		L
d.	kana			1 W	L

The ranking in (9) explains why FOOT-BINARITY is satisfied by vowel lengthening in cases of the unstressable word syndrome: lengthening is the only way to achieve a binary foot while respecting NON-FINALITY, and HEADEDNESS(PrWd) demands that some foot be present.

Kager's P-OT analysis also accounts for why iambic lengthening occurs. He assumes a constraint UNEVEN-IAMB that is violated by any foot except a light-heavy one. Iambic lengthening is the result of ranking UNEVEN-IAMB over DEP-μ, as in tableau (10). PARSE-SYLLABLE, which favors more complete footing, must also dominate DEP-μ, because simply positing fewer feet would be a way of avoiding the necessity of lengthening their head syllables.

replaced by a constraint against a sequence of two unfooted syllables, PARSE-2 (Kager 1994), which *is* surface-unviolated in Hixkaryana. But this move leads to an obvious loss of generality: there are now two foot binarity constraints, FOOT-BINARITY itself and PARSE-2. This loss of generality is not at all atypical of efforts to recode violable constraints as inviolable (McCarthy 2002: 16–17).

(10) /nemokotono/ → *(ne'mo:)(ko'to:)no* in P-OT

	/nemokotono/	FT-BIN	NON-FIN	HEAD (PrWd)	UNEVEN-IAMB	PARSE-SYLL	DEP-μ
a. →	(ne'mo:)(ko'to:)no					1	2
b.	(ne'mo)(ko'to)no				2 W	1	L
c.	(ne'mo:)kotono					3 W	1 L

Tableau (10) shows UNEVEN-IAMB ranked below NON-FINALITY(foot) and HEADEDNESS(PrWd) because *('ka:)na* violates it but **(ka'na:)* and **kana* do not.

Finally, this P-OT analysis accounts for why lengthening is not allowed in stranded penults, such as the third syllable of **(a'ʧo:)('wo:)wo*. The crucial ranking puts UNEVEN-IAMB, which according to Kager *('wo:)* violates, over PARSE-SYLLABLE. This rules out the creation of a monosyllabic foot to achieve fuller foot parsing, as shown in tableau (11).

(11) /aʧowowo/ → *(a'ʧo:)wowo* in P-OT

	/aʧowowo/	FT-BIN	NON-FIN	HEAD (PrWd)	UNEVEN-IAMB	PARSE-SYLL	DEP-μ
a. →	(a'ʧo:)wowo					2	1
b.	(a'ʧo:)('wo:)wo				1 W	1 L	2 W
c.	(a'ʧo:)('wo)wo	1 W			1 W	1 L	1
d.	(a'ʧo:)(wo'wo:)		1 W			L	2 W

In P-OT, the interaction between operations (in GEN) and constraints (in EVAL) is unidirectional: the operations apply, singly and in all combinations, and then the most harmonic result is chosen. From (9)–(11), it might seem that this is the only way to solve the problem of how operations and constraints interact in Hixkaryana and perhaps in CLIs generally. As we will now show, however, parallelism is not essential when constraints are violable.

3.4 Cross-level interactions in Harmonic Serialism

Because it inherits OT's violable constraints, HS offers a successful and fully formalized solution to the problem of how constraints and operations interact in a derivation – a problem that could not be solved in earlier serial theories that combined rules with *in*violable constraints.

(For the background necessary to follow this argument, see McCarthy (this volume).)

We will illustrate this claim with an HS analysis of Hixkaryana. GEN includes two operations that are relevant to this analysis: footing and vowel lengthening. The vowel lengthening operation was defined in (5). Following Pruitt (2010), the footing operation is defined so as to parse any pair of adjacent unfooted syllables as a headed binary foot or any single unfooted syllable as a headed unary foot. The full candidate set includes the unchanged faithful candidate and the various results of applying any one of the operations once. For simplicity, we will ignore candidates with obviously fatal constraint violations, such as candidates with disyllabic trochaic feet.

We will also replace Kager's ad hoc constraint UNEVEN-IAMB with the standard constraint SWP, which was mentioned in Section 3.2. SWP is violated by any stressed light syllable, and thus it accounts for the widespread process of lengthening under stress, including iambic lengthening.[6] In contrast, UNEVEN-IAMB disfavors even (ˈCV:) iambs relative to (CVˈCV:) ones, wrongly predicting the existence of languages that epenthesize a vowel to change (ˈCV:) into (VˈCV:).[7]

Because the mapping /kana/ → *(ˈka:)na* involves two distinct operations in GEN, stress assignment and lengthening, it has to pass through an intermediate step. That intermediate step is *(ˈka)na*, with a degenerate foot. For *(ˈka)na* to win at step 1, NON-FINALITY and HEADEDNESS(PrWd) have to dominate FOOT-BINARITY and SWP, as shown in (12). (The reason for ranking FOOT-BINARITY over SWP will emerge in (16) and (18).)

6 The constraint SWP uniformly favors heavy stressed syllables in both iambic and trochaic feet. This predicts that stressed syllable lengthening should not be unique to iambic languages. Although Hayes (1995: 82ff.) argues that stressed syllable lengthening is most robustly attested in iambic languages, this generalization does not appear to be categorical. For discussion of trochaic lengthening, see for example Hyde (2007), McGarrity (2003), and Revithiadou (2004).

7 Kager's analysis of Hixkaryana will not work if UNEVEN-IAMB is replaced with the better-motivated constraint SWP. The problem is clear from tableau (11): if UNEVEN-IAMB were replaced by SWP, then **(aˈʧo:)(ˈwo:)wo* in (11b) would wrongly win. The HS analysis does not have this liability, as we explain later in this section.

(12) Step 1 of /kana/ → *('ka:)na* in HS

	/kana/	NON-FIN	HEAD (PrWd)	FT-BIN	SWP	DEP-μ
a. →	('ka)na			1	1	
b.	kana		1 W	L	L	
c.	(ka'na)	1 W		L	1	
d.	ka:na		1 W	L	L	1 W

The ranking in (12) is similar but not identical to the ranking required for the P-OT analysis in (9). The main difference is this: FOOT-BINARITY is surface-unviolated in Hixkaryana, so it is undominated in P-OT, which knows no other output level of representation than surface structure. In HS, however, FOOT-BINARITY must be dominated because it is crucially violated at the first step of the /kana/ → *('ka)na* → *('ka:)na* derivation. This is therefore an instance of what we referred to in the introduction as *violation of the surface-true.*

At step 2, the candidate *('ka:)na* becomes available, because it is one operation away from the step 1 output/step 2 input *('ka)na*. It satisfies FOOT-BINARITY and SWP at the expense of violating lower-ranking DEP-μ:

(13) Step 2 of /kana/ → *('ka:)na* in HS

	('ka)na	NON-FIN	HEAD (PrWd)	FT-BIN	SWP	DEP-μ
a. →	('ka:)na					1
b.	('ka)na			1 W	1 W	L

This derivation converges at step 3 because no footing or lengthening operation will produce any further improvement:

(14) Step 3 of /kana/ → *('ka:)na* in HS — Convergence

	('ka:)na	NON-FIN	HEAD (PrWd)	FT-BIN	SWP	DEP-μ
a. →	('ka:)na					
b.	('ka:)('na)	1 W			1 W	
c.	('ka:)na:					1 W

The derivation of /nemokotono/ also involves footing and lengthening steps. At step 1, the first two syllables are parsed into a foot, satisfying HEADEDNESS(PrWd). (Footing proceeds from left-to-right because of the constraint ALIGN-L(ft, wd) (McCarthy and Prince 1993a), not shown here.)

(15) Step 1 of /nemokotono/ → *(ne'mo:)(ko'to:)no* in HS

	/nemokotono/	NON-FIN	HEAD (PrWd)	FT-BIN	PARSE-SYLL	SWP	DEP-μ
a. →	(ne'mo)kotono				3	1	
b.	nemokotono		1 W		5 W	L	

This is another instance of violation of the surface-true because the winning candidate violates SWP, a constraint that is never violated by surface forms of this language.

At step 2 of this derivation, a second foot is built next to the first one, further improving performance on PARSE-SYLLABLE. Adding a second foot is preferred to lengthening the stressed vowel in the foot already built because PARSE-SYLLABLE dominates SWP:

(16) Step 2 of /nemokotono/ → *(ne'mo:)(ko'to:)no* in HS

	(ne'mo)kotono	NON-FIN	HEAD (PrWd)	FT-BIN	PARSE-SYLL	SWP	DEP-μ
a. →	(ne'mo)(ko'to)no				1	2	
b.	(ne'mo)kotono				3 W	1 L	
c.	(ne'mo:)kotono				3 W	L	1 W

Once again, the surface-unviolated constraint SWP is violated at an intermediate derivational step.

At step 3, one of the stressed vowels lengthens. The derivation produces the same result regardless of which one lengthens first. (This is therefore a *convergent tie* — see Section 2.4 of McCarthy (this volume).) In tableau (17), we take the arbitrary option of lengthening the stressed vowel on the left first:

(17) Step 3 of /nemokotono/ → *(ne'mo:)(ko'to:)no* in HS

	(ne'mo)(ko'to)no	NON-FIN	HEAD (PrWd)	FT-BIN	PARSE-SYLL	SWP	DEP-μ
a. →	(ne'mo:)(ko'to)no				1	1	1
b.	(ne'mo)(ko'to)no				1	2 W	L

The other stressed vowel lengthens at step 4, and then the derivation converges at step 5.

The derivation of /atʃowowo/ proceeds as follows: /atʃowowo/ → *(a'tʃo)wowo* → *(a'tʃo:)wowo*. Because the step 1 tableau involves the same constraint interactions as (15), we will go straight to step 2, which is where the most interesting candidates are found:

(18) Step 2 of /atʃowowo/ → *(aˈtʃo)wowo* in HS

	(aˈtʃo)wowo	NON-FIN	HEAD (PrWd)	FT-BIN	PARSE-SYLL	SWP	DEP-μ
a. →	(aˈtʃoː)wowo				2		1
b.	(aˈtʃo)(ˈwo)wo			1 W	1 L	2 W	L
c.	(aˈtʃo)(woˈwo)	1 W			L	2 W	L

At step 2, FOOT-BINARITY blocks the creation of the degenerate foot *(ˈwo)*. Because further foot parsing is foreclosed by FOOT-BINARITY and NON-FINALITY, iambic lengthening occurs instead, in response to the ranking of SWP above DEP-μ. Finally, the derivation converges at step 3.

As promised, this analysis illustrates the HS solution to the problem of how an active constraint can come to be violated in a derivation. Instead of relaxing serialism and allowing operations to apply simultaneously, as in P-OT or the repair-strategy analysis of Mester (1994), HS relies on OT's ranked and violable constraints. In OT, from the fact that FOOT-BINARITY is active in forcing lengthening in *(ˈkaː)na*, it does not follow that it is inviolable. The potential for violability allows for the creation of a degenerate foot if that is required by a higher ranked constraint, like NON-FINALITY and HEADEDNESS(PrWd) in this analysis. The challenging conspiratorial interaction between blocking and repair is also dealt with straightforwardly in this serial theory. The constraint ranking produces the effect of the conspiracy by choosing which operation provides the best way of satisfying FOOT-BINARITY in a given context: by not footing at all when HEADEDNESS(PrWd) is already satisfied, as in *(aˈtʃoː)wowo*, and by lengthening when NON-FINALITY and HEADEDNESS(PrWd) have forced the creation of a degenerate foot, as in *(ˈkaː)na*.

This analysis also explains why the lengthening repair is available for violations of SWP and HEADEDNESS(PrWd), but not in **(aˈtʃoː)(ˈwoː)wo*. Because PARSE-SYLLABLE dominates SWP, (CVˈCV) iambs are created in the course of the derivation, but they later change to (CVˈCVː) because SWP itself dominates DEP-μ. And because HEADEDNESS(PrWd) dominates FOOT-BINARITY, CVCV words are parsed with a degenerate foot (ˈCV)CV, which later becomes (ˈCVː)CV because FOOT-BINARITY itself dominates DEP-μ. There is no lengthening in **(aˈtʃoː)(ˈwoː)wo* because there is nothing to repair: *(aˈtʃoː)(ˈwo)wo* is not optimal earlier in the derivation, because PARSE-SYLLABLE is ranked below FOOT-BINARITY. Except when higher-ranking HEADEDNESS(PrWd) forces them, violations of FOOT-BINARITY never arise in the course of the derivation.

In HS, temporary violation of the surface-unviolated constraints SWP and FOOT-BINARITY is controlled by the grammar in the same way that P-OT controls violation of surface-violated constraints. Significantly, the HS analysis accounts for these facts without the dubious constraint UNEVEN-IAMB (violated by *(ˈwo)* as well as *(neˈmo)* and *(koˈto)*) that is required in the P-OT analysis.

Finally, the Hixkaryana analysis illustrates another of the core properties of HS highlighted in the introduction: *full availability of structural operations*. In tableaux (16) and (18), for example, lengthening a stressed vowel competes against building another foot. Because PARSE-SYLLABLE dominates SWP, foot-building is favored until no more feet can be built. Operations at different levels of the hierarchy are always available in GEN, and their applicability is controlled as usual by the constraint hierarchy. Thus, structure-building in HS is not strictly bottom-up (or top-down, for that matter). This has relevance to Prince and Smolensky's argument against bottom-up constructionism in Tongan, for which see Elsman (this volume).[8]

Cross-level interactions are a special case of the general problem of formalizing the interaction between constraints and operations. There is nothing in the theory that would lead us to expect that footing could not create a violation of a constraint, or that lengthening could not apply afterwards to resolve it. By solving the problem of formalizing temporary violation and repair, HS also solves this problem of constraint/operation interaction, and maybe all others. Parallelism is not required.

3.5 Violation of the surface-true

The analysis of Hixkaryana in Section 3.4 relies on an inherent property of HS: derivations may involve temporary violation of surface-unviolated constraints. Two other putative arguments for parallelism in the literature can also be reanalyzed by relying on this aspect of HS: the interaction of stress and syllable weight in Latin (Mester 1994: 15); and the interaction of syllabification and segmental processes in Lardil (Prince and Smolensky 1993/2004: 148–149).

8 The prosodic hierarchy (Selkirk 1980) imposes a limited amount of bottom-upness. For example, it is a standard assumption in prosodic hierarchy theory that feet are constructed out of syllables and that feet cannot have unfilled daughter nodes. If this is taken to be an inviolable constraint, then no foot can be built until at least one syllable has been.

3.5.1 Latin stress and syllable weight

Like Hixkaryana, Latin has CLIs between vowel length and metrical structure. The processes involved are known as iambic and cretic shortening, because the underlying forms of the words they affect look like the iambic (LH) and cretic (HLH) feet of Greek meter. Both processes are primarily associated with pre-Classical, Plautine Latin and are usually assumed to reflect a more colloquial speech style than Classical Latin.

In iambic shortening, a final long vowel in a LH disyllable is shortened:

(19) Iambic shortening[9]

Underlying	Surface	
amo:	ˈa.mo	'I love'
puta:	ˈpu.ta	'believe!'
wolo:	ˈwo.lo	'I want'
wiri:	ˈwi.ri	'men'
homo:	ˈho.mo	'human'

HH disyllables, such as ˈ*lau.do*: 'I praise', do not undergo shortening.

In cretic shortening, a final long vowel in a HLH trisyllable is shortened:

(20) Cretic shortening

Underlying	Surface	
di:kito:	ˈdi:.ki.to	'say! (fut.)'
impera:	ˈim.pe.ra	'rule!'
mentio:	ˈmen.ti.o	'mention
de:sino:	ˈde:.si.no	'I cease'

LLH trisyllables, such as ˈ*stu.de.o*: 'I study', do not undergo shortening.

Mester (1994) argues that both of these shortening processes improve metrical parsing.[10] He argues that the Latin foot is a maximally and minimally bimoraic trochee, (ˈLL) or (ˈH). Shortening in /puta:/ → *(ˈpu.ta)* yields a foot that is neither too small like **(ˈpu)ta*: nor too large like **(ˈpu.ta:)*. Cretic shortening is a response to this same imperative, under the assumption that *(ˈdi:)(ˌki.to)* has a (ˌLL) foot after the primary stress.

Prince and Smolensky (1993/2004: Section 4.5) show how Mester's ideas can be expressed in P-OT. They begin with the basic stress pattern,

9 Iambic shortening also affects words with final closed syllables: *(ˈputat)* can be scanned in Plautine verse as LL rather than LH.

10 Mester's analysis is not without its critics; see Fortson (2008) and Lahiri *et al.* (1999).

which puts stress on a heavy penult, else the antepenult, in words of sufficient length:

(21) Latin stress pattern

Weight Pattern	Main-stress foot	
...ˈLLσ	sus(ˈpi.ki)a	'suspicions (nom./acc.)'
...ˈHLσ	deː.klaː(ˈraː)ti.o	'declaration (nom.)'
...LˈHσ	a(ˈmiː)ke	'friend (voc.)'
...HˈHσ	ser(ˈmoː)ne	'speech (abl.)'

The foot is obviously trochaic. It is attracted to the right edge of the word, but subject to the requirement that the final syllable not be footed. This means that NON-FINALITY(ˈft), which is violated if the main-stress foot is final in the prosodic word, dominates ALIGN-R(ˈσ, word), which is fully satisfied only by final stress:

(22) NON-FINALITY(ˈft) ≫ ALIGN-R(ˈσ, word)

	/suspikia/	NON-FIN(ˈft)	ALIGN-R(ˈσ)
a. →	sus(ˈpi.ki)a		2
b.	sus.pi(ˈki.a)	1 W	1 L

Iambic shortening shows that NON-FINALITY(ˈft) is itself crucially dominated. Underlying /amoː/ becomes *(ˈa.mo)* because there is no other allowable way of parsing this word into a well-formed trochaic foot. The candidate *(ˈa.moː)* in (24b) fatally violates the Weight-to-Stress Principle of Prince (1990), which we formulate here as a constraint on foot structure, mimicking the notion of quantity-sensitivity in Hayes (1980):

(23) WSP-FOOT
Assign a violation mark for any heavy syllable in the weak branch of a foot.

The candidate *(ˈa)moː* in (24c) is ruled out by FOOT-BINARITY, because the *(ˈa)* foot is monomoraic. The next candidate, *a(ˈmoː)* in (24d), has main stress on the final syllable in contravention of a different NON-FINALITY constraint, NON-FINALITY(ˈσ).[11] Finally, footless (24e) is out because of the requirements that every lexical word correspond to

11 Prince and Smolensky (1993/2004: 62) have a single non-finality constraint that is violated once for a word-final main-stressed foot or syllable, and twice for both. In keeping with the standard practice in later work, we have split it into two constraints.

a prosodic word (Lx≈Pr) and that every prosodic word contain a head foot (Headedness(PrWd) (Selkirk 1995)). Ranking these markedness constraints above Non-Finality(ˈft) and the faithfulness constraint Max-μ produces the desired result:

(24) Iambic shortening in P-OT (after Prince and Smolensky 1993/2004: Section 4.5)

/amoː/	WSP-Foot	Ft-Bin	Non-Fin(ˈσ)	Lx≈Pr Hd(Pr)	Non-Fin(ˈft)	Max-μ
a. → (ˈa.mo)					1	1
b. (ˈa.moː)	1 W				1	L
c. (ˈa)moː		1 W			L	L
d. a(ˈmoː)			1 W		1	L
e. a.moː				1 W	L	L

Prince and Smolensky's analysis of cretic shortening is similar, except that shortening affects a foot that is assumed to follow the main stress in cretic words: /diːkitoː/ → (ˈdiː)(ˌki.to). The presence of this foot is an effect of Parse-Syllable, which rules out *(ˈdiː)ki.toː* and *(ˈdiː)ki(ˌtoː)* in (25c,d). As in the case of iambic shortening, failure to shorten in *(ˌki.toː)* violates WSP-Foot (see (25b)):

(25) Cretic shortening in P-OT (after Prince and Smolensky 1993/2004: Section 4.5)

/diːkitoː/	WSP-Foot	Parse-Syll	Max-μ
a. → (ˈdiː)(ˌki.to)			1
b. (ˈdiː)(ˌki.toː)	1 W		L
c. (ˈdiː)ki.toː		2 W	L
d. (ˈdiː)ki(ˌtoː)		1 W	L

To complete this analysis of Cretic shortening, Prince and Smolensky require one additional constraint and ranking. A constraint against (ˈHL) trochees, dubbed RhHrm (for rhythmic harmony) by Prince and Smolensky (1993/2004: 70–71), also dominates Max-μ:

(26) Cretic shortening in P-OT — RhHrm ≫ Max-μ

/diːkitoː/	RhHrm	Max-μ
a. → (ˈdiː)(ˌki.to)		1
b. (ˈdiː.ki)(ˌtoː)	1 W	L

This will suffice for now as a summary of the P-OT analysis. Turning now to HS, we will apply the same analytic strategy seen in the analysis of Hixkaryana: foot parsing creates a foot with marked quantitative structure that is made less marked by changing quantity at a later stage of the derivation. In the case of Latin, the (ˈLH) trochee that is initially created violates WSP-FOOT, which is satisfied later in the derivation when MAX-μ is violated.

We will enter the derivation at what might be called step 2, after syllabification. Tableau (27) shows step 2 of the derivation of /amoː/. GEN can build any single foot or shorten a vowel, but not both. Failure to build a foot, as in (27d), is ruled out by LX≈PR and HEADEDNESS(PrWd).[12] Building a foot on just the final or initial syllable, as in (27b,c), violates FOOT-BINARITY or NON-FINALITY(ˈσ). The crucial move in (27) is ranking WSP-FOOT below FOOT-BINARITY and NON-FINALITY(ˈσ). This ranking allows *(ˈa.moː)* to win at step 2, setting the stage for shortening at step 3.

(27) Step 2 from /amoː/: NONFIN(ˈσ), FOOT-BINARITY ≫ WSP-FOOT

a.moː	NON-FIN(ˈσ)	FT-BIN	LX≈PR HD(Pr)	NON-FIN(ˈft)	WSP-FOOT	ALIGN-R(ˈσ)	MAX-μ
a. → (ˈa.moː)				1	1	1	
b. a(ˈmoː)	1 W			1	L	L	
c. (ˈa)moː		1 W		L	L	1	
d. a.moː			1 W	L	L	L	

This is a case of violation of the surface-true – Latin has no surface (ˈLH) feet, because WSP-FOOT is surface-unviolated.

Shortening occurs at step 3 because WSP-FOOT dominates MAX-μ, as in the P-OT analysis:

(28) Step 3 from /amoː/: WSP-FOOT ≫ MAX-μ

(ˈamoː)	NON-FIN(ˈft)	WSP-FOOT	ALIGN-R(ˈσ)	MAX-μ
a. → (ˈamo)	1		1	1
b. (ˈamoː)	1	1 W	1	L

Shortening occurs at this step because all of the high-ranking foot-parsing constraints have been satisfied. This allows EVAL to turn its

12 Candidate (27d) also violates PARSE-SYLLABLE. That is not the reason why it loses to *(ˈa.moː)*, however, because NON-FINALITY(ˈft) dominates PARSE-SYLLABLE, as shown by (29c).

attention to the quantitative constraint WSP-FOOT. The result of shortening is that *(ˈamo)* now conforms with expectations about Latin surface forms, and indeed the derivation converges at the next step.

The HS derivation of cretic shortening includes two steps of foot construction before shortening occurs: /deːsinoː/ → *deː.si.noː* → *(ˈdeː)si.noː* → *(ˈdeː)(ˌsi.noː)* → *(ˈdeː)(ˌsi.no)*. At the first post-syllabification step, step 2, the heavy antepenult is parsed into a monosyllabic foot that receives main stress. This candidate wins because the candidates with more complete footing, (29b,c), violate RHHRM or NON-FINALITY(ˈft), both of which dominate PARSE-SYLL:

(29) Step 2 from /deːsinoː/ — NON-FINALITY(ˈft) ≫ PARSE-SYLL

	deː.si.noː	NON-FIN(ˈσ)	RHHRM	NON-FIN(ˈft)	WSP-FOOT	ALIGN-R(ˈσ)	PARSE-SYLL
a. →	(ˈdeː)si.noː					2	2
b.	(ˈdeː.si)noː		1 W			2	1 L
c.	deː(ˈsi.noː)			1 W	1 W	1 L	1 L
d.	deː.si(ˈnoː)	1 W		1 W		L	2

Candidate (29c) involves a ranking disjunction – NON-FINALITY(ˈft) *or* WSP-FOOT dominates ALIGN-R(ˈσ, word) and PARSE-SYLLABLE – but this disjunction will be partly resolved in a moment, when we see that PARSE-SYLLABLE has to dominate WSP-FOOT.

At step 3, the options for further parsing of *(ˈdeː)si.noː* include all of the ways of adding a foot to the last two syllables, plus shortening one of the vowels or doing nothing. Shortening at this point would gratuitously violate MAX-μ, while doing nothing violates PARSE-SYLLABLE. To get the desired parse *(ˈdeː)(ˌsi.noː)* at this step, PARSE-SYLLABLE has to dominate WSP-FOOT:

(30) Step 3 from /deːsinoː/ — PARSE-SYLLABLE ≫ WSP-FOOT

	(ˈdeː)si.noː	PARSE-SYLL	WSP-FOOT
a. →	(ˈdeː)(ˌsi.noː)		1
b.	(ˈdeː)si.noː	2 W	L
c.	(ˈdeː)si(ˌnoː)	1 W	L

As in the derivation of *(ˈamo)*, the surface unviolated constraint WSP-FOOT is violated at this intermediate step.

Next, at step 4, input *(ˈdeː)(ˌsi.noː)* undergoes shortening of the final vowel because WSP-FOOT dominates MAX-μ:

(31) Step 4 from /deːsinoː/

	(ˈdeː)(ˌsi.noː)	WSP-Foot	Max-μ
a. →	(ˈdeː)(ˌsi.no)		1
b.	(ˈdeː)(ˌsi.noː)	1 W	L

The derivation converges at step 5.

WSP-Foot is surface-unviolated in this variety of Latin, and it is therefore undominated in the P-OT analysis. In the HS analysis, however, it is violated at an intermediate step of the iambic and cretic shortening derivations. This is reflected in a difference in ranking: WSP-Foot has to be dominated by Non-Finality(ˈσ), Foot-Binarity, and Parse-Syllable in the HS analysis, but these four constraints are surface-unviolated and therefore unrankable in P-OT. Thus, although (ˈLH) trochees are absent from surface forms, they emerge and are later eliminated by shortening in the course of the derivation. This is therefore an instance of violation of the surface-true.

In terms of empirical coverage, the analysis in HS presented here and the analysis in P-OT in Prince and Smolensky (1993/2004) are essentially equivalent,[13] but there is an interesting difference between the constraints required for each analysis. With the constraints discussed thus far, the P-OT analysis cannot account for why /deːsinoː/ shortens the final vowel, yielding *(ˈdeː)(ˌsi.no)*, instead of shortening the antepenultimate vowel, yielding **(ˈde.si)(ˌnoː)*. Prince and Smolensky resolve this tie by calling on the constraint PkProm, which favors having a heavy main-stressed syllable. More recent developments in OT offer a couple of other options: some sort of contiguity constraint favoring peripheral over medial shortening (Kenstowicz 1994b, McCarthy and Prince 1995, 1999); or a positional faithfulness constraint discouraging shortening in the main-stressed syllable (Beckman 1998).

This issue does not arise at all in the HS analysis, however. As is apparent from tableaux (29)–(31), **(ˈde.si)(ˌnoː)* is never even a candidate, much less a serious challenger to the intended winner. It is not a candidate because it would require an intermediate step with a (ˈHL)

13 The P-OT and HS analyses make different predictions about foot parsing in syllables that precede the main stress. For example, when the main stress is preceded by two heavy syllables, as in *audiːˈtoːte* 'hear! (fut. pl.)', the P-OT analysis predicts *(ˌau)(ˌdiː)(ˈtoː)te* while the HS analysis predicts *(ˌaudiː)(ˈtoː)te*. Little or nothing is known for sure about pretonic secondary stress in Latin, and existing conjectures (e.g., initial secondary stress if main stress is on the peninitial (Allen 1973: 190–191)) are unable to settle empirical questions of such subtlety.

trochee, *(ˈdeː.si)*. Undominated RHHRM ensures that such feet never win even at intermediate derivational steps. The HS analysis therefore has an advantage over the P-OT analysis: it can explain, and need not stipulate, which syllable is affected by cretic shortening.

3.5.2 *Lardil syllabification and augmentation*

Prince and Smolensky (1993/2004: 148–149) develop an argument in favor of a parallel architecture for OT from the interaction of syllabification and augmentation in Lardil.[14] Lardil limits its coda consonants to coronal sonorants, plus labial and dorsal nasals that share place with the following onset. Lardil has a process of augmentation that affects unsuffixed CVC roots, making them disyllabic for word-minimality reasons. If the final consonant of the CVC root is a sonorant, augmentation adds *Ca*, where C is a stop that is homorganic with the preceding consonant (32a). Roots ending in obstruents augment by adding only *a*, with the root-final consonant syllabified as an onset (32b). The *a* augment is also used with ɾ-final roots (32c), because the language prohibits clusters of ɾ followed by a coronal (except lamino-alveolars) (Hale and Nash 1997: 250).

(32) Lardil augmentation (all words nominative case, data from Hale *et al.* (1981))
a. CV augmentation after sonorants (except ɾ)

/wun/	wun.ta	'rain'
/maɽ/	maɽ.ʈa	'hand'
/ɽil/	ɽil.ta	'neck'
/kaŋ/	kaŋ.ka	'speech'

b. V augmentation after obstruents

/pat/	pa.ta	'west'
/jak/	ja.ka, *jak.ka	'fish'

c. V augmentation after ɾ

/teɾ/	te.ɾa, *teɾ.ta	'thigh'

Why do Prince and Smolensky see these data as challenging for serialism? They consider a serial theory in which 'Syllabification' and 'Augmentation' are rules whose order must be fixed, and they argue that Syllabification must precede Augmentation to distinguish the forms in which V is added from those that get CV: 'Augmentation inserts an onset only when the stem-final consonant is *already* parsed as a coda' (p. 148). Thus, the serial derivation as they conceive it must go something like this:

14 They credit a 1992 manuscript by Robert Kirchner.

(33) Serial derivation after Prince and Smolensky (1993/2004: 148–149)

	/wun/	/jak/
Syllabification	$[\text{wun}]_\sigma$	$[\text{ja}]_\sigma\text{k}$
Augmentation	$[\text{wun}]_\sigma[\text{ta}]_\sigma$	$[\text{ja}]_\sigma[\text{ka}]_\sigma$

Prince and Smolensky's argument against serialism relies on the observation that *ŋ* codas must be followed by *k*, so *ŋ*'s [dorsal] place is licensed by association with the following onset: 'Now consider /kaŋ/. When Syllabification applies, the *ŋ* is no more syllabifiable as a coda than the *k* of /yak/ [our /jak/] — *ŋ* can only be a coda when linked to a following onset. The situation is obviously not improved by trying to allow Augmentation to precede Syllabification ... In short, Syllabification and Augmentation are mutually interdependent: each "triggers" the other.'

Although this is a valid argument against the serial theory implicit in (33), HS is not such a theory. The theory in (33) has inviolable constraints – because free-standing *ŋ* is never a licit coda in surface forms, it cannot be parsed as a coda in intermediate *kaŋ* – and it presupposes that epenthesis and (re-)syllabification can never occur in parallel. This is not HS. In HS, as in OT generally, constraints are violable, and even surface-unviolated constraints may be violated in the course of a derivation.

A further point of difference between HS, as we understand it here, and the serial theory implied by (33) is that research in HS, as in OT generally, takes the nature of GEN to be an empirical hypothesis. Although we may speak of HS's GEN 'performing only one operation at a time', the question of what a single operation can do is an empirical one. Another way to put this question is to ask exactly how much parallelism is needed, or how much must a single operation in GEN be allowed to do. An example of this sort of reasoning can be found in McCarthy's (this volume) argument that resyllabification must occur in parallel with syllable-altering processes like epenthesis or syncope.

Both of these matters – violation of the surface-true and the question of GEN's power – are important in understanding Lardil augmentation in HS. A typical HS derivation for Lardil will look like this:

(34) HS derivation of *wun.ta*

/wun/ → wun → (ˈwun) → (ˈwu.na) → (ˈwun.Ca) → (ˈwun.ta)

The symbol *C* stands for a placeless plosive. At step 1 of this derivation, /wun/ is syllabified as the single syllable *wun*, and at step 2 (shown in tableau (35)) it is parsed into a foot to satisfy HEADEDNESS(PrWd), at the expense of violating FOOT-BINARITY. (Lardil codas are non-moraic.)

(35) Step 2 of /wun/ → ˈ*wun.ta*

	wun	Hd(Pr)	Ft-Bin
a. →	(ˈwun)		1
b.	wun	1 W	L

Augmentation occurs at step 3 because Foot-Binarity dominates Dep, as shown in tableau (36) (though see Moore-Cantwell (this volume) for a different view of such word-minimality effects).[15] This tableau also shows that Foot-Binarity dominates the constraint Align-R(MWord, σ), which is the trigger of *Ca* augmentation in Prince and Smolensky's analysis. This constraint is violated by intermediate *(ˈwu.na)* because the rightmost consonant in the morphological word /wun/ is not also rightmost in some syllable.

(36) Step 3 of /wun/ → ˈ*wun.ta*

	(ˈwun)	Ft-Bin	Align-R	Dep
a. →	(ˈwu.na)		1	1
b.	(ˈwun)	1 W	L	L

Step 4 of this derivation brings it into conformity with Align-R(MWord, σ) by epenthesizing a placeless plosive consonant C into onset position, pushing the *n* back into the coda where it belongs. The epenthesized consonant is placeless because of the undominated faithfulness constraint Dep(place), which is violated if any more fully specified consonant is epenthesized, as in candidates (37b,c,d).[16] It is a plosive by emergence of the unmarked: plosives minimize sonority in the onset, as required by the margin hierarchy (Prince and Smolensky 1993/2004: 151ff.).

15 The quality of the epenthetic vowel is a result of emergence of the unmarked (McCarthy and Prince 1994). Epenthetic *a* maximizes sonority in the nucleus, as required by the peak hierarchy (Prince and Smolensky 1993/2004: 151ff.).

16 Dep(place) or its equivalent is also needed in any P-OT analysis. Flap-final /teɾ/ augments as ˈ*te.ɾa*, not ˈ*teɾ.ta*, because of an undominated constraint against ɾ+coronal clusters. Dep(place) is needed to rule out **teɾ.pa* and **teɾ.ka*, which are phonotactically impeccable.

(37) Step 4 of /wun/ → ˈ*wun.ta*

	(ˈwu.na)	Dep (place)	Align-R	Dep
a. →	(ˈwun Ca) with [cor] linked to n			1
b.	(ˈwu na) with [cor] linked to n		1 W	L
c.	(ˈwun ta) with [cor] linked to n and [cor] linked to t	1 W		1
d.	(ˈwun ka) with [cor] linked to n and [dors] linked to k	1 W		1

At the last pre-convergence step of the derivation, the [coronal] feature of *n* spreads into the following placeless C. This occurs because a markedness constraint favoring consonants with place (call it Have-Place) dominates any faithfulness constraint that militates against spreading place:

(38) Step 5 of /wun/ → ˈ*wun.ta*

	(ˈwun.Ca)	Have-Place	"No-Spread"
a. →	(ˈwun ta) with one [cor] linked to both n and t		1
b.	(ˈwun Ca) with [cor] linked to n	1 W	L

Have-Place must be ranked below Align-R(MWord, σ), else it would block epenthesis of the placeless plosive C at step 4.

Augmentation with *a* rather than *Ca* occurs when the root-final consonant is an obstruent. Obstruent codas are illicit because of the constraint in (39), which is based on typological observations by Zec (1995) and others about the preference for high sonority in codas:

(39) *Coda(obst)
Assign a violation mark for every [–sonorant] consonant in coda position.

Independent evidence for *Coda(obst) in Lardil comes from the observation that a word-final coronal obstruent /t/ or /ʈ/ changes into the corresponding rhotic /ɾ/ OR /ɽ/ (Hale 1973: 426 fn. 32): /jaɽput/ → *jaɽpuɾ* 'snake, bird'.[17]

Now, suppose the derivation of /pat/ → *ˈpa.ta* has proceeded through step 3, yielding *(ˈpa.ta)*. At step 4, shown in tableau (40), *(ˈpat.Ca)* is a candidate, but it loses because *CODA(obst) dominates ALIGN-R(MWORD, σ). This is therefore the convergence step for the /pat/ → *ˈpa.ta* derivation.

(40) Step 4 of /pat/ → *ˈpa.ta*

	(ˈpa.ta)	*CODA(obst)	ALIGN-R	DEP
a. →	(ˈpa.ta)		1	
b.	(ˈpat.Ca)	1 W	L	1 W

Now we are ready to tackle /kaŋ/ → *ˈkaŋ.ka*, which is the basis of Prince and Smolensky's argument for parallelism. In Lardil surface forms, coda *ŋ* must always be followed by onset *k* – although Lardil has nonhomorganic clusters like *nk*, it does not allow *ŋt*, and there are alternations showing that it deletes word-final *ŋ*: /wuŋkunuŋ/ → *wuŋkunu* 'queen-fish' (cf. *wuŋkunuŋin* 'id. non-future accusative'). Except for [coronal], consonantal place features have to be licensed by association with onset position:

(41) *CODA(lab/dors)
Assign a violation mark for every instance of the features [labial] or [dorsal] not associated with a consonant in onset position.

A place feature that is licensed in this way can also be associated with a preceding coda without violating this constraint (Goldsmith 1990: 123–128, Ito 1989), so coda *ŋ* is allowed only if followed by onset *k*. They share a single [dorsal] place feature that is associated with the onset and thereby avoids violating *CODA(lab/dors).

Although *CODA(lab/dors) is surface-unviolated, violation in the course of the derivation is also possible in HS, and that is how /kaŋ/ → *ˈkaŋ.ka* is derived. Suppose that *CODA(lab/dors) is crucially dominated by PARSE-SEGMENT, which is violated once by every unsyllabified segment. It follows that /kaŋ/ will be syllabified as the single syllable *kaŋ* at step 1, as shown in tableau (42). This tableau pits fully-syllabified (42a)

17 The change of /t/ is optional, according to Klokeid (1976: 39). There is variation between *t* and *ɾ* word-finally, according to Hale and Nash (1997).

against partially syllabified (42b), which loses because of its violation of top-ranked PARSE-SEGMENT. Candidate (42c) takes a different path entirely, deleting a segment rather than parsing a syllable. This too is a non-starter because of PARSE-SEGMENT.

(42) Step 1 of /kaŋ/ → ˈ*kaŋ.ka*

/kaŋ/	PARSE-SEG	HEAD	FT-BIN	*CODA (lab/dors)	DEP	MAX-C
a. → σ / | \ k a ŋ		1		1		
b. σ / | k a ŋ	1 W	1		L		
c. ka	2 W	1		L		1 W

In tableau (42), all candidates violate HEADEDNESS(PrWd). This is unavoidable at step 1 if it is assumed that GEN cannot build syllables and feet at the same time, and feet cannot be built before syllables (see note 8). At step 2, satisfying this top-ranked constraint takes precedence over other markedness violations. Although there is also a violation of *CODA(lab/dors) to worry about, the ranking of HEADEDNESS(PrWd) over *CODA(lab/dors) gives priority to building a foot rather than repairing the coda *ŋ*. Tableau (43) certifies this argument formally.

(43) Step 2 of /kaŋ/ → ˈ*kaŋ.ka*

σ / | \ k a ŋ	PARSE-SEG	HD(Pr)	FT-BIN	*CODA (lab/dors)	DEP	MAX-C
a. → Ft | σ / | \ k a ŋ			1	1		
b. σ / | \ k a ŋ		1 W	L	1		
c. σ / \ k a		1 W	L	L		1 W

From now on, we will return to the 'flat' representation of syllable and foot structure used in earlier tableaux like (35).

The input/faithful candidate at step 3 violates two active markedness constraints, FOOT-BINARITY and *CODA(lab/dors). Epenthesizing the augment vowel deals with both, making the foot binary and (temporarily) moving the *ŋ* out of coda position:

(44) Step 3 of /kaŋ/ → *ˈkaŋ.ka*

	(ˈkaŋ)	PARSE-SEG	HD(Pr)	FT-BIN	*CODA (lab/dors)	DEP	MAX-C
a. →	(ˈka.ŋa)					1	
b.	(ˈkaŋ)			1 W	1 W	L	
c.	(ˈka)			1 W		L	1 W

What happens next depends on the ranking of ALIGN-R(MWORD, σ). Previously, in tableau (36), we saw that it is crucially dominated by the surface-unviolated constraint FOOT-BINARITY. But it must itself dominate another surface-unviolated constraint, *CODA(lab/dors). This ranking forces *(ˈka.ŋa)* to map to *(ˈkaŋ.Ca)* at step 4:

(45) Step 4 of /kaŋ/ → *ˈkaŋ.ka*

	(ˈka.ŋa)	PARSE-SEG	HD(Pr)	FT-BIN	ALIGN-R	*CODA (lab/dors)	DEP	MAX-C
a. →	(ˈkaŋ.Ca)					1	1	
b.	(ˈka.ŋa)				1 W	L	L	

Finally, at step 5, the [dorsal] feature of the *ŋ* spreads rightward to supply a place feature for the placeless plosive C. This is ensured by the ranking in (38) as well as by *CODA(lab/dors), which requires that the [dorsal] feature be licensed by association with an onset. Deletion of *ŋ* from the coda, as in (46c), is non-viable because C remains placeless.

(46) Step 5 of /kaŋ/ → *ˈkaŋ.ka*

	(ˈkaŋ.Ca)	*CODA (lab/dors)	HAVE-PLACE	MAX-C	"NO-SPREAD"
a. →	(ˈkaŋ.ka)				1
b.	(ˈkaŋ.Ca)	1 W	1 W		L
c.	(ˈka.Ca)		1 W	1 W	L

This derivation then converges at step 6.

Due diligence requires us to show that this analysis of /kaŋ/ → *ˈkaŋ.ka* does not adversely affect *ŋ* deletion in polysyllables like /wuŋkunuŋ/ → *wuŋkunu*. In the derivation of *ˈkaŋ.ka*, epenthesis of *a* occurs not to license the *ŋ* but rather to satisfy FOOT-BINARITY; licensing of *ŋ* is a mere side effect. When FOOT-BINARITY is not at stake, unlicensed *ŋ* is deleted because DEP dominates MAX-C – a ranking that we have shown (but not used) throughout the analysis of Lardil. With this ranking, *wuŋkunu* beats **wuŋkunuŋa*, and **wuŋkunuŋka* never gets a chance to be a candidate.[18]

Lardil has another potential argument for parallelism: flap-final /teɾ/ augments as *ˈte.ɾa*, not **ˈteɾ.ta*. Hale attributes this to a surface-true prohibition against clusters of ɾ+coronal. The question is how the constraint *ɾ+coronal can rule out the candidate **teɾ.Ca* at step 4 of the derivation.

Following Chomsky and Halle (1968: 336–337) and the literature on underspecification theory (e.g., Kiparsky 1982, Pulleyblank 1986), we might propose that a markedness constraint is violated by a form if and only if its structural description is nondistinct from that form. A structural description and a form are non-distinct if and only if the form does not contradict any of the requirements in the structural description. The clusters *ɾt* and *ɾC* are nondistinct from each other and from the structural description of *ɾ+coronal. But the clusters *ɾp* and *ɾk* are distinct from the structural description of *ɾ+coronal, and so they are allowed. Considerations of this sort have not played much of a role in P-OT because P-OT's markedness constraints evaluate only fully-specified surface representations. But HS allows for the possibility of underspecified intermediate representations, such as *teɾ.Ca*. It is therefore no surprise that HS should adopt a criterion for constraint applicability that recalls the pre-OT derivational literature.

Alternatively, GEN could be given the power to epenthesize a segment and spread place onto it in a single step. This approach conflates steps

18 Although ALIGN-R(MWORD, σ) dominates *CODA(lab/dors), it must not prevent *CODA(lab/dors) from favoring deletion of *ŋ* in /wuŋkunuŋ/ → *wuŋkunu*. Prince and Smolensky's (1993/2004: 127) definition of ALIGN-R(MWORD, σ) reads like this: 'The final edge of a Morphological Word corresponds to the final edge of a syllable.' Under this definition, ALIGN-R(MWORD, σ) would wrongly block deletion in /wuŋkunuŋ/ as well as correctly trigger epenthesis in /kaŋ/. In correspondence theory (McCarthy and Prince 1995, 1999), ANCHOR constraints decouple these two senses of alignment. One type of ANCHOR constraint says that the underlying word-final consonant must have an output correspondent. This constraint is dominated by *CODA(lab/dors), as shown by /wuŋkunuŋ/ → *wuŋkunu*. The other type of ANCHOR constraint says that if there is an output correspondent of an underlying word-final consonant, then it must be syllable-final. This is the sense of ANCHOR that we impute to ALIGN-R(MWORD, σ) in the analysis of Lardil.

4 and 5 in (45) and (46) into a single step, and it allows *ˈte.ɾa* and **ˈteɾ.ta* to compete directly. Ideally, we would resolve the question of which approach is right, but we have exhausted the Lardil evidence bearing on this question. General techniques of OT and HS typology are applicable (Wolf and McCarthy 2009), but applying them now would divert us from our goals.

Returning to our main point, we have argued that Lardil exemplifies violation of the surface-true. The surface-unviolated constraint *CODA(lab/dors) is violated in the course of the derivation of *ˈkaŋ.ka* from /kaŋ/. Constraint ranking provides a principled account of when and how this violation arises and when and how it is eliminated, not only in /kaŋ/ → *ˈkaŋ.ka* but also in /wuŋkunuŋ/ → *wuŋkunu*. Prince and Smolensky's argument for parallelism is really an argument against a serial theory with inviolable constraints. It is a valid objection to rule+-constraint theories like those discussed in Section 3.2, but it is not a valid objection to HS.

Prince and Smolensky's argument also shows the inadequacy of bottom-up serial theories. The serial derivation in (33) is one in which syllabification strictly precedes foot parsing, so the augmentation step is absolutely respectful of the difference between codas (e.g., in *[wun]$_\sigma$*) and non-codas (e.g., in *[ja]$_\sigma$k*) emerging from the syllabification step. As shown by augmentation-triggered resyllabification in tableau (36), HS is not strictly bottom-up. Indeed, that is the topic of the next section.

3.6 Harmonic Serialism is not bottom-up

Some arguments for parallelism in the literature rely on the assumption that serial theories are inherently bottom-up: the interaction of phonological phrasing and cliticization in English (McCarthy 2002: 146–149); the interaction of stress and sonorant syllabification in English (Pater 2000: 248–250); and the interaction of stress and syllabification in Tongan (Prince and Smolensky 1993/2004: 33–36). We discuss the two English cases in this section; for Tongan, see Elsman (this volume). As we will argue, HS has full availability of structural operations at every step of the derivation; thus, it is not bottom-up.

The prosodic structure assigned to a function word often depends on the larger context in which it finds itself. In English, for example, monosyllabic prepositions like *to* are reduced phrase-medially (47a) but unreduced phrase-finally (47b):

(47) Function word stress in English
(a) Unstressed phrase-medially
John spoke tŏ Bill. (pronounced *tə*)
(b) Stressed phrase-finally
Who did John speak tó? (pronounced *tú*)
Who did John speak tó yesterday?

The general rule is that monosyllabic function words (other than object pronouns) are stressed and consequently unreduced at the end of a phonological phrase (PPh), but they are otherwise unstressed and reduced.

Following Selkirk (1995), we assume that an unstressed, reduced function word like *tŏ* is not a prosodic word (PrWd) like $[\textit{John}]_{\text{PWd}}$ or $[\textit{spoke}]_{\text{PWd}}$. Instead, English function words are normally parsed as immediate constituents of the phonological phrase:

(48) Structure of reduced function word (after Selkirk 1995)
$\{[\text{John}]_{\text{PWd}}\}_{\text{PPh}}$ $\{[\text{spoke}]_{\text{PWd}}$ tŏ $[\text{Bill}]_{\text{PWd}}\}_{\text{PPh}}$

As in this example, throughout this section we will use brackets to delimit prosodic words and braces for phonological phrases, omitting the subscripts. See Selkirk (1995) for evidence that this is the correct structure, rather than attachment of *tŏ* to the preceding or following prosodic word.

Selkirk's analysis relies on a combination of alignment constraints and constraints on the prosodic hierarchy. Among the alignment constraints are two that require a lexical word and a prosodic word to begin and end together:

(49) ALIGN-L(LexWd, PWd)
Assign a violation mark for every lexical word whose left edge does not coincide with the left edge of a prosodic word.

(50) ALIGN-R(LexWd, PWd)
Assign a violation mark for every lexical word whose right edge does not coincide with the right edge of a prosodic word.

These constraints are violated by structures where a function word is incorporated into a preceding or following prosodic word: [*spoke tŏ*], [*tŏ Bill*].

The constraints on the prosodic hierarchy include these two:

(51) PWDCON
Assign a violation mark for every prosodic word that does not contain a lexical word.

(52) EXHAUSTIVITY(PPh)
Assign a violation mark for every constituent of type lower than prosodic word that is immediately dominated by PPh.

PWDCON is violated by any stressed function word like $[\text{tó}]_{PWd}$. EXHAUSTIVITY(PPh) is a no-skipping constraint on the levels of the prosodic hierarchy: because the phonological phrase immediately dominates the prosodic word in the hierarchy, a structure is marked if it contains an instance of phonological phrase immediately dominating some lesser constituent, such as the foot or syllable. The syllable *tŏ* in (48) violates this constraint.

A P-OT grammar that will produce the desired output in (48) is given in (53). The candidates represent the four logically possible dispositions of *to*, according to Selkirk (1995): as a 'free clitic' that is immediately dominated by phonological phrase (53a); as a free-standing prosodic word (53b); as a proclitic incorporated into the following prosodic word (53c); and as an enclitic incorporated into the preceding prosodic word (53d). Each of the four constraints in (49)–(52) is violated by one of these candidates. The winner is (53a) because it violates the lowest-ranking constraint, EXHAUSTIVITY(PPh).

(53) P-OT analysis of *(John) spoke to Bill* (after Selkirk 1995)

	spoke to Bill	AL-L (LexWd, PWd)	AL-R (LexWd, PWd)	PWD CON	EXH (PPH)
a. →	{[spoke] tŏ [Bill]}				1
b.	{[spoke] [tó] [Bill]}			1 W	L
c.	{[spoke] [tŏ Bill]}	1 W			L
d.	{[spoke tŏ] [Bill]}		1 W		L

Thus far, nothing has been said about the distribution of phonological phrases. In Selkirk (1995), the boundaries of phonological phrases are determined by the following constraint:

(54) ALIGN-R(Lex^{max}, PPh)
Assign a violation mark for every maximal projection of a lexical category whose right edge does not coincide with the right edge of a phonological phrase.

This constraint is unviolated in English surface forms. It is responsible for the single phonological phrase in ... {*spoke to Bill*}, and it is also responsible for the pair of phonological phrases in ... {*speak to*} {*yesterday*}.

Phrase-internally, a function word is parsed as an immediate constituent of a phonological phrase, as shown in (53), but this option is ruled out when the function word is phonological phrase-final. The responsible constraint is one that requires every phonological phrase to end in a prosodic word:

(55) ALIGN-R(PPh, PWd)
Assign a violation mark for every phonological phrase whose right edge does not coincide with the right edge of a prosodic word.

With ALIGN-R(LexWd, PWd) also prohibiting a phonological phrase-final function word from encliticizing onto the preceding prosodic word, the only option left is to parse the function word as a prosodic word on its own. This is a violation of PWDCON, which must therefore be ranked below ALIGN-R(Lex^{max}, PPh), ALIGN-R(PPh, PWd) and ALIGN-R(LexWd, PWd):

(56) P-OT analysis of *(Who did you) speak to yesterday?* (after Selkirk 1995)

speak to yesterday	AL-R (Lex^{max}, PPh)	AL-R (PPh, PWd)	AL-R (LexWd, PWd)	PWD CON	EXH (PPh)
a. → {[speak] [tó] } {[yesterday] }				1	
b. {[speak] tŏ [yesterday] }	1 W			L	1 W
c. {[speak] tŏ } {[yesterday] }		1 W		L	1 W
d. {[speak tŏ] } {[yesterday] }			1 W	L	

The interaction illustrated in tableau (56) is a top-down effect. Whether a function word is parsed as a prosodic word depends on where it is situated in the phonological phrase, which is the next level of structure above the prosodic word. McCarthy (2002) uses this top-down effect to argue for P-OT over HS. The essence of the argument is this: phonological phrases are constructed out of prosodic words, so the prosodic words cannot be built after the phonological phrases. But (56) shows that prosodic words can't be built without knowing where the boundaries of phonological phrases are. (That is why the crucial competitors in (56) include candidates that differ from one another in *both* prosodic word and phonological phrase structure.) There is no serial ordering of prosodic word construction and phonological phrase construction that can meet both of these requirements, so the prosodic words and phonological phrases need to be built simultaneously, by parallel rather than serial optimization.

This argument for P-OT goes through only under the assumption that the serial alternative is *strictly* bottom-up – specifically, it argues against a theory that builds prosodic word structure, then builds phonological phrase structure, but is unable to revise the prosodic word structure to reflect the changed conditions occasioned by the appearance of phonological phrase structure. This assumption is neither necessary nor even plausible in HS, however, as we now demonstrate.

We will first lay out the derivations and then explain how the grammar produces them. The derivation of *(John) spoke to Bill* proceeds as follows:

(57) Derivation of (John) spoke to Bill

Step 1	[spoke] to Bill (*or* spoke to [Bill])
Step 2	[spoke] to [Bill]
Step 3	{[spoke] to [Bill]}
Step 4	{[spoke] to [Bill]} – *Convergence*

At steps 1 and 2, prosodic word structure is built. The grammar does not specify whether *spoke* or *Bill* is parsed into a prosodic word first, but by the end of step 2 both prosodic words have been created. (This is another example of a convergent tie – cf. tableau (17).) At step 3, every prosodic word that the grammar requires has been built, and the only remaining harmony-improving operation is parsing the VP into a phonological phrase. The output of step 4 is identical with the output of step 3, so the derivation terminates.

A grammar that will produce this derivation is given in (58)–(61). It includes all of the constraints and rankings that are in Selkirk's P-OT analysis, plus two additional rankings that will be explained shortly. At step 1, the options include building a prosodic word (58a), doing nothing (58b), and building a phonological phrase (58c). Doing nothing is no option at all, because (58b) is harmonically bounded by the winner. Building a phonological phrase first (58c) is ruled out by ranking ALIGN-L/R(LexWd, PWd) above ALIGN-R(Lex^{max}, PPh). This is the first of the two rankings that are required in the HS analysis but are not required (though consistent with) the P-OT analysis:

(58) Step 1 of HS analysis of *(John) spoke to Bill*

	spoke to Bill	AL-L (LexWd, PWd)	AL-R (LexWd, PWd)	AL-R (Lex^{max}, PPh)	AL-R (PPh, PWd)	PWD CON	EXH (PPh)
a. →	[spoke] to Bill	1	1	1			
b.	spoke to Bill	2 W	2 W	1			
c.	{spoke to Bill}	2 W	2 W	L	1 W		3 W

We are ignoring the convergent tie between (58a) and *spoke to [Bill].*

At step 2, the other prosodic word is built in response to the same constraint interaction as step 1:

(59) Step 2 of HS analysis of *(John) spoke to Bill*

	[spoke] to Bill	AL-L (LexWd, PWd)	AL-R (LexWd, PWd)	AL-R (Lex^{max}, PPh)	AL-R (PPh, PWd)	PWD CON	EXH (PPh)
a. →	[spoke] to [Bill]			1			
b.	[spoke] to Bill	1 W	1 W	1			
c.	{[spoke] to Bill}	1 W	1 W	L	1 W		2 W

At step 3, the choice is between building a phonological phrase (60a), doing nothing (60b), or parsing the function word *to* as a prosodic word. The highest-ranking constraint that distinguishes among these candidates is ALIGN-R(Lex^{max}, PPh), and it favors building a phonological phrase.

(60) Step 3 of HS analysis of *(John) spoke to Bill*

	[spoke] to [Bill]	AL-L (LexWd, PWd)	AL-R (LexWd, PWd)	AL-R (Lex^{max}, PPh)	AL-R (PPh, PWd)	PWD CON	EXH (PPh)
a. →	{[spoke] to [Bill]}						1
b.	[spoke] to [Bill]			1 W			L
c.	[spoke] [to] [Bill]			1 W		1 W	L

Finally, at step 4, there are two main options, doing nothing (61a) or parsing *to* as a prosodic word (61b). As in Selkirk's analysis, ranking PWDCON over EXHAUSTIVITY(PPh) disfavors the prosodic word parse of *to* in phrase-medial position, and so the derivation converges.

(61) Step 4 of HS analysis of *(John) spoke to Bill*

{[spoke] to [Bill]}	AL-L (LexWd, PWd)	AL-R (LexWd, PWd)	AL-R (Lex^{max}, PPh)	AL-R (PPh, PWd)	PWD CON	EXH (PPh)
a. → {[spoke] to [Bill]}						1
b. {[spoke] [to] [Bill]}					1 W	L

In Selkirk's P-OT analysis, although function words are not normally parsed as prosodic words because PWDCON dominates EXHAUSTIVITY(PPh), a phonological phrase-final function word is parsed as a prosodic word because ALIGN-R(PPh, PWd) dominates PWDCON. Exactly the same interaction occurs in the HS analysis, but its effect is intrinsically ordered after construction of the phonological phrase. We take up the derivation of *(Who did John) speak to yesterday?* at step 3, when *speak* and *yesterday* have already been parsed as prosodic words. The winning candidate parses the prosodic word *yesterday* as a phonological phrase. It harmonically bounds all of its competitors, as tableau (62) shows.

(62) Step 3 of HS analysis of *(Who did John) speak to yesterday?*

[speak] to [yesterday]	AL-L (LexWd, PWd)	AL-R (LexWd, PWd)	AL-R (Lex^{max}, PPh)	AL-R (PPh, PWd)	PWD CON	EXH (PPh)
a. → [speak] to {[yesterday]}			1			
b. [speak] to [yesterday]			2 W			
c. {[speak] to} [yesterday]			1	1 W		1 W
c. [speak] [to] [yesterday]			2 W		1 W	

At step 4, there is a critical conflict among several constraints. The winner in (63a) violates both ALIGN-R(PPh, PWd) and EXHAUSTIVITY(PPh). The losers satisfy both of these constraints, but at the expense of violating higher-ranking ALIGN-R(Lex^{max}, PPh). The ranking of ALIGN-R(Lex^{max}, PPh) above ALIGN-R(PPh, PWd) required in this tableau is the second of the two rankings that are necessary in the HS analysis but unnecessary in (though consistent with) Selkirk's P-OT analysis.

(63) Step 4 of HS analysis of *(Who did John) speak to yesterday?*

[speak] to {[yesterday]}	AL-L (LexWd, PWd)	AL-R (LexWd, PWd)	AL-R (Lex^{max}, PPh)	AL-R (PPh, PWd)	PWD CON	EXH (PPh)
a. → {[speak] to} {[yesterday]}				1		1
b. [speak] to {[yesterday]}			1 W	L		L
c. [speak] [to] {[yesterday]}			1 W	L	1 W	L

Step 5 is the point in the derivation where the HS analysis shows that it is capable of producing top-down effects. The winner of step 4 has a phonological phrase, *{[speak] to}*, that does not end in a prosodic word. When prosodic words were first built at the earliest steps of the derivation, there was no reason to create a prosodic word *[to]*, and there was a very good reason not to: PWDCON. At that derivational stage, there was no phonological phrase structure yet, and so the winning candidates vacuously satisfied ALIGN-R(PPh, PWd). But now, at step 5, there is phonological phrase structure present and the markedness of *{[speak] to}* is apparent. Because ALIGN-R(PPh, PWd) dominates PWDCON, the prosodic structure of *to* is revised:

(64) Step 5 of HS analysis of *(Who did John) speak to yesterday?*

{[speak] to} {[yesterday]}	AL-L (LexWd, PWd)	AL-R (LexWd, PWd)	AL-R (Lex^{max}, PPh)	AL-R (PPh, PWd)	PWD CON	EXH (PPh)
a. → {[speak] [to]} {[yesterday]}					1	
b. {[speak] to} {[yesterday]}				1 W	L	1 W

The derivation converges at step 6.

The interaction in tableau (64) is a top-down effect. The presence of the phonological phrase is forcing the creation of prosodic words where none was required previously. This derivation is certainly not strictly bottom-up, and it therefore constitutes a demonstration that HS does not entail strict bottom-upness. In a strictly bottom-up derivation, decisions about the parse into prosodic words would be made once and for all before any phonological phrase is constructed. One might imagine constructing a derivational theory of prosodic structure that is strictly bottom-up in this sense, but HS is not such a theory because it has *full availability of structural operations*. With all structural operations available throughout the derivation, as long as there are markedness constraints that mention two levels of the prosodic hierarchy – such as ALIGN-R(PPh, PWd) – there can be top-down as well as bottom-up effects.

McCarthy's (2002) argument for P-OT, based on Selkirk's (1995) analysis, is really an argument against strict bottom-up derivations. But, as we have shown, this is a straw man: HS is not strictly bottom-up, and indeed it is capable of analyzing the English function word data with exactly the same constraints that Selkirk used.

Pater (2000: 248–250) presents a P-OT analysis of the sonorant destressing phenomenon in English, and he argues that it too demonstrates the need for parallelism. To keep the discussion manageable, we focus on a representative subset of the data: words of the form #H̀Xσ́ ..., where H denotes a heavy initial syllable and X is the syllable whose stressedness is the point of the analysis.

When X has an obstruent coda, it is regularly stressed:

(65) #H̀**X**σ́... words, X=CVO

ˌtɪmˌ**bʌk**ˈtuː	Timbuctoo
ˌdiːˌ**lɛk**ˈteːʃn̩	delectation
ˌsɪnˌ**tæg**ˈmærɪk	syntagmatic
ˌɪnˌ**dɪg**ˈneːʃn̩	indignation

When X would be expected to have a sonorant coda, however, it is unstressed and the would-be coda is parsed as the nucleus:

(66) #H̀**X**σ́... words, X=CN̩

ˌgɔɹ**gn̩**ˈzoːlə	gorgonzola
ˌmoː**zm̩**ˈbiːk	Mozambique
ˌsɑj**ml̩**ˈteːnijəs	simultaneous
ˌpɛn**sl̩**ˈveːnijə	Pennsylvania

Neither generalization is exceptionless (for which see Burzio 2007, Elfner 2007, Ross 1972), but we follow Pater in assuming that this is a grammatically controlled pattern that must be accounted for in any analysis.

In Pater's P-OT analysis, the behavior exemplified in (66) is a consequence of two constraints. One, *CLASH-HEAD, rules out *ˌ*gɔɹ*ˌ*gan*ˈ*zoːlə* because of the stress clash between *gan* and the main-stressed syllable *zoː*.[19] The other constraint is, in our terms, WSP-WORD, which is violated by *ˌ*gɔɹgan*ˈ*zoːlə* because *gan* is a heavy syllable that is unstressed and unfooted (cf. (23)). Both of these constraints dominate *NUC/SON,

19 We arbitrarily assume that the underlying vowel of the medial syllable of *gorgonzola* is /ɑ/.

which is violated by the sonorant consonant nucleus of the syllable *gn̩*. Tableau (67) presents this ranking argument.

(67) *gorgonzola* in Pater's (2000) P-OT analysis

	/gɔɹganzoːlə/	*Clash-Head	WSP-Word	*Nuc/Son
a. →	ˌgɔɹgn̩ˈzoːlə			1
b.	ˌgɔɹˌganˈzoːlə	1 W		L
c.	ˌgɔɹganˈzoːlə		1 W	L

The examples in (65) work differently, however, because *Clash-Head is itself crucially dominated by *Nuc/Obst, which ensures that *ˌtɪmbk̩ˈtuː* is not a viable option:

(68) *Timbuctoo* in Pater's (2000) P-OT analysis

	/tɪmbʌktuː/	*Nuc/Obst	WSP-Word	*Clash-Head	*Nuc/Son
a. →	ˌtɪmˌbʌkˈtuː			1	
b.	ˌtɪmbk̩ˈtuː	1 W		L	
c.	ˌtɪmbʌkˈtuː		1 W	L	

The argument for parallelism goes like this (Pater 2000: 250):

> In this treatment of the asymmetry between sonorant-final and obstruent-final syllables, it appears to be crucial that syllabification and stress assignment be evaluated in parallel, rather than established and evaluated in sequence. Whether a syllable in a pretonic sequence is unstressed depends in part upon whether the syllable-final consonant can be parsed as a nucleus. Whether a sonorant is parsed as a nucleus in turn depends upon whether it is unstressed. This sort of interdependence between the well-formedness of stress and syllable structure is awkward to express in a theory in which syllabification derivationally precedes stress placement...

Like the argument for parallelism discussed in the previous section, this is an argument against a particular kind of serial theory and not against serial theories generally. It is an argument against strict bottom-up serialism, in which all syllabification operations precede all stress-assignment operations. Bottom-up serialism allows stress to depend on syllabification, but it does not allow syllabification to depend on stress. It is not possible to construct a strictly bottom-up serial analysis that mimics the analysis of *gorgonzola* in (67).

But HS is not bottom-up serialism. In HS, syllabification and stress assignment operations are available at every step of the derivation. If

ˌgɔɹganˈzoːlə is the input to some intermediate step of the derivation, then GEN can produce the candidates *ˌgɔɹgn̩ˈzoːlə* and *ˌgɔɹˌganˈzoːlə*, one of which has changed the syllabification and the other of which has changed the stress. This means that HS can in principle have exactly the same candidate competition as the P-OT analyses in (67) and (68). Thus, there is no prima facie case here for P-OT over HS.

3.7 Limits on cross-level interactions in Harmonic Serialism

P-OT and HS analyses of the same data may place different demands on CON. Because HS maps underlying to surface forms via a gradual, harmonically improving derivation, it will sometimes require more finely differentiated constraints than P-OT. Because any theory of CON has typological consequences, this point of difference offers a way of falsifying a HS analysis using techniques that are standard in OT (see, e.g., McCarthy 2008: chapter 5).

To make this concrete, consider the interaction of apocope and vowel shortening in Yawelmani (Archangeli 1984, Kisseberth 1970, Kuroda 1967, Newman 1944, and many others). Yawelmani has a process of *i*-epenthesis that breaks up unsyllabifiable consonant clusters:

(69) Epenthesis in Yawelmani

/ʔilk-hin/	ʔi.lik.hin	'sing (nonfuture)'
/lihm-hin/	li.him.hin	'run (nonfuture)'
cf.		
/ʔilk-al/	ʔil.kal	'sing (dubitative)'
/lihm-al/	lih.mal	'run (dubitative)'

It also has a process that shortens long vowels in closed syllables:

(70) Closed syllable shortening in Yawelmani

/laːn-hin/	lan.hin	'hear (nonfuture)'
/ṣaːp-hin/	ṣap.hin	'burn (nonfuture)'
cf.		
/laːn-al/	laː.nal	'hear (dubitative)'
/ṣaːp-al/	ṣaː.pal	'burn (dubitative)'

Together, these processes ensure that Yawelmani has no surface CVCC or CVːC syllables.

A process of apocope deletes final short vowels. When apocope would create a final CV:C syllable, vowel shortening occurs. We might also expect to see epenthesis called on when apocope would create a final CVCC syllable. In fact, though, apocope is blocked in that situation.

(71) Apocope in Yawelmani

/taxa:-kʔa/	ta.xakʔ	'bring!'
/taxa:-mi/	ta.xam	'having brought'
cf.		
/xat-kʔa/	xat.kʔa	'eat!'
	not *xa.tikʔ	
/xat-mi/	xat.mi	'having eaten'
	not *xa.tim	

In P-OT, the difference in (71) can be attributed to a difference in the ranking of two faithfulness constraints relative to the markedness constraint that demands apocope, *V#. *V# dominates MAX-μ to account for *ta.xakʔ* (tableau (72)), but it must itself be dominated by DEP-V to account for **xa.tikʔ* (tableau (73)). In both cases, undominated *SUPERHEAVY rules out the CV:C and CVCC syllables that would result from apocope alone:

(72) *V# ≫ MAX(μ) in Yawelmani (P-OT analysis)

	/taxa:-kʔa/	*SUPER	*V#	MAX-μ	MAX-V
a. →	ta.xakʔ			2	1
b.	ta.xa:.kʔa		1 W	L	L
c.	ta.xa:kʔ	1 W		1 L	1

(73) DEP-V ≫ *V# in Yawelmani (P-OT analysis)

	/xat-kʔa/	*SUPER	DEP-V	*V#	MAX-μ	MAX-V
a. →	xat.kʔa			1		
b.	xatkʔ	1 W		L	1 W	1 W
c.	xa.tikʔ		1 W	L	1 W	1 W

This constraint set is insufficient for a HS analysis of Yawelmani, however. The HS factorial typology of this constraint set over the inputs in (72) and (73) was computed using OT-Help (Staubs *et al.* 2010), under the assumption that deletion, shortening, and epenthesis each require their own derivational step (though resyllabification does not). This typology includes only the languages in (74), and Yawelmani is not among them.

(74) A factorial typology in HS

taxa:k$^?$a	xatk$^?$a
taxa:k$^?$	xatk$^?$
taxak$^?$	xatk$^?$
taxa:k$^?$	xatik$^?$
taxak$^?$	xatik$^?$

The reason why Yawelmani is not in this typology is that *SUPERHEAVY, as the only constraint against both CVCC and CV:C syllables, is simply too general. At the first step of the derivation of /taxa:-k$^?$a/ (after initial syllabification), apocope needs to be allowed, yielding *ta.xa:k$^?$*, which will undergo vowel shortening at the next step. At the first step of the derivation of /xat-k$^?$a/, however, apocope needs to be blocked, so the derivation immediately converges on faithful *xat.k$^?$a*. The constraint *SUPERHEAVY is obviously inadequate to this task: if ranked above *V#, it will wrongly block apocope in both **ta.xa:.k$^?$a* and *xat.k$^?$a*, and if ranked below *V#, it will wrongly allow apocope in both *ta.xa:k$^?$* (→ *ta.xak$^?$*) and *xatk$^?$*. At the first step of the derivation, then, there is no way of knowing that the mapping /taxa:-k$^?$a/ → *ta.xa:k$^?$* should be allowed because at the next step the violation of *SUPERHEAVY will be eliminated by shortening, while the mapping /xat-k$^?$a/ → *xatk$^?$* should be blocked by the same constraint.

This problem arises because positing a CON with just *SUPERHEAVY entails that CV:C and CVCC syllables are always equally marked. Now suppose we include *COMPLEX-CODA in CON. This constraint introduces an additional bias against CVCC syllables, and thereby eliminates any need for doing shortening and apocope in parallel. At step 1, shown in tableaux (75) and (76), *V# compels violation of low-ranking *SUPERHEAVY, but it is unable to compel violation of high-ranking *COMPLEX-CODA.

(75) Step 1 of /taxa:-k$^?$a/ → ta.xa:k$^?$ → ta.xak$^?$ (HS analysis)

	/taxa:-k$^?$a/	*COMP-CODA	*V#	*SUPER	MAX-V	MAX-μ
a. →	ta.xa:k$^?$			1	1	1
b.	ta.xa:.k$^?$a		1 W	L	L	L

(76) Step 1 of /xat-k$^?$a/ → xat.k$^?$a (HS analysis)

	/xat-k$^?$a/	*COMP-CODA	*V#	*SUPER	MAX-V	MAX-μ
a. →	xat.k$^?$a		1			
b.	xatk$^?$	1 W	L	1 W	1 W	1 W

The derivation of *xat.kʔa* has already converged. The derivation of *ta.xakʔ* continues for one more step before the convergence step:

(77) Step 2 of /taxa:-kʔa/ → ta.xa:kʔ → ta.xakʔ (HS analysis)

	ta.xa:kʔ	*COMP-CODA	*V#	*SUPER	MAX-V	MAX-μ
a. →	ta.xakʔ					1
b.	ta.xa:kʔ			1 W		L

The upshot is that the derivations of *ta.xakʔ* and *xat.kʔa* can be distinguished by the markedness of the intermediate form that would result from apocope: *ta.xa:kʔ* is allowed because *SUPERHEAVY is ranked below *V#, but *xatkʔ* is not because *COMPLEX-CODA is ranked above *V#. As we have so often seen in HS analyses, a surface-unviolated constraint (here *SUPERHEAVY) is violated at an intermediate step of the derivation of *ta.xakʔ*.

We have shown that the HS analysis of Yawelmani requires both *COMPLEX-CODA and *SUPERHEAVY, while the P-OT analysis requires only *SUPERHEAVY. Is this merely an unwelcome expedient to which HS is forced by its assumptions, or does even P-OT need both of these constraints? Data from Sudanese Arabic show that both constraints are indeed required in both the parallel and serial theories. This language has a syncope process that can produce surface CVC or CV:C syllables (78a), but is blocked when a CVCC syllable would result (78b).

(78) Syncope in Sudanese Arabic (Hamid 1984: 82ff.)

a.	/fihim-u/	ˈfihmu	'understood (m. pl.)'	
	/jika:tib-u/	jiˈka:dbu	'correspond (m. pl.)'	
	/ma:sik-a/	ˈma:ska	'holding (f. sg.)'	
b.	/jaktub-u/	ˈjaktibu	'write(m. pl.)'	*ˈjagdbu
	/jitardʑim-u/	jiˈtardʑimu	'translate (m. pl.)'	*jiˈtardʑmu

It is impossible in P-OT or HS to analyze this language without a markedness constraint that disfavors *ˈ*jagdbu* but not *jiˈka:dbu*. That constraint is *COMPLEX-CODA, which is therefore indispensible even in P-OT.

In OT, constraints are hypotheses about CON that are subject to empirical verification through language typology. Typology therefore provides a check on constraint proliferation. As Yawelmani illustrates, harmonic improvement in HS derivations may require a richer constraint set than a P-OT analysis of the same facts. But language typology offers a straightforward way of testing proposed constraints, with results that could in principle falsify a proposed HS analysis and even undermine the theory in which it is embedded.

As in P-OT, the logic of language typology can be used in HS to discriminate among competing hypotheses about how some data should be analyzed. McCarthy and Prince (1993b) present an analysis of word-final vowel shortening in Axininca Campa[20] that appears to require parallelism. A HS reanalysis is possible, but not the most obvious one.

Stress in Axininca Campa is generally quantity-sensitive.[21] In sequences of light syllables the pattern is left-to-right iambic; every other syllable beginning with the second is stressed (examples in (79a)), and this pattern is restarted after a stress-attracting heavy syllable. Stress nearly always avoids the final syllable, however. Disyllables generally receive initial stress, as in (79b), while polysyllabic words ending in an even number of light syllables vary between shifting the last stress to the penult and omitting the stress altogether, (79c). McCarthy and Prince (1993b) analyze the penult stress cases with a final trochee.

(79) Axininca stress (Payne, Payne and Santos 1982: 188–189, 193)

a.	tʃʰoˈrina	'species of palm'
	iˈtʃʰikaˈkina	'he has cut me'
b.	ˈsari	'macaw'
	ˈkito	'shrimp'
c.	kiˈmiˈtaka ~ kiˈmitaka	'perhaps'
	hoˈtiˈtana ~ hoˈtitana	'he let me in'

Because the distinction between main and secondary stress is unimportant for our purposes here, we have not indicated it in the transcriptions.

On the surface, syllables with a long vowel or diphthong always receive stress, but this requirement is satisfied somewhat heterogeneously when it conflicts with a constraint prohibiting stress on the final syllable. When the final syllable underlyingly contains a long vowel, it surfaces as short and unstressed, as the examples in (80) show.[22] But

20 This language is now known as Ajyˈininka Apurucayali.

21 The system is more complicated than heavy/light, however. Payne *et al.* (1982) propose a weight/prominence continuum to account for the relative likelihood of syllable types attracting or repelling stress.

22 Examples like *sima/no-sima-ni* 'fish'/'my fish' show that this process is indeed shortening rather than lengthening before a suffix. Examples like *mi:* 'otter' show that shortening is blocked in monosyllables – a typical word-minimality effect (McCarthy and Prince 1993b: 164).

when the final syllable contains a diphthong, it surfaces faithfully and receives stress.

(80) Final long vowels shorten (Payne 1981: 119–121)

/sampa:/	ˈsampa	‘balsa’	cf. no-sampa:-ti ‘my balsa’
/sawo:/	ˈsawo	‘case’	cf. no-sawo:-ti ‘my case’
/cʰimi:/	ˈcʰimi	‘ant’	cf. no-cʰimi:-ti ‘my ant’

(81) Final diphthongs stressed (Payne, Payne and Santos 1982: 187–188)

noˈpai	‘my sugarcane’
ampoˈkai	‘we will come back’
kiˈtiʃiˈtakoˈtai	‘we came in the morning’
ˈa:ˈtai	‘we will go’

McCarthy and Prince’s (1993b) P-OT analysis accounts for the differential treatment of final long vowels and diphthongs in terms of differing degrees of unfaithfulness. WSP, which requires all heavy syllables to be stressed, and NON-FINALITY, which requires that final syllables *not* be stressed, are in conflict when a diphthong or a long vowel is in the final syllable. In the long vowel case (tableau (82)), both markedness constraints are satisfied by shortening the vowel (i.e., deleting a mora), so WSP and NON-FINALITY dominate MAX-μ. In the diphthong case (tableau (83)), a violation of MAX-V would be needed for simultaneous satisfaction of the markedness constraints. Instead, we actually find that a violation of NON-FINALITY is tolerated in order to avoid violation of MAX-V and WSP. Thus, WSP and MAX-V both dominate NON-FINALITY.[EX]

(82) Final long vowels shortened in P-OT

	/sawo:/	WSP	MAX-V	NON-FIN	MAX-μ
a. →	(ˈsawo)				1
b.	(ˈsawo:)	1 W			L
c.	(saˈwo:)			1 W	L

(83) Final diphthongs stressed in P-OT

	/nopai/	WSP	MAX-V	NON-FIN	MAX-μ
a. →	(noˈpai)			1	
b.	(ˈnopa)		1 W	L	1 W
c.	(ˈnopai)	1 W		L	

This is a CLI between a moraic process (vowel shortening) and stress. When a final heavy syllable can be shortened, stress will surface on a different syllable or be omitted, but when a final heavy syllable cannot be shortened because it contains a diphthong, it must surface with the stress. The ranking of NON-FINALITY above MAX-μ and below MAX-V determines how WSP will be satisfied in final syllables: by shortening if possible, but otherwise by stressing.

When we try to restate this analysis derivationally, however, we run into a problem. Under the ranking in (82) and (83), the first step of the derivation from /sawo:/ will produce the output *(saˈwo:)*. This is shown in tableau (84).

(84) Step 1 from /sawo:/

	/sawo:/	WSP	MAX-V	NON-FIN	MAX-μ
a. →	(saˈwo:)			1	
b.	(ˈsawo:)	1 W		L	

Throughout, we have assumed that GEN cannot shorten a vowel and move a stress together, in a single step. It follows, then, that the next step after *(saˈwo:)* will include a candidate with shortening *(saˈwo)* and a candidate with stress shift *(ˈsawo:)*, but not a candidate with both *(ˈsawo:)*. Unfortunately, neither *(saˈwo)* nor *(ˈsawo:)* is more harmonic than *(saˈwo:)* – the former merely adds a violation of MAX-μ, and the latter swaps a violation of NON-FINALITY for a violation of higher-ranking WSP. Therefore, the derivation will immediately converge on the wrong surface form, **(saˈwo:)*.

One obvious but typologically unsound line of attack is to separate WSP into two constraints, one requiring that long vowels be stressed and the other requiring that diphthongs be stressed. If WSP-DIPHTHONG is ranked above NON-FINALITY, then /nopai/ will correctly map to *no(ˈpai)* at step 1:

(85) Step 1 of /nopai/ with WSP-DIPHTHONG and WSP-V:

	/nopai/	WSP-DIPH	NON-FIN	WSP-V:
a. →	(noˈpai)		1	
b.	(ˈnopai)	1 W	L	

And if WSP-V: is ranked below NON-FINALITY, then /sawo:/ will map to *(ˈsawo:)* at step 1, which will change to *(ˈsawo)* at step 2 before converging.

(86) Step 1 of /sawoː/ with WSP-DIPHTHONG and WSP-V:

	/sawoː/	WSP-DIPH	NON-FIN	WSP-Vː
a. →	(ˈsawoː)			1
b.	(saˈwoː)		1 W	L

Although this analysis initially looks promising, it is typologically shaky. Two languages are known to exist in which long vowels and diphthongs have different weight in stress assignment, and in both cases the long vowels are treated as heavier than the diphthongs. (The languages are Kara and Maori (de Lacy 1997, Gordon 2006: 123).) If this is a valid typological generalization, then it must mean that WSP-DIPHTHONG never dominates WSP-V. The ranking in (85) and (86) is inconsistent with that.

Although typological considerations militate against one imaginable account of the Axininca facts, they also point the way toward a better one. Buckley (1998) presents extensive cross-linguistic evidence that NON-FINALITY cannot explain all cases where final long vowels are prohibited, so a constraint against final length *tout court* is needed in CON.[23] For example, Italian regularly lengthens stressed vowels in open syllables (*kapiˈtaːno* 'captain'), but not if the stressed syllable is word-final (*kafˈfe* 'coffee'). The dispreference for word-final long vowels extends even to the sandhi process *raddopiamento sintattico*, in which lengthening of final stressed syllables is accomplished by consonant gemination rather than vowel lengthening: *kafˈfenˈneːro* 'black coffee'.

If CON includes a constraint against final long vowels, then the *raison d'être* for McCarthy and Prince's analysis vanishes. The point of their analysis is to use NON-FINALITY and the faithfulness difference between shortening a long vowel and simplifying a diphthong to explain why final long vowels are unstressed and short while final diphthongs are stressed and unchanged. This explanation is unnecessary if final long vowels are marked regardless of stress.

As the Axininca Campa and Arabic examples have illustrated, analyses of CLIs in HS may require finer differentiation of markedness constraints than analyses of the same data in P-OT. But these richer constraint sets (and ultimately HS itself) are falsifiable using techniques of typological analysis that are standard in OT (see, e.g., McCarthy 2008: chapter 5), and that have now been automated for HS with the OT-Help software package (Staubs *et al.* 2010).

23 On the functional basis of final vowel length neutralization, see Myers and Hansen (2007).

3.8 Conclusion

Cross-level interactions are a good place to look for differences between parallel and serial theories. It is no surprise that they have figured in almost all arguments for parallelism,[24] nor that they present some of the severest challenges to rule+constraint theories.

Cross-level interactions have certain properties that can challenge derivational theories of constraint satisfaction. In many cases, they require surface-true constraints to be violated in the course of the derivation. They also show that derivations cannot always build structure in a strictly monotonic, bottom-up fashion.

We have argued that Harmonic Serialism has these characteristics. Because it is simply a derivational version of Optimality Theory, it has violable constraints. There is nothing in Harmonic Serialism that requires surface-unviolated constraints to be unviolated throughout the derivation; indeed, it is hard to imagine how this requirement could even be stipulated within the theory. Further, there is nothing in Harmonic Serialism that requires structure to be built from the bottom up; rather, every structure-building or -altering operation in GEN is available in principle at every step of the derivation.

Perhaps the most important result to emerge from this study is this. Serialism is nothing more than a vague description of a large class of theories that posit intermediate derivational steps. Critiques of serialism necessarily deal with specifics, and those specifics are not something that is known *a priori*. Different serial theories have very different properties, and facts that are a problem for one are very likely not a problem for all. The arguments against serialism in the OT literature are, as we have shown, arguments against specific serial theories that have very little to do with Harmonic Serialism.

24 CLIs are not the only source of arguments for parallelism in the OT literature. One other prominent argument is based on reduplicative overapplication (McCarthy and Prince 1995, 1999); for a HS approach, see McCarthy, Kimper, and Mullin (2012); for a general critique of the argument, see among others Inkelas and Zoll (2005) and Kiparsky (2010). Another prominent argument comes from phonologically-conditioned allomorphy (Mascaró 1996, Mester 1994, Tranel 1996, among others); for a HS approach, see van Oostendorp (2009) and cf. Wolf (2008); for recent discussion of the issue and references to related work, see Nevins (2011).

References

Allen, W. S. (1973) *Accent and Rhythm*. Cambridge: Cambridge University Press.

Archangeli, Diana (1984) *Underspecification in Yawelmani Phonology and Morphology*. Doctoral dissertation, MIT, Cambridge, MA.

Beckman, Jill (1998) *Positional Faithfulness*. Doctoral dissertation, University of Massachusetts Amherst, Amherst, MA. Available on Rutgers Optimality Archive, ROA-234.

Broselow, Ellen (1982) On predicting the interaction of stress and epenthesis. *Glossa* 16: 115–132.

Buckley, Eugene (1998) Iambic lengthening and final vowels. *International Journal of American Linguistics* 64: 179–223.

Burzio, Luigi (2007) Phonology and phonetics of English stress and vowel reduction. *Language Sciences* 29: 154–176.

Calabrese, Andrea (2009) Markedness theory vs. phonological idiosyncrasies in a realistic model of language. In Charles Cairns and Eric Raimy (eds), *Contemporary Views on Architecture and Representations in Phonological Theory*, 261–304. Cambridge, MA: MIT Press.

Chomsky, Noam and Morris Halle (1968) *The Sound Pattern of English*. New York: Harper and Row.

de Lacy, Paul (1997) *Prosodic Categorisation*. Master's thesis, University of Auckland, Auckland, New Zealand.

Derbyshire, Desmond (1979) *Hixkaryana*. Amsterdam: North-Holland.

Elfner, Emily. (2007) Moraic faithfulness: Evidence from Blackfoot and English. Unpublished paper, University of Massachusetts Amherst, Amherst, MA. Handout of paper presented at the 15th Manchester Phonology Meeting, University of Manchester, Manchester, England, May 26. Available at http://emilyelfner.sites.olt.ubc.ca/files/2014/10/elfner_2007_morafaith_mfm_handout.pdf

Fortson IV, Benjamin W. (2008) *Language and Rhythm in Plautus*. Berlin: Walter de Gruyter.

Goldsmith, John (1990) *Autosegmental and Metrical Phonology*. Oxford and Cambridge, MA: Blackwell.

Gordon, Matthew Kelly (2006) *Syllable Weight: Phonetics, Phonology, Typology*. New York: Routledge.

Gouskova, Maria (2003) *Deriving Economy: Syncope in Optimality Theory*. Doctoral dissertation, University of Massachusetts Amherst, Amherst, MA. Available on Rutgers Optimality Archive, ROA-610.

Hale, Kenneth (1973) Deep-surface canonical disparities in relation to analysis and change: An Australian example. In Thomas Sebeok (ed.), *Current Trends in Linguistics*, 401–458. The Hague: Mouton.

Hale, Kenneth, Ann Farmer, David Nash and Jane Simpson. (1981) A preliminary dictionary of Lardil. Unpublished paper, MIT, Cambridge, MA.

Hale, Kenneth and David Nash (1997) Damin and Lardil phonotactics. In Darrell Tryon and Michael Walsh (eds), *Boundary Rider: Essays in Honour of Geoffrey O'Grady*, 247–259. Canberra: Australian National University. Pacific Linguistics C-136.

Hamid, Abdel Halim M. (1984) *A Descriptive Analysis of Sudanese Colloquial Arabic Phonology*. Doctoral dissertation, University of Illinois, Urbana, IL.
Hayes, Bruce (1980) *A Metrical Theory of Stress Rules*. Doctoral dissertation, MIT, Cambridge, MA. Published by Garland Press, New York, 1985.
Hayes, Bruce (1995) *Metrical Stress Theory: Principles and Case Studies*. Chicago: The University of Chicago Press.
Hyde, Brett (2007) Non-finality and weight-sensitivity. *Phonology* 24: 287–334.
Inkelas, Sharon and Cheryl Zoll (2005) *Reduplication: Doubling in Morphology*. Cambridge; New York: Cambridge University Press.
Ito, Junko (1989) A prosodic theory of epenthesis. *Natural Language and Linguistic Theory* 7: 217–259.
Kager, René. (1994) Ternary rhythm in alignment theory. Unpublished paper, Utrecht University, Utrecht. Available on Rutgers Optimality Archive, ROA-35.
Kager, René (1995) The metrical theory of word stress. In John A. Goldsmith (ed.), *The Handbook of Phonological Theory*, 367–402. Cambridge, MA, and Oxford, UK: Blackwell.
Kager, René (1999) *Optimality Theory*. Cambridge: Cambridge University Press.
Kenstowicz, Michael (1994a) *Phonology in Generative Grammar*. Oxford: Blackwell.
Kenstowicz, Michael (1994b) Syllabification in Chukchee: A constraints-based analysis. In Alice Davison, Nicole Maier, Glaucia Silva and Wan Su Yan (eds), *Proceedings of the Formal Linguistics Society of Mid-America 4*, 160–181. Iowa City: Department of Linguistics, University of Iowa. Available on Rutgers Optimality Archive, ROA-30.
Kiparsky, Paul (1973) Phonological representations. In O. Fujimura (ed.), *Three Dimensions of Linguistic Theory*, 3–136. Tokyo: TEC.
Kiparsky, Paul (1982) Lexical phonology and morphology. In I. S. Yang (ed.), *Linguistics in the Morning Calm*, 3–91. Seoul: Hanshin.
Kiparsky, Paul (2010) Reduplication in Stratal OT. In Linda Uyechi and Lian Hee Wee (eds), *Reality Exploration and Discovery: Pattern Interaction in Language and Life*, 125–142. Stanford, CA: CSLI Publications.
Kisseberth, Charles (1970) On the functional unity of phonological rules. *Linguistic Inquiry* 1: 291–306.
Klokeid, Terry (1976) *Topics in Lardil Grammar*. Doctoral dissertation, MIT, Cambridge, MA.
Kuroda, S.-Y. (1967) *Yawelmani Phonology*. Cambridge, MA: MIT Press.
Lahiri, Aditi, Tomas Riad and Haike Jacobs (1999) Diachronic prosody. In Harry van der Hulst (ed.), *Word Prosodic Systems in the Languages of Europe*, 335–422. Berlin: De Gruyter.
Liberman, Mark and Alan Prince (1977) On stress and linguistic rhythm. *Linguistic Inquiry* 8: 249–336.
Mascaró, Joan (1996) External allomorphy as emergence of the unmarked. In Jacques Durand and Bernard Laks (eds), *Current Trends in Phonology: Models and Methods*, 473–483. Salford, Manchester: European Studies Research Institute, University of Salford.
McCarthy, John J. (2002) *A Thematic Guide to Optimality Theory*. Cambridge: Cambridge University Press.
McCarthy, John J. (2008) *Doing Optimality Theory*. Malden, MA, and Oxford, UK: Blackwell.

McCarthy, John J., Wendell Kimper and Kevin Mullin (2012) Reduplication in Harmonic Serialism. *Morphology* 22: 173–232.

McCarthy, John J. and Alan Prince (1986/1996) *Prosodic Morphology 1986*. New Brunswick, NJ: Rutgers University Center for Cognitive Science. Available at http://ruccs.rutgers.edu/pub/papers/pm86all.pdf.

McCarthy, John J. and Alan Prince (1993a) Generalized Alignment. In Geert Booij and Jaap van Marle (eds), *Yearbook of Morphology*, 79–153. Dordrecht: Kluwer. Available on Rutgers Optimality Archive, ROA-7.

McCarthy, John J. and Alan Prince (1993b) *Prosodic Morphology: Constraint Interaction and Satisfaction*. New Brunswick, NJ: Rutgers University Center for Cognitive Science. Available on Rutgers Optimality Archive, ROA-482.

McCarthy, John J. and Alan Prince (1994) The emergence of the unmarked: Optimality in prosodic morphology. In Mercè Gonzàlez (ed.), *Proceedings of the North East Linguistic Society* 24, 333–379. Amherst, MA: GLSA Publications. Available on Rutgers Optimality Archive, ROA-13.

McCarthy, John J. and Alan Prince (1995) Faithfulness and reduplicative identity. In Jill Beckman, Laura Walsh Dickey and Suzanne Urbanczyk (eds), *University of Massachusetts Occasional Papers in Linguistics* 18, 249–384. Amherst, MA: GLSA Publications. Available on Rutgers Optimality Archive, ROA-103.

McCarthy, John J. and Alan Prince (1999) Faithfulness and identity in Prosodic Morphology. In René Kager, Harry van der Hulst and Wim Zonneveld (eds), *The Prosody-Morphology Interface*, 218–309. Cambridge: Cambridge University Press.

McGarrity, Laura W. (2003) *Constraints on Patterns of Primary and Secondary Stress*. Doctoral dissertation, Indiana University, Bloomington, IN. Available on Rutgers Optimality Archive, ROA-651.

Mester, Armin (1994) The quantitative trochee in Latin. *Natural Language and Linguistic Theory* 12: 1–61.

Myers, Scott (1991) Persistent rules. *Linguistic Inquiry* 22: 315–344.

Myers, Scott and Benjamin B. Hansen (2007) The origin of vowel length neutralization in final position: Evidence from Finnish speakers. *Natural Language and Linguistic Theory* 25: 157–193.

Nevins, Andrew (2011) Phonologically conditioned allomorph selection. In Colin Ewen, Elizabeth Hume, Marc van Oostendorp and Keren Rice (eds), *The Companion to Phonology*, 2357–2382. Malden, MA, and Oxford, UK: Wiley-Blackwell.

Newman, Stanley (1944) *Yokuts Language of California*. New York: Viking Fund.

Paradis, Carole (1988a) On constraints and repair strategies. *The Linguistic Review* 6: 71–97.

Paradis, Carole (1988b) Towards a theory of constraint violations. *McGill Working Papers in Linguistics* 5: 1–43.

Pater, Joe (2000) Nonuniformity in English secondary stress: The role of ranked and lexically specific constraints. *Phonology* 17: 237–274. Available on Rutgers Optimality Archive, ROA-107.

Payne, David (1981) *The Phonology and Morphology of Axininca Campa*. Arlington, TX: The Summer Institute of Linguistics and University of Texas at Arlington.

Payne, David, Judith Payne and Jorge Santos (1982) *Morfología, Fonología y Fonética del Asheninca del Apurucayali (Campa – Arawak Preandino).* Yarinacocha, Peru: Summer Institute of Linguistics.

Prince, Alan (1980) A metrical theory for Estonian quantity. *Linguistic Inquiry* 11: 511–562.

Prince, Alan (1990) Quantitative consequences of rhythmic organization. In M. Ziolkowski, M. Noske and K. Deaton (eds), *Parasession on the Syllable in Phonetics and Phonology*, 355–398. Chicago: Chicago Linguistic Society.

Prince, Alan and Paul Smolensky (1993/2004) *Optimality Theory: Constraint Interaction in Generative Grammar.* Malden, MA, and Oxford: Blackwell.

Pruitt, Kathryn (2010) Serialism and locality in constraint-based metrical parsing. *Phonology* 27: 481–526.

Pulleyblank, Douglas (1986) *Tone in Lexical Phonology*. Dordrecht: D. Reidel.

Revithiadou, Anthi (2004) The Iambic/Trochaic Law revisited: Lengthening and shortening in trochaic systems. *Leiden Working Papers in LInguistics* 1: 37–62.

Ross, John Robert (1972) A reanalysis of English word stress. In Michael Brame (ed.), *Contributions to Generative Phonology*, 229–323. Austin, TX: University of Texas Press.

Selkirk, Elisabeth (1980) Prosodic domains in phonology: Sanskrit revisited. In Mark Aronoff and Mary-Louise Kean (eds), *Juncture*, 107–129. Saratoga, CA: Anma Libri.

Selkirk, Elisabeth (1995) The prosodic structure of function words. In Jill Beckman, Laura Walsh Dickey and Suzanne Urbanczyk (eds), *University of Massachusetts Occasional Papers in Linguistics 18: Papers in Optimality Theory*, 439–470. Amherst, MA: GLSA Publications.

Sommerstein, Alan (1974) On phonotactically motivated rules. *Journal of Linguistics* 10: 71–94.

Sommerstein, Alan (1977) *Modern Phonology*. London: Arnold.

Staubs, Robert, Michael Becker, Christopher Potts, Patrick Pratt, John J. McCarthy and Joe Pater (2010) OT-Help 2.0 [computer program]. Available at http://web.linguist.umass.edu/~OTHelp/.

Tranel, Bernard (1996) French liaison and elision revisited: A unified account within Optimality Theory. In Claudia Parodi, Carlos Quicoli, Mario Saltarelli and Maria Luisa Zubizarreta (eds), *Aspects of Romance Linguistics*, 433–455. Washington, DC: Georgetown University Press. Available on Rutgers Optimality Archive, ROA-15.

van Oostendorp, Marc. (2009) How I learned to stop worrying and love the derivation. Unpublished paper, Meertens Institute, Amsterdam. Handout of talk presented at Manchester Phonology Meeting 17. Downloaded 10 August 2012 from http://www.vanoostendorp.nl/pdf/090528manchester.pdf.

Wolf, Matthew (2008) *Optimal Interleaving: Serial Phonology-Morphology Interaction in a Constraint-Based Model.* Doctoral dissertation, University of Massachusetts Amherst, Amherst, MA. Available on Rutgers Optimality Archive, ROA-996.

Wolf, Matthew and John J. McCarthy (2009) Less than zero: Correspondence and the null output. In Curt Rice (ed.), *Modeling Ungrammaticality in Optimality Theory.* London: Equinox Publishing. Earlier version available on Rutgers Optimality Archive, ROA-722.

Zec, Draga (1995) Sonority constraints on syllable structure. *Phonology* 12: 85–129.

Chapter 4

Parallelism vs. serialism, or constraints vs. rules? Tongan stress and syllabification revisited[1]

Minta Elsman

4.1 Introduction

When comparing linguistic theories, it is not sufficient to merely declare one theory superior to the other based on the extent of its empirical coverage. The last, and perhaps most crucial step in the evaluation process, is to determine the *specific mechanisms* of the losing theory (e.g. structural representations, constraint definitions, operations inventory, etc.) that make it unable to provide a principled explanation for the phenomenon in question. This step is critical to the overall progress of linguistic theory, as multiple diverse theories often share common mechanisms, and information about the disadvantages of a particular mechanism can inform the development of new theoretical frameworks.

This paper revisits a classic evaluation of competing theories found in Chapter 3 of Prince and Smolensky (1993/2004). In this chapter, Prince and Smolensky introduce a parallel Optimality-Theoretic analysis of a stress and syllabification interaction that has been well documented in Tongan. They compare their parallel OT account with an earlier derivational, rule-based analysis (Poser 1985), and conclude that the parallel OT analysis accounts for the Tongan data in a principled way that is consistent with other empirical observations about Tongan, while the

1 This research was supported by grant BCS-0813829 from the National Science Foundation to the University of Massachusetts, Amherst.

derivational analysis relies on rules that lack independent empirical motivation and are theoretically inconsistent with the basic principles of markedness. Crucially, the mechanism that Prince and Smolensky identify as giving OT the advantage in accounting for the Tongan data is its parallel architecture. They argue that because OT is parallel while the rule-based analysis is derivational, the Tongan data support the superiority of a parallel framework over a derivational one.

In this paper, I challenge Prince and Smolensky's claim that the Tongan data distinguish between derivational and parallel frameworks. Instead, I argue that the Tongan data distinguish between rule-based and constraint-based frameworks, and constitute evidence in favor of the latter. In support of this claim, I offer a Harmonic Serialist analysis of the Tongan data. Harmonic Serialism (Prince and Smolensky 1993/2004, McCarthy 2007, McCarthy 2010b) inherits from parallel OT the idea of a set of violable, ranked constraints, but unlike OT (and like rule-based theories), it assumes a derivational architecture. In this paper, I show that the HS analysis of Tongan is no less theoretically principled or empirically supported than the OT analysis and that therefore, the Tongan data do not signal the superiority of parallel frameworks over derivational ones. Instead, given that OT and HS (which account equally well for the Tongan data) are constraint-driven, while rule-based theories (which have difficulty handling the Tongan data) are not, what can be learned from the Tongan case is that constraint-based frameworks are preferable to rule-based systems. The discussion in this paper serves to illustrate the importance of correctly identifying the specific mechanism that promotes or inhibits the ability of a particular theory to account for a given empirical phenomenon.

4.2 Data, Previous Analyses, and Critique

In support of their argument for the superiority of parallel OT over rule-based theories, Prince and Smolensky cite the interaction of stress and syllabification in Tongan. The crucial facts (first observed in Churchward 1953) are as follows. Vowel length is contrastive in Tongan, and primary stress falls on the penultimate syllable (1), or the final syllable if it is heavy (2). However, in words consisting of a long penult followed by a short final, the long penult is split into two identical short vowels, the latter of which bears stress (3):

(1) Penultimate Stress on Light Syllables

	a. mohéŋa	'bed'
	b. fetúʔu	'starry'
	c. hahá ke	'east'

(2) Final Stress on Heavy Syllables

a.	kumá:	'rat'
b.	kotokó:	'to cackle'
c.	haŋé:	'to be like'

(3) Long Penult followed by Short Final

a.	huúfi	'to open officially'
b.	fakahaáʔi	'to show'

The crucial pattern that must be accounted for in the data set above is the fact that when a penultimate long vowel precedes a final light syllable, it is broken up into two short vowels, with stress falling on the second vowel.

The rule-based analysis of Tongan that Prince and Smolensky select for their critique is found in Poser 1985. Poser relies on three ordered rules to derive the patterns in (1–3). First, an initial round of syllabification parses all long vowels as heterosyllabic sequences of identical short vowels, producing only light syllables. Then, a trochaic foot is built at the right edge of the word, such that stress falls on the penultimate syllable. Finally, a rule of syllable fusion applies, which merges each pair of adjacent identical vowels into a single long vowel, except where a foot boundary intervenes.[2] The process is illustrated below for words of all light syllables (4a), words with heavy final syllables (4b), and words with heavy penults (4c):

(4) Stress and Syllabification in Tongan (Poser 1985):

	a.	b.	c.
Input:	/moheŋa/	/kotoko:/	/hu:fi/
Syllabification:	mo.he.ŋa	ko.to.ko.o	hu.u.fi
Stress:	mo(hé.ŋa)	ko.to(kó.o)	hu(ú.fi)
Fusion:	N/A	ko.to(kó:)	blocked by ft boundary
Output:	[mohéŋa]	[kotokó:]	[huúfi]

Prince and Smolensky identify empirical and theoretical deficiencies in Poser's analysis, and attribute both to its serial architecture, which

2 This is a slightly improved version of Poser's analysis. In the original analysis, fusion is blocked if one of the vowels is stressed. However, the assumption that stress blocks fusion derives the wrong output for words with a final long vowel: /kotoko:/ → ko.to.ko.o → *[kotokó.o].

requires syllables to be constructed before feet can be built or stress can be assigned. Their primary focus is Poser's stipulation of a default V.V syllabification of long vowels in Tongan. They observe that on an empirical level, Poser provides no independent motivation for such a syllabification process in Tongan. Moreover, they argue that the default V.V syllabification of long vowels assumed in Poser's account cannot be correct, since long vowels in Tongan regularly surface as tautosyllabic, and are only 'split' to ensure that the stressed foot aligns with the right edge of the word. Prince and Smolensky demonstrate that this pattern can be captured under a parallel optimality-theoretic analysis, in which ONSET, which prefers tautosyllabic parse of long vowels, is dominated by EDGEMOST, a constraint requiring the stressed foot to align with the right edge of the word:

(5) Long Vowels 'Split' for Optimal Foot Alignment[3]

(a)

	/kotoko:/	EDGEMOST	ONSET
a. →	koto(kó:)		
b.	koto(kó.o)		1 W

(b)

	/hu:fi/	EDGEMOST	ONSET
a. →	hu(úfi)		1
b.	(hú:)fi	1W	L

Prince and Smolensky observe that Poser's analysis is unable to capture the fact that V.V syllabification arises only when foot alignment is at stake, and they attribute this deficiency to the fact that within a serial framework, structure-building processes cannot apply in parallel, such that 'there is no way to anticipate the specific demands of main-stressing when the syllable level is being constructed' (1993/2004: 30). Thus, to ensure correct syllabification of words with penultimate long vowels, Poser must assume a default disyllabic parse for *all* long vowels, despite the evidence to the contrary.

Prince and Smolensky's objection to Poser's analysis extends to the theoretical implications of assuming V.V as the default syllabification of long vowels in Tongan. Since a tautosyllabic parse of long vowels appears to be the unmarked case both in Tongan and crosslinguistically, Poser's analysis requires 'levels of derivation in which the syllabification of words is highly marked ... with heterosyllabic sequences of identical

3 To complete the analysis of /hu:fi/, an undominated foot form constraint such as *HL or FOOT BINARITY-μ must be added to rule out the form *(hú:fi), which satisfies both EDGEMOST and ONSET.

vowels' (1993/2004: 30). Thus, Poser's analysis essentially trivializes markedness by ignoring it wherever necessary:

> Poser's analysis shows that [bottom-up prosodic structure-building] is empirically inconsistent with ... the idea that initial syllabification is governed by markedness considerations. By this, only the least marked options are initially available to syllabify underlying forms. Indeed, we might argue that [bottom-up prosodic structure-building] is *antithetical* to the idea. Since the burden of establishing well-formedness falls on the later rules that function as repair strategies [i.e., syllable fusion], it is always possible to allow great divergences from well formedness at earlier stages. It becomes necessary to do so in cases like Tongan, where the generality of early rules of construction can be maintained only by letting them ignore basic markedness principles. (1993/2004: 30)

In summary, Prince and Smolensky argue that Poser's assumption of a default V.V syllabification of long vowels in Tongan, which he must necessarily adopt as a consequence of working under a derivational approach, is empirically unmotivated and theoretically inconsistent with the generally accepted role of formal markedness. They therefore conclude that the Tongan stress and syllabification pattern constitutes evidence against Poser's analytical framework, and by extension, against all derivational approaches. The following section refutes this conclusion, showing that in a derivational framework in which phonological processes are motivated by violable, ranked constraints, the initial V.V syllabification of long vowels emerges as the optimal way to eliminate violations of a highly-ranked (and independently motivated) markedness constraint.

4.3 An HS Analysis of Tongan Stress and Syllabification

4.3.1 Syllabification

As noted elsewhere in this volume, a defining characteristic of HS is that the function GEN is restricted to producing candidates that differ from the input by a single change, or operation. Following previous work in HS (Pruitt 2008, Elfner 2009) each of the following operations constitutes a single change:

(a) insertion of a single syllable node
(b) deletion of a single syllable node
(c) construction of a single stressed foot

The operations listed above can all be considered 'discrete', in the sense that each operation can apply only once per iteration, and no two operations may apply concurrently. In contrast, (re)association of a segment with a syllable is a 'for-free' operation in HS (McCarthy 2008a), in the sense that it may apply: (a) multiple times within a single iteration; and (b) concurrently with any of the discrete operations listed above.[4] As a consequence, GEN can build a maximum of only one syllable per iteration, but the size and shape of that syllable (e.g. CV, CVC, CVCC, CV:, CV:C, etc) is restricted only by: (a) the available input segments; and (b) the constraint ranking that determines the syllable inventory for the language under analysis.

Within this framework, syllable node insertion and foot building are discrete operations. As a result, it is impossible to simultaneously build a syllable and parse it into a foot. This in turn entails that syllables must be constructed before feet, since feet are built on syllables. Therefore, although the analysis proposed here differs from Poser's in that it replaces ordered rules with violable and interacting constraints, it nonetheless retains the characteristic of building syllables before feet. In order to ensure that syllables are constructed before feet, it is necessary to rank PARSESEGMENT over PARSESYLLABLE. In fact, this requirement holds for all HS-based analyses of prosodic structure, given that the reverse ranking precludes any possibility of either syllabification or foot building, since syllabification creates unfooted syllables, in violation of PARSESYLLABLE. This is illustrated in the tableau below, where bold font indicates segments that have been parsed into syllables:

(6) Argument for PARSESEGMENT over PARSE SYLLABLE

	/moheŋa/	PARSESEG	PARSESYLL
a. →	**mo**heŋa	4	1
b.	moheŋa	6 W	L

Like Poser's analysis, the current account proposes that long vowels are initially parsed as sequences of identical short vowels. If, as Prince and Smolensky claim, ONSET is responsible for the general tautosyllabic parsing of long vowels in Tongan surface forms, then to derive initial V.V syllabification within the constraint-based framework of HS, some constraint demanding V.V syllabification must outrank ONSET. This

4 For an argument in favor of treating segmental association with a syllable as a discrete operation within a serial constraint-based architecture, see Pater (2012).

constraint must meet two conditions. First, because it interacts with ONSET to determine syllabification, and because syllabification precedes footing, this constraint must be able to evaluate candidates that lack foot structure. Second, there should be independent evidence that this constraint is active in Tongan, and that it dominates ONSET.

A constraint that meets these conditions is Weight-to-Stress (WTS), which assigns a violation mark to every heavy syllable that does not bear stress, without reference to feet or foot structure. That this constraint is highly ranked in Tongan is evidenced by the fact that long vowels always bear stress, regardless of their position in the word (Churchward 1953):

(7) Evidence for Undominated WTS in Tongan

a. kà:ká:	'to cheat'
b. mà:lò:ló:	'to rest'
c. mà:lóhi	'strong'
d. ù:kúma	'patient'

Because WTS is never violated in Tongan, while ONSET is (i.e., penultimate long vowels split into a V.V sequence), it must be the case that WTS occupies a higher stratum than, and therefore dominates, ONSET.

Before illustrating the interaction of WTS and ONSET in Tongan syllabification, it is necessary to formalize the underlying representation of long vowels. Although the lexical representation of vowel length is often formalized as a single vowel segment linked to two morae (Hyman 1985, Zec 1988, Hayes 1989), arguments have also been made for treating long vowels as underlyingly bisegmental (e.g., Selkirk 1991). The analysis proposed here adopts the bisegmental representation for expositional expedience only; the proposed analysis can be made compatible with the moraic theory of length by the addition of constraints that evaluate the moraic level of representation.

The constraint ranking in (8) derives V.V syllabification of long vowels in Tongan:

(8) Tongan Constraint Ranking
WTS >> PARSESEGMENT >> PARSESYLLABLE, ONSET

The tableau below illustrates how the ranking in (8) selects as optimal a disyllabic parse of an input such as /huu/ ('to go in'). In the first iteration, candidate (b), which exhaustively parses the input as a CV: syllable, is ruled out by a fatal WTS violation. Note that it is not possible to avoid this violation by stressing the heavy syllable, since syllable node insertion and foot building are discrete operations, and cannot both apply within a single iteration. Of the remaining candidates, (a) incurs the fewest violations of PARSESEGMENT, and is therefore selected as optimal, becoming the input to the second iteration. The input to the second iteration contains a single, unparsed segment, in violation of PARSESEGMENT. This violation is repaired by inserting a new syllable node, and associating the unparsed segment with the newly-inserted node (candidate a). Associating the unparsed segment with the syllable node inherited from Iteration 1 (candidate b) is ruled out by a fatal violation of WTS. Failure to parse the segment (candidates c, d) incurs a fatal violation of PARSESEGMENT.

(9) WTS >> PARSESEG >> ONSET forces V.V Syllabification:

Iteration 1

	/huu/	WTS	PARSESEG	PARSESYLL	ONSET
a. →	**hu**.u		1	1	
b.	**hu:**	1 W	L	1	
c.	huu		3 W	L	

Iteration 2

	hu.u	WTS	PARSESEG	PARSESYLL	ONSET
a. →	**hu.u**			2	1
b.	**hu:**	1 W		1 L	L
c.	**(hú)** u		1 W	L	
d.	**hu**.u		1 W	1 L	L

The ranking above syllabifies words of all short vowels as in (10a), words with long final vowels as in (10b), and words with long penultimate vowels as in (10c):

(10) Syllabification Derived Via Ranking WTS>> PARSESEG >>ONSET:
a. /moheŋa/ → mo.he.ŋa
b. /kotokoo/ → ko.to.ko.o
c. /huufi/ → hu.u.fi

The HS analysis of Tongan presented here follows Poser's original analysis in proposing that long vowels are initially assigned a heterosyllabic parse. Since it is this aspect of Poser's analysis that leads Prince and Smolensky to dismiss it on the grounds that its theoretical assumptions lack an empirical basis in the Tongan data, it is appropriate to stop here and ask whether in adopting this same proposal, the HS analysis is also guilty of implementing an empirically baseless assumption. The answer is no. Unlike Poser's analysis, in which the initial disyllabic parse of long vowels is stipulated without any independent empirical support, in the HS analysis proposed here, this syllabification is forced by a ranking of WTS>>ONSET, which receives direct empirical support from the fact that Tongan categorically prohibits unstressed long vowels, but tolerates onsetless vowels in certain prosodic contexts.

4.3.2 Stress Assignment and Syllable Fusion

As in Poser's rule-based analysis, in the HS account proposed here, initial syllabification is followed by primary stress assignment at the right edge of the word. Within the constraint-based framework of HS, primary stress assignment and alignment can be motivated by the following constraint (adapted from Pruitt 2011):

(11) ALIGNHD-R
Assign a violation mark for each syllable that does not have a primary stress foot to its right, and is itself not a constituent of a primary stress foot[5].

Similarly to PARSESYLLABLE, this constraint penalizes the construction of syllables in the absence of foot structure. Thus, consistent with the requirement that syllables be constructed before feet (Section 4.3.1), ALIGNHD-R must also rank below PARSESEG.

The tableau in (12) illustrates primary stress assignment for the fully syllabified input mo.he.ŋa. Candidates (b) and (c) are both ruled out by violations of ALIGNHD-R, as each contains at least one syllable that is neither a constituent of the primary stress foot, nor has the primary stress foot to its right. Thus in this iteration, ALIGNHD-R selects candidate (a) as optimal, because it is the only candidate in which each syllable is either a constituent of the primary stress foot, or has the primary stress foot to its right.

5 Under Pruitt's definition, ALIGNHD-R is a single constraint that requires both (a) the *presence* of primary stress and (b) the *alignment* of primary stress with the right edge of the word. This constraint is intended as an improvement on standard OT approaches to deriving edge-sensitive primary stress assignment (one constraint demanding the presence of primary stress, and two others demanding alignment to the left and right word edges), which predict unattested languages in which words of three or more syllables vacuously satisfy both alignment constraints by not realizing primary stress (e.g., ALIGN-R(HD, WD), ALIGN-L(HD,WD) >> HAVEPRIMARYSTRESS).

(12) Primary Stress Assignment by ALIGNHD-R

	mo.he.ŋa	WTS	PARSESEG	ALIGNHD-R	PARSESYLL	ONSET
a. →	**mo(hé.ŋa)**				1	
b.	**(mó.he)ŋa**			1 W	1	
c.	**mo.he.ŋa**			3 W	3 W	

The tableau in (13) illustrates primary stress assignment for the fully syllabified input ko.to.ko.o, which contains a final long vowel parsed as a disyllabic sequence. Here again, candidates (b) and (c) are both ruled out by violations of ALIGNHD-R. At this stage in the derivation, the fusion of V.V sequences favored by ONSET continues to be suboptimal, because creating an unstressed heavy syllable fatally violates WTS (candidate d). Thus, vowel fusion is delayed to a subsequent iteration:

(13) Primary Stress Assignment Precedes Vowel Fusion for ko.to.ko.o

	ko.to.ko.o	WTS	PARSESEG	ALIGNHD-R	PARSESYLL	ONSET
a. →	**ko.to(kó.o)**				2	1
b.	**(kó.to)ko.o**			2 W	2	1
c.	**ko.to.ko.o**			4 W	4 W	1
d.	**ko.to.ko:**	1 W		3 W	3 W	L

Finally, the tableau in (14) shows how primary stress is assigned to the fully syllabified input hu.u.fi, which contains a penultimate long vowel parsed as a disyllabic sequence. As was the case in the previous tableau, the ranking WTS >> ONSET rules out vowel fusion (candidate d). Of the remaining candidates, ALIGNHD-R selects (a) as optimal, because it is the only candidate in which each syllable is either a constituent of the primary stress foot, or has the primary stress foot to its right.

(14) Primary Stress Assignment Precludes Vowel Fusion for hu.u.fi

	hu.u.fi	WTS	PARSESEG	ALIGNHD-R	PARSESYLL	ONSET
a. →	**hu(ú.fi)**				1	1
b.	**(hú.u)fi**			1 W	1	1
c.	**hu.u.fi**			3 W	3 W	1
d.	**hu:fi**	1 W		2 W	2 W	L

Crucially for this input, the optimal footing is one that eliminates the possibility of vowel fusion in any subsequent iteration: because the adjacent vowels are separated by a foot boundary, fusion is ruled out by the principle of SYLLABLE INTEGRITY (Hayes 1995; Prince and Smolensky

1993/2004), which states that GEN cannot produce structures in which a foot boundary is contained within a syllable.

The placement of main stress illustrated Tableaux 12–14 above yields the attested output forms for words containing only short vowels (mo.he.ŋa → [mo(hé.ŋa)]), and for words containing a long penult followed by a short final (hu.u.fi → [hu(ú.fi)]). However, main stress placement cannot be the final step for words with final long vowels (ko.to.ko.o → ko.to(kó.o)), since word-final long vowels surface with a monosyllabic parse (*kotokó:*).

To derive the final output for forms such as ko.to.(kó.o), this analysis follows Poser's in implementing syllable fusion, which is motivated by ONSET. For concreteness, syllable 'fusion' is defined here as deletion of a syllable node (a discrete operation), accompanied by for-free re-association of stranded segmental material with an adjacent syllable, as illustrated in (15):

(15) Syllable 'Fusion' as Syllable Deletion and Segment Re-association

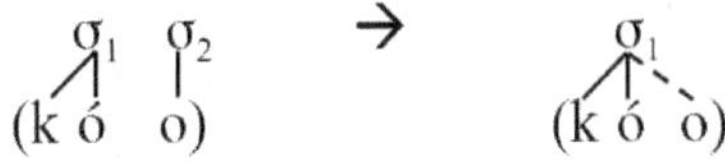

It is crucial to note that the input to the operation in (15) already contains a stressed syllable (*kó.o*). As a result, re-association of the vowel *o* with the preceding CV syllable does not incur a WTS violation, since the resulting heavy syllable inherits stress from the input.[6] The tableau below illustrates this final step in the derivation of [ko.to.kó:]

(16) Fusion No Longer Blocked by WTS

	ko.to(kó.o)	WTS	PARSESEG	ALIGNHD-R	PARSESYLL	ONSET
a. →	**ko.to(kó:)**				2	
b.	**ko.to(kó.o)**				2	1 W

To summarize, under the HS analysis proposed here, the syllable fusion demanded by ONSET becomes optimal only *after* a prosodic

6 Crucially, in order to satisfy ONSET without incurring a fatal violation of WTS, fusion must be achieved via deletion of the unstressed syllable node (σ_2), and the reassociation of its segmental material with the stressed syllable node (σ_1). Deletion of the stressed syllable node entails deletion of the stress itself, in which case fusion produces a heavy, unstressed syllable, in violation of WTS.

environment has been created in which fusion can apply without violating WTS.[7]

At this point, it is useful once again to contrast the proposed HS analysis with Poser's rule-based account, focusing this time on the extent to which each analysis maintains theoretical consistency with what Prince and Smolensky call 'the basic principles of markedness.' As observed earlier, Prince and Smolensky take issue with the fact that Poser's analysis produces marked structures (identical V.V sequences) before unmarked structures (long vowels). In adopting the same order of operations, does the HS analysis proposed here similarly flout the notion of markedness? The answer, once again, is no. Under the HS analysis proposed here, the optimal output of each iteration is the least-marked prosodic structure (where markedness is determined by the constraint ranking) that is possible given the input to that iteration (from which GEN can only produce candidates that differ by a single discrete operation). In the case of Tongan, the markedness of identical V.V sequences relative to long vowels depends on both the structure of the input and the constraint ranking. When the input contributes a stress that can protect a long vowel from penalization by WTS, then the long vowel is *less* marked than a V.V sequence, because it satisfies ONSET. However, when no stress is available to protect a potential long vowel from penalization by WTS, the long vowel is *more* marked than the V.V sequence, given

7 Since fusion of identical vowels is only permitted when one of the vowels is stressed, in order to derive attested words with multiple stressed long vowels, it is necessary to assume that Tongan has iterative footing independent of the requirements of WTS, as the mapping below illustrates:

(i) /maalooloo/ → ma.a.lo.o.lo.o → (mà.a)(lò.o)(ló.o) → (mà:)(lò:)(ló:) → [mà:lò:ló:] 'to rest.'

This assumption cannot be tested on native Tongan words, since morphologically simple words do not exceed three syllables, and morphologically complex words (compound words, words with polysyllabic prefixes) have been argued to contain multiple prosodic domains, each with its own primary stress (Feldman 1978, Schütz 2001). However, a study of stress placement in loanwords from English (Zuraw *et al.* 2010) shows that loanwords of four or more syllables are assigned one secondary stress on the first or second syllable of the word. That this reflects more than simple faithfulness to the English input is shown by the fact that assignment of secondary stress applies to epenthetic syllables:

(ii) (sì.kụ)(é.a) 'square'

This evidence suggests that the native grammar of Tongan does require secondary stress assignment independently of WTS requirements, as predicted by the HS analysis proposed here.

the (independently-motivated) ranking WTS >> ONSET. Thus, unlike Poser's rule-based analysis, the HS analysis proposed here adheres to basic markedness principles: when faithfulness is irrelevant (as it is in these Tongan examples), at each stage in the derivation, the optimal output is always the least marked form.

4.4 Discussion

The harmonic serialist analysis laid out in Section 4.3 derives the Tongan stress and syllabification pattern through the following steps:

(a) an initial round of V.V syllabification (WTS >> PARSESEG >> ONSET)
(b) the construction of a stressed foot at the right edge of the word (ALIGNHD-R)
(c) the fusion of identical tautopodal vowels (motivated by ONSET and protected from WTS by the input stress).

Under this analysis, the construction of prosodic structure in Tongan follows the same steps proposed in Poser's original rule-based analysis, and adheres to a derivational approach in which syllables are constructed before feet. If it is the case that the failure of Poser's rule-based analysis to provide an empirically justified and theoretically principled account of the Tongan data stems from its derivational architecture, then the HS analysis presented in Section 4.3, which is also derivational in nature, and proceeds through exactly the same steps, should suffer from the same flaws that Prince and Smolensky identify in Poser's rule-based analysis. But it does not. Under the HS account, the initial round of V.V syllabification is motivated by a high ranking of WTS in Tongan, which is evidenced by the fact that long vowels always surface with stress. Moreover, under the HS analysis presented here, the production of identical V.V sequences before long vowels is consistent with the principles of markedness: given their respective inputs and the gradual nature of GEN, each of these structures is the least marked output candidate available in the iteration that produces it.

The analysis and discussion presented in this paper show that as far as Tongan is concerned, serialism is not the problem. Both parallel Optimality Theory and Harmonic Serialism provide empirically motivated and theoretically principled accounts of the phenomenon in question. But if a serial architecture is not an impediment to a sound analysis of the Tongan stress and syllabification pattern, then why does Poser's rule-based analysis have such difficulty in accounting for the Tongan data? The answer is clear. The formal mechanism that unites the OT and HS analyses, and distinguishes both from Poser's account, is the implementation of violable, interacting constraints in place of categorical rules.

On an empirical level, the use of constraints allows both OT and HS to relate the 'splitting' of long vowels in Tongan to broader generalizations about prosodic well-formedness requirements in Tongan. In the OT analysis, the requirement that each syllable have an onset is less important than the requirement that the head foot align with the right edge of the word (EDGEMOST >> ONSET). Under the HS account, the requirement that each syllable have an onset is less important than the requirement that heavy syllables be stressed (WTS >> ONSET). In contrast, under the rule-based analysis provided by Poser, Tongan simply has two conflicting syllabification rules (V.V syllabification and vowel fusion) that are related only by an arbitrary rule ordering.

At the theoretical level, adoption of a constraint set is a way of implementing relative markedness, which allows the OT and HS analyses to treat identical V.V sequences as more marked than long vowels in some environments, and as less marked than long vowels in others. Under the OT account, an identical V.V sequence is more marked than a long vowel, except in penultimate position preceding a light syllable (where a long vowel would pull the head foot away from the right edge). In the HS account, an identical V.V sequence is more marked than a long vowel, unless the long vowel is unstressed. In contrast, Poser's rule-based account must simply stipulate V.V syllabification as the default case in Tongan, despite the fact that this syllabification pattern is marked cross-linguistically, and appears to be the exceptional case in Tongan as well. The HS analysis presented here thus constitutes an improvement on Poser's original rule-based analysis, addressing Prince and Smolensky's critique by showing how a reversal in the relative markedness of conflicting syllabification algorithms (V.V. vs. V:) emerges as a natural consequence in a constraint-based framework in which GEN is restricted to producing one change at a time, such that a mapping that produces a potentially marked structure (e.g. a long vowel) is only optimal when the input provides the context that licenses that structure (e.g., stress).

This paper has compared the respective abilities of OT, HS, and rule-based frameworks to provide a sound analysis of Tongan stress and syllabification. This cross-theory comparison has demonstrated that in the single case of Tongan, the choice between a parallel and serial architecture is irrelevant, while the choice between a constraint-based and a rule-based framework is absolutely crucial. It is important to note that this one case study is in no way meant to serve as some sort of argument that parallel Optimality Theory and Harmonic Serialism are rendered equivalent by virtue of the fact that both make use of violable, interacting constraints. Indeed, the majority of papers in this volume are dedicated to arguing that HS is better-equipped than parallel OT to handle certain phonological phenomena, and it is undoubtedly the case that counterarguments will be forthcoming. What this case study *is* intended to show is that when one theory outperforms another, it is crucial to correctly identify the formal difference that gives one theory the advantage over the other. In the case of Tongan, the superiority of Prince and Smolensky's OT analysis over Poser's rule-based account stems not from the use of parallel architecture, but rather from the implementation of violable constraints in place of categorical rules. This claim is supported by the fact that the derivational, constraint-based HS analysis of Tongan stress and syllabification presented in this paper shares the advantages of its parallel OT counterpart, while suffering none of the flaws found in its derivational, rule-based correlate.

References

Churchward, C. Maxwell (1953) *Tongan Grammar*. London and New York: Oxford University Press.

Elfner, Emily (2009) Syllabification and stress-epenthesis interaction in Harmonic Serialism. Unpublished paper, University of Massachusetts Amherst, Amherst, MA. Available on Rutgers Optimality Archive, ROA-1047.

Feldman, Harry (1978) Some notes on Tongan phonology. *Oceanic Linguistics* 17: 133–139.

Hayes, Bruce (1989) Compensatory lengthening in moraic phonology. *Linguistic Inquiry* 20: 253–306.

Hayes, Bruce (1995) *Metrical Stress Theory: Principles and Case Studies*. Chicago: The University of Chicago Press.

Hyman, Larry (1985) *A Theory of Phonological Weight*. Dordrecht: Foris.

McCarthy, John J. (2007) Restraint of analysis. In Sylvia Blaho, Patrik Bye and Martin Krämer (eds), *Freedom of Analysis*, 203–231. Berlin and New York: Mouton de Gruyter.

McCarthy, John J. (2010a) Studying Gen. *Journal of the Phonetic Society of Japan* 13: 3–12.

McCarthy, John J. (2010b) An introduction to Harmonic Serialism. *Language and Linguistics Compass* 4: 1001–1018.

Pater, Joe (2012) Serial Harmonic Grammar and Berber syllabification. In Toni Borowsky, Shigeto Kawahara, Takahito Shinya and Mariko Sugahara (eds), *Prosody Matters: Essays in Honor of Lisa Selkirk*. London: Equinox Publishing.

Poser, William J. (1985) Cliticization to NP and Lexical Phonology. *The Proceedings of the West Coast Conference on Formal Linguistics* 4, 262–272.

Prince, Alan and Paul Smolensky (1993/2004) *Optimality Theory: Constraint Interaction in Generative Grammar*. Malden, MA, and Oxford, UK: Blackwell.

Pruitt, Kathryn. (2008) Iterative foot optimization and locality in stress systems. Unpublished paper, University of Massachusetts Amherst, Amherst, MA. Available on Rutgers Optimality Archive, ROA-999.

Pruitt, Kathryn. (2011) Primary stress in GEN and CON. Unpublished paper, University of Massachusetts Amherst, Amherst, MA. Presentation to Phonology Grant Group.

Schütz, Albert J. (2001) Tongan accent. *Oceanic Linguistics* 40: 303–323.

Selkirk, Elisabeth (1991) A two-root theory of length. In E. Dunlap and Jaye Padgett (eds), *University of Massachusetts Occasional Papers in Linguistics 14: Papers in Phonology*. Amherst, MA: GLSA Publications.

Zec, Draga (1988) *Sonority Constraints on Prosodic Structure*. Doctoral dissertation, Stanford University.

Zuraw, Kie, Kathleen O'Flynn and Kaeli Ward. (2010) Marginal contrasts in Tongan loanword phonology. Unpublished paper, UCLA, Los Angeles. Presentation at Linguistics Colloquium, University of Massachusetts Amherst.

Chapter 5
Serial restrictions on feature/stress interactions[1]

Robert Staubs

5.1 Introduction

The most crucial distinction between Harmonic Serialism and Optimality Theory is the property of *gradualness*. Harmonic Serialism (Prince and Smolensky 1993/2004, McCarthy 2000, 2008a, 2008b) requires that outputs be relatable with their inputs via a series of individual optimally-improving steps. Parallel Optimality Theory (Prince and Smolensky 1993/2004) imposes no such requirement. This type of restriction creates predictions for HS that differ from parallel OT in two ways: first, HS predicts some languages to exist that crucially exclude non-global optimization; second, it *excludes* languages that *require* global optimization. This latter property, the restrictiveness of HS, is the one relevant to this paper: HS constrains the range of predicted languages to those that do not exhibit arbitrarily large degrees of 'lookahead'.

I will focus on interactions between segmental features and stress, showing that in Harmonic Serialism, positional markedness constraints yield desirable interactions between segments and stress such as sonority-driven stress and vowel reduction but do not yield certain pathological

1 Many thanks to the participants in the UMass Phonology Grant Group (Summer–Fall 2010), the audience at NELS 41, John J. McCarthy, Claire Moore-Cantwell, Joe Pater, Paul Smolensky, Colin Wilson, and Kie Zuraw for helpful discussion. This material is based upon work supported by the National Science Foundation under Grant No. S121000000211 to the author and BCS-0813829 to the University of Massachusetts Amherst. Opinions expressed do not necessarily reflect those of any of these parties. All remaining mistakes are my own.

languages predicted by Optimality Theory, noted by de Lacy (2002). Such predicted languages would use arbitrarily chosen segmental features to condition the placement of stress, in contrast to the restricted and abstract manner in which stress is actually observed to interact with segmental information. I will focus principally on constraints making reference to sonority, but do not intend this reference to be exclusive.

This paper begins with an abbreviated overview of Harmonic Serialism (see McCarthy this volume for a more thorough overview). The discussion of positional markedness in the theory follows, along with the problems for Optimality Theory and the serial solution. I describe necessary restrictions on the form of positional markedness constraints and the concomitant expansion of the set of representations to include recursive foot structure. I finally outline the scope of the solution and conclude.

5.2 Harmonic Serialism overview

Harmonic Serialism (HS) is a serial version of parallel Optimality Theory (OT). In Optimality Theory, an input yields infinitely many candidate forms via the relation Gen. Of these, the output is the candidate that is maximally harmonic according to the constraint ranking. Harmonic Serialism modifies this architecture by modifying the role of Gen. Rather than producing all possible modifications to the input form, a Harmonic Serialist Gen only produces those forms accessible from the input via a *single* step. For each production of Gen, the most harmonic candidate is chosen as the input to another iteration. The derivation converges when two successive steps yield the same optimum.

The Harmonic Serialist interpretation of Gen imposes additional requirements for typological prediction. In Optimality Theory, typological predictions are determined fully by the universal constraints in Con. Gen imposes few additional restrictions because it provides all logically conceivable linguistic structures. For Harmonic Serialism, however, the case is somewhat different. Con is still vitally important, but now so is Gen (McCarthy this volume). Rather than generating everything, Gen only applies a single change. It is therefore important to consider what constitutes a single change. The operations that together make up Gen are therefore also crucial for making typological predictions. The constituency of Gen is an empirical question accessible to the researcher, just as the constituency of Con is in either HS or OT.

5.3 Positional markedness in Harmonic Serialism

5.3.1 General analysis

Positional markedness effects are described similarly in both Optimality Theory and Harmonic Serialism: both frameworks use positional markedness constraints. Such constraints are of the following form:

(1) Positional markedness schema
Assign a violation for every structure X in a position Y.

In OT, these constraints give rise to a variety of typologically desirable interactions between levels of the prosodic hierarchy. McCarthy, Pater, and Pruitt (this volume) discuss cross-level interactions more generally. In the simple case, nothing new needs to be said of HS. Such effects fall out simply. However, OT also makes undesirable predictions not made in HS. OT accounts predict, for example, that arbitrary segmental features may affect stress assignment.

To exemplify both the desirable and undesirable cases, we will examine constraints that govern the sonority of vowels. The particular family of constraints we will look at penalizes vowels of high sonority in the dependent of a foot, as in e.g. de Lacy (2002).

For our purposes, we will use the following abbreviated sonority scale:

(2) Sonority scale ($>_s$ means 'is of greater sonority than')
a $>_s$ e, o $>_s$ i, u $>_s$ ə

Such constraints penalize vowels in foot-dependent position over a certain sonority threshold. An example (one which I will use throughout this paper) is the following constraint, which penalizes unstressed vowels of sonority greater than [i, u]:

(3) *UNSTRESSED [a, e, o]
Assign a violation for every vowel in the set [a, e, o] in the dependent of a foot.

This type of constraint is motivated by two symmetric interactions between stress and segmental information: sonority-driven stress and vowel reduction. The former reflects an influence of segmental material on stress, the latter an influence of stress on possible segments. With parallel optimization in OT, both phenomena may be described using a single type of constraint.

Languages with sonority-driven stress prefer stressing high-sonority vowels even at the expense of breaking an otherwise general stress pattern. For example, consider the following partial language. This language has trochaic stress except when the second vowel is sonorous and the first is not:

(4) Example of sonority-driven stress
(a) píki
(b) póki
(c) péko
(d) pikó

For sonority-driven stress, we simply require that such a constraint dominate the normally active stress-placement constraints of the language and that it *not* dominate faithfulness to vowel quality:

(5) No sonority effects

piki	IDENT	*UNSTRESSED[a,e,o]	TROCHEE
a. → (píki)			
b. (pikí)			1 W
c. (píkə)	1 W		

(6) Sonority-driven stress

piko	IDENT	*UNSTRESSED[a,e,o]	TROCHEE
a. (píko)		1 W	L
b. → (pikó)			1
c. (píkə)	1 W		L

In (5), no high-sonority vowels are present and thus no sonority effects emerge. In (6), the constraint against sonorous unstressed vowels outranks the constraint on the regular stress pattern, ruling out candidates such as (6a) which surface with high-sonority unstressed vowels. The sonority constraint is ranked under IDENT, ruling out candidates such as (6c) with reduction.

Such an analysis is immediately adaptable to Harmonic Serialism. The same rankings hold:

(7) No sonority effect
Step 1

piki	IDENT	*UNSTRESSED[a,e,o]	TROCHEE
a. → (píki)			
b. (pikí)			1 W

Step 2

(píki)	IDENT	*UNSTRESSED[a,e,o]	TROCHEE
a. → (píki)			
b. (píkə)	1 W		

(8) Sonority-driven stress
Step 1

piko	IDENT	*UNSTRESSED[a,e,o]	TROCHEE
a. (píko)		1 W	L
b. → (pikó)			1

Step 2

(pikó)	IDENT	*UNSTRESSED[a,e,o]	TROCHEE
a. → (pikó)			1
b. (pəkó)	1 W		1

This derivational account of a schematic language splits the single step of parallel OT into two, but one of these steps is just the (near-trivial) convergence step. The decision of whether or not to stress a particular location is still accomplished in one operation and thus the analysis does not change substantially. Although the overall derivation is serial, the comparison of candidates in Harmonic Serialism remains stepwise parallel. That is, the optimization compares many candidates at a time even though those candidates can differ from their input by at most one operation from GEN. Therefore, the derivation can still decide to stress the high-sonority vowels. This results in a single step sufficing to replicate the effect of parallel OT for such forms.

The situation is quite similar for vowel reduction. In languages with vowel reduction, *UNSTRESSED constraints motivate the selection of less sonorous vowels in dependent position. For example, consider a language with strict trochees and reduction:

(9) Example of vowel reduction
(a) píki
(b) póki
(c) pékə (underlying /peko/)
(d) píkə (underlying /piko/)

Such a language may be analyzed with a simple reranking. If TROCHEE is undominated and IDENT is dominated by *UNSTRESSED[a, e, o] we obtain the pattern in OT:

(10) No sonority effects

	piki	TROCHEE	*UNSTRESSED[a,e,o]	IDENT
a. →	(píki)			
b.	(pikí)	1 W		
c.	(píkə)			1 W

(11) Sonority-based reduction

	piko	TROCHEE	*UNSTRESSED[a,e,o]	IDENT
a.	(píko)		1 W	L
b.	(pikó)	1 W		L
c. →	(píkə)			1

As in (5), the tableau in (10) does not show the stress shift, simply because there are no high-sonority vowels in the input. (11) differs from (6) in that TROCHEE is high-ranked, giving priority to a canonically trochaic foot form. This rules out sonority-driven stress in (11b), leaving only (11c) – reduction – as a repair for the sonorous vowel. As before, the adaptation of this analysis to Harmonic Serialism is trivial. An identical ranking will suffice and the derivation proceeds intuitively:

(12) No sonority effects
Step 1

	piki	TROCHEE	*UNSTRESSED[a,e,o]	IDENT
a. →	(píki)			
b.	(pikí)	1 W		

Step 2

	(píki)	TROCHEE	*UNSTRESSED[a,e,o]	IDENT
a. →	(píki)			
b.	(píkə)			1 W

(13) Sonority-based reduction
Step 1

	piko	TROCHEE	*UNSTRESSED[a,e,o]	IDENT
a. →	(píko)		1	
b.	(pikó)	1 W	L	

Step 2

	(píko)	TROCHEE	*UNSTRESSED[a,e,o]	IDENT
a.	(píko)		1 W	L
c. →	(píkə)			1

In the first step, syllables are obligatorily parsed into feet in the assumed ranking. Following this, reduction is performed if necessary.

5.3.2 Overprediction with positional markedness in Optimality Theory

We have thus far seen that the typologically motivated strategies for avoiding *UNSTRESSED[a, e, o] are both easily analyzed in both OT and HS. Harmonic Serialism progresses through intermediate ill-formed steps to arrive at the same global optima as OT – for example, parsing sonorous segments as dependents in a vowel-reducing language. Such cases therefore do not distinguish parallel and serial theories.

De Lacy (2002: 94–96) identifies a type of pathological language predicted to exist by Parallel OT. In such a language, there is a general tendency for vowel reduction but a specific prohibition against reducing some particular type of vowel. This type of vowel is preferentially stressed to avoid a faithfulness violation, even if the feature referenced is not actually observed to draw stress typologically. The features seen attracting stress, such as sonority and weight, are directly related to prosodic well-formedness. This is not true of any arbitrary features controlled by faithfulness constraints; thus, we do not expect arbitrary features to drive stress in this way. I will show that Harmonic Serialism allows an account of sonority effects that does not also predict these languages to exist.

To exemplify this pathology, consider the IDENT(round), a constraint mandating faithfulness to vowel roundness. With this constraint in CON, OT predicts that there should exist languages that always have trochaic stress unless an iamb would result in stressing a sonorous, otherwise stressless, round vowel:

(14) Example of undesirable prediction
(a) píki
(b) póki
(c) píkə (underlying /pike/)
(d) pekó
(e) pikó

Note the crucial case of (14c). When a sonorous but unrounded vowel is in second position, it does not draw stress. This is therefore not a simple sonority-driven stress system. Instead, such a language is of an essentially disjunctive type: reduction if the vowel is sonorous and not round (14c), sonority-driven stress if the vowel is sonorous and round (14d, e), and trochaic stress otherwise (14a, b). Such patterns are easily obtained by ranking IDENT(round) and *UNSTRESSED[a, e, o] above the stress-placement constraints and all other IDENT below:

(15) Reduction

	pike	*UNSTRESSED [a,e,o]	IDENT (round)	TROCHEE	IDENT (other)
a.	(píke)	1 W			L
b. →	(píkə)				1
c.	(piké)			1 W	L

(16) Sonority-driven stress

	piko	*UNSTRESSED [a,e,o]	IDENT (round)	TROCHEE	IDENT (other)
a.	(píko)	1 W		L	
b.	(píkə)		1 W	L	
c. →	(pikó)			1	

In (15), reduction of [e] is compelled by a ranking similar to the one seen in (10, 11) for general reduction: *UNSTRESSED and TROCHEE dominate IDENT(other), motivating the repair of sonorous vowels, but preventing sonority-driven stress. However, the situation with IDENT(round) is different: the faithfulness constraint dominates TROCHEE. This partial order was seen in (5, 6). Thus, sonority-driven stress and reduction coexist in this predicted language.

Such a language shows an unobserved interaction between segments and stress. Rather than merely showing sonority-driven stress, it shows specifically *roundness*-driven stress. The only sonorous vowels that will attract stress are the ones that incur IDENT(round) violations in reduction

– that is, the round vowels. In a similar way, sonority constraints combined with different instantiations of IDENT can create stress systems driven by arbitrary vowel features; all that is required is a Faithfulness constraint that isolates the feature and is violated by reduction.

This prediction is not made for these candidates under Harmonic Serialism. The prediction itself is intrinsically linked to OT's access to a full candidate set – OT can see the full consequences of individual changes by comparing candidates that differ by arbitrarily many changes. HS is limited by gradualness, restricting its search to the immediately optimal. See Moore-Cantwell (this volume) for related cases in which HS avoids unwanted OT predictions concerning stress.

To make an explicit analysis on these lines, I must adopt a concrete proposal for GEN. In terms of GEN's segmental component, I only consider a single 'reduction' operation. This reduces any vowel to [ə]. Other reduction operations in Harmonic Serialism would share the properties discussed here. For the stress-manipulating component, for our present purposes I adopt the proposal of Pruitt (2010): GEN contains only operations that build mono- or disyllabic right- or left-headed feet.

Under this proposal, GEN contains operations manipulating stress and sonority but not operations that change both at once. This means that we are left only with derivations like the following:[2]

(17) Sonority-driven stress

Step 1

pike	*UNSTRESSED [a,e,o]	IDENT (round)	TROCHEE	IDENT (other)
a. (píke)	1 W		L	
b. → (piké)			1	

Step 2

(piké)	*UNSTRESSED [a,e,o]	IDENT (round)	TROCHEE	IDENT (other)
a. → (piké)			1	

In the first step, parsing must occur. The ranking prioritizes removing unstressed sonorous vowels and thus the stress is shifted to [e] – sonority-driven stress.

2 All typological claims about Harmonic Serialism were confirmed using OT-Help 2.0 (Staubs, *et al.* 2010), software for computing typological predictions in HS.

(18) Sonority-driven stress
Step 1

	piko	*UNSTRESSED [a,e,o]	IDENT (round)	TROCHEE	IDENT (other)
a.	(píko)	1 W		L	
b. →	(pikó)			1	

Step 2

	(pikó)	*UNSTRESSED [a,e,o]	IDENT (round)	TROCHEE	IDENT (other)
a. →	(pikó)			1	

The picture is not much different with an [o] to stress. No competitive reduction candidates are entertained in Step 1. In this first step, GEN cannot produce forms like [(píkə)] that require the application of two operations.

(19) Vowel reduction (sonority-driven stress irrelevant)
Step 1

	peko	*UNSTRESSED [a,e,o]	IDENT (round)	TROCHEE	IDENT (other)
a. →	(péko)	1			
b.	(pekó)	1		1 W	

Step 2

	(péko)	*UNSTRESSED [a,e,o]	IDENT (round)	TROCHEE	IDENT (other)
a.	(péko)	1 W	L		
b. →	(pékə)		1		

When both vowels are high-sonority, one must be stressed and the other must be reduced. The choice is necessarily made without reference to the features of the vowels, because the parse happens in Step 1 before any IDENT violations are relevant.

The shift to serial evaluation with a restricted GEN causes this ranking to produce a language with sonority-driven stress and vowel reduction, lacking an arbitrary sensitivity to particular feature values.

The unwanted type of effect predicted in OT is eliminated for these candidates in Harmonic Serialism. HS can never compare a candidate with trochaic stress and reduction of a non-round vowel with a candidate with iambic stress and reduction of a round vowel. Under the assumed GEN, these candidates are more than two operations apart and thus can never coexist within a single HS tableau. One HS tableau can only ever contain candidates that are at most two operations apart, each candidate being one step away from the input. HS thus cannot look ahead in a derivation and therefore cannot make a tradeoff between a temporary loss (aberrant stress) and a long-term gain (faithfulness). It therefore converges on the local optimum, avoiding the unobserved global optimum.

Another way to view this solution is in terms of ordering. The constraint *UNSTRESSED is violated exclusively in the dependents of feet. Vowel reduction must be a response to a violation of *UNSTRESSED and must therefore follow the placement of some sonorous syllable in a foot dependent. Therefore, footing is naturally ordered before reduction just by the definitions of the constraints involved (see McCarthy 2008 and Elfner this volume for more on such ordering results).

This prediction is independent of possible rankings. However, it *is* dependent on GEN, which must not include any single operation that changes sonority and stress simultaneously. However, we should feel reasonably confident in excluding such an operation without strong evidence in its favor; sonority-changing operations and stress-assigning operations are both independently motivated, so to also include an operation which merges the two would not be parsimonious. With the type of constraints specified, this condition is sufficient.

Harmonic Serialism replicates Optimality Theory's desirable predictions in the domain of positional markedness, and specifically sonority/stress interactions. These effects relate disparate levels of the prosodic hierarchy but do not in fact require parallel evaluation of all candidates. Instead, stepwise optimization of options suffices. Harmonic Serialist derivations can achieve the necessary power by proceeding through intermediate ill-formed steps.

Optimality Theory overpredicts with positional markedness constraints, predicting languages wherein arbitrary features drive stress. Such languages are not predicted in Harmonic Serialism due to gradualness. These pathological languages require a choice between candidates that differ in both stress and segmental features – impossible in Harmonic Serialism. This domain provides a case in which Harmonic Serialism and Optimality Theory are equally capable of capturing a type of interaction between stress and segments but for which OT is too powerful overall.

5.4 Restrictions on positional markedness constraints

5.4.1 Distinguishing positions

Throughout, I have been assuming a definition of positional markedness constraints in which they make reference exclusively to foot dependent position. It is important to consider whether this restriction is a necessary one, as other types of constraints have also been proposed. For example, de Lacy (2002) proposes sonority/stress constraints that are violated for high-sonority segments in any 'unstressed' position, not just in foot-dependent position.

The flagship case for distinguishing sonority/stress constraints in foot-dependent position from a general 'unstressed' position is Dutch semi-formal reduction. In this register of Dutch, reduction targets some vowels in any unstressed position and others only in the dependents of feet (Kager 1989, van Oostendorp 1995). In particular, /e/ and /a/ always reduce, while /i/ and /o/ only do so if in the dependent of a foot.

(20) Reduction pattern of Dutch semi-formal register
 (a) /a/ reduces in any unstressed position
  (i) (kàrə)(mɛ́l) *karamel* 'caramel'
  (ii) rə(dɛ́is) *radijs* 'radish'
 (b) /e/ reduces in any unstressed position
  (i) sə(nát) *senaat* 'senate'
  (ii) (ékə)(nòmə)(trí) *econometrie* 'econometrics'
 (c) /o/ reduces in foot dependents only
  (i) to(mát) *tomaat* 'tomato'
  (ii) (pɛ̀lə)(tón) *peloton* 'platoon'
 (d) /i/ reduces in foot dependents only
  (i) mi(nút) *minuut* 'minute'
  (ii) (rɛ̀lə)(kwí) *relikwie* 'relic'

In Optimality Theory, this phenomenon calls out for a division in the relevant positional markedness constraints. De Lacy (2002) proposes that there are actually two families of constraints like *Unstressed – one that makes reference to any unstressed syllable and one that refers only to foot dependents.

(21) *Unstressed-FootDependent[a, e, o]
 Assign a violation for every vowel in the set [a, e, o] in the dependent of a foot.

(22) *Unstressed-All[a, e, o]
 Assign a violation for every vowel in the set [a, e, o] lacking stress.

Before this point, I have been using only the first definition – the one which calls for the relevant vowel to be in the dependent of a foot. If this were the only kind of sonority/stress constraint available, however, how are we to make sense of Dutch semi-formal reduction? Arguments for a division along these lines seem sound. Clearly, we must distinguish different kinds of 'unstressed' positions. Given this, the most natural mechanism we are allowed in Optimality Theory (or Harmonic Serialism) is positional markedness constraints. Therefore, the restriction made above – namely, that positional markedness constraints only refer to unstressed position as foot dependent position – cannot be true in general. We must seemingly abandon it and allow both kinds of constraint. However, this does not come without cost – these other constraints reintroduce the kinds of pathologies discussed above.[3]

Our earlier ranking did not produce a pathological, roundness-driven stress system when *UNSTRESSED-FT was used. This is not the case with *UNSTRESSED-ALL. Such a constraint allows for a distinction between the reducibility of vowels to be made based on faithfulness *prior to stress assignment*. *UNSTRESSED-FT avoided the pathology in HS precisely because serial evaluation imposes an order with this constraint: violations are only incurred after stress is assigned, therefore stress assignment must always precede reduction due to this constraint. With this restriction removed, it follows that we can now create an instance of the pathological a roundness-driven stress system.

(23) Vowel reduction

Step 1

	pike	*UNSTRESSED-ALL [a,e,o]	IDENT (round)	TROCHEE	IDENT (other)
a.	(píke)	1 W			L
b.	(piké)			1 W	L
c. →	pikə				1

Step 2

	pikə	*UNSTRESSED-ALL [a,e,o]	IDENT (round)	TROCHEE	IDENT (other)
a. →	(píkə)				
b.	(pikə́)			1 W	

3 Particular thanks are due to Colin Wilson for stressing the importance of this point.

Step 3

(píkə)	*UNSTRESSED-ALL [a,e,o]	IDENT (round)	TROCHEE	IDENT (other)
a. → (píkə)				

In the first step, [e] is reduced to satisfy *UNSTRESSED. No intermediate step of parsing is needed for the violation to be incurred and thus its satisfaction can demand segmental change.

(24) Sonority-driven stress
Step 1

piko	*UNSTRESSED-ALL [a,e,o]	IDENT (round)	TROCHEE	IDENT (other)
a. (píko)	1 W		L	
b. → (pikó)			1	
c. pikə		1 W		

Step 2

(pikó)	*UNSTRESSED-ALL [a,e,o]	IDENT (round)	TROCHEE	IDENT (other)
a. → (pikó)			1	

With an [o] in the input, reduction incurs a violation of IDENT(round). Stress is attracted to the vowel instead to avoid this outcome. The contrast between (23) and (24) shows that in this language, stress is attracted to rounding. Thus a change in the formal specification of *UNSTRESSED has reintroduced the pathology.

With *UNSTRESSED-FT, stress is naturally ordered before reduction for reasons discussed above. This is not the case for *UNSTRESSED-ALL. This problem arises due to how violations are assigned. Unstressed high-sonority vowels violate such constraints prior to the creation of any relevant prosodic structure. If these violations are resolved, any sort of change that can follow can resolve them. This includes both sonority-changing operations (reduction) and shifts in stress. In combination, the pathology can emerge.

A contrast is needed between different types of unstressed positions, but the relevant constraints must not be violated immediately upon syllable creation. One option is thus to make *both* kinds of constraints refer to *built* structure rather than a lack of structure. I thus propose a solution in two parts: an extension of GEN to include adjunction to the foot and a redefinition of *UNSTRESSED-FT and *UNSTRESSED-ALL to make use of this extension.

5.4.2 *Ordering through foot-adjunction*

The ordering effect achieved by *UNSTRESSED-FT comes about due to its requirement that some foot must exist prior to any violation. If there is no foot, there can be no foot dependents and therefore no violations. Further, the effect is local: it is only the syllable encompassed by the foot itself that can cause violations – syllables elsewhere in the word cannot cause them. This effect is replicable with a form of *UNSTRESSED-ALL if the notion of the foot dependent is extended.

This extension is possible if we allow syllables to be adjoined to the foot. Previously, GEN was assumed to only construct feet with one head syllable and at most one dependent syllable. We can expand the representational space by introducing an operation of (Chomsky) adjunction to GEN: a syllable can be adjoined to a foot to create a larger (possibly super-binary) foot. This type of structure is shown in (25). In this example, a trochaic foot has an additional syllable adjoined to the right. This kind of adjunction creates a recursive foot structure, making it similar to the superfoot, colon, or internally-layered foot (e.g. McCarthy 1982, van Oostendorp 1995, Jensen 2000, Caballero 2008, Kager 2012, Martínez-Paricio 2012, 2014).

(25) Adjoined foot structure

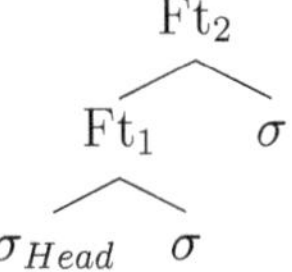

If GEN is supplemented in this way, the constraint set can refer to two different kinds of dependents. In particular, I propose that instead of constraints on foot dependents and constraints on all unstressed syllables, all such constraints make reference to foot dependents: either specifically 'unadjoined' ones or any dependents at all:

(26) *UNSTRESSED-UNADJOINED[a, e, o]
Assign a violation for every vowel in the set [a, e, o] in the dependent of a foot that is sister to the head of the foot.

(27) *UNSTRESSED-FOOTDEPENDENT[a, e, o]
Assign a violation for every vowel in the set [a, e, o] in the dependent of a foot in any capacity.

Although these recursive feet are constructed gradually, the distinction between unadjoined dependents and other dependents is interpretable based on the prosodic structure itself – no reference is needed to previous steps of the derivation. Martínez-Paricio (2012) relates similar kinds of distinctions in recursive foot structure to notions of minimal and maximal projection of the foot. This account is compatible with these ideas of the representation, in particular as used to analyze Dutch by Kager and Martínez-Paricio (2014). Broadly speaking, we only need the following two properties for an HS theory to avoid the pathology: (1) two degrees of reducibility must be distinguished based on locality to stress and (2) metrical structure must be built as the derivation proceeds, such that candidates differing in both stress and reduction cannot compete. In an approach using something like recursive feet, this means that all 'unstressed' syllables must be the dependents of (some sort of) feet, and some syllables are dependents in more than one way.

With these constraints and the adjunction operation, languages like the Dutch semi-formal register can be analyzed. The reduction in 'dependents' only is a result of *UNSTRESSED-UNADJOINED, affecting only the syllables which are direct sisters to the stressed syllable. These feet are created by the initial Pruitt-type foot creation operation. Reduction in general 'unstressed' position is accounted for by the remaining *UNSTRESSED-FOOTDEPENDENT. These syllables are due to both initial foot creation as well as adjunction. Any syllables that appear to reduce in 'stray' position must actually first adjoin to a nearby foot (motivated by some constraint). Prior to adjunction, there are no violations of the *UNSTRESSED constraint. After their adjunction, violations of *UNSTRESSED are incurred and may be resolved by reduction. This approach to Dutch reduction mirrors van Oostendorp's (1995) account in terms of superfeet. Importantly, although this approach changes representations, it also requires something like Harmonic Serialist evaluation. The idea of derivational order does not exist in OT, and therefore it cannot be leveraged to prohibit certain types of languages.

This setup does not imply that there is no such thing as stray syllables – parsing is not necessarily exhaustive. Adjunction may be desirable in many situations, but will also create marked structures (such as non-binary feet) which may violate higher-ranked constraints. This proposal is therefore not a wholesale return to stray syllable adjunction (Liberman and Prince 1977, Hayes 1981). What this method *does* imply, however, is that reduction (because of lack of stress) can never occur in stray syllable position. Adjunction must always precede it. This is a desirable result for the reason outlined above: if this is not the case, we necessarily predict the existence of systems in which stress is driven by arbitrary features.

As discussed above, the crucial property of a theory that avoids this pathology is that sonority violations do not come before foot creation. This property holds of recursive feet: any syllable that is not incorporated into a foot simply does not enter into the computation of reduction. It is only when a syllable is placed in a dependent, either by initial footing or by adjunction, that sonority violations occur. The desired separation is achieved, eliminating cases such as (23, 24) and thereby avoiding the pathology.

This useful property, avoiding violation before stress, could potentially be achieved in other ways. For example, a reviewer proposes that this could be understood as the gradual adjunction of syllables to the prosodic word (directly and indirectly through feet). Another alternative would be to base the evaluation of positional markedness constraints in part on a previous step, similarly to Jesney's approach to positional faithfulness (Jesney 2008). The foot-dependent approach is superior to these chiefly in simplicity: the status of syllables that fail to parse into the prosodic word is not well-understood in current approaches, and a Jesney-like approach introduces another dimension of complexity in that constraints can 'look back' to previous steps.

5.4.3 Beyond sonority

In the preceding discussion, I have dealt exclusively with constraints on the relationship between the sonority of vowels and stress. I have demonstrated how such constraints in a parallel framework predict languages that use arbitrary features to drive stress and proposed a serial solution to this pathological prediction. However, such arbitrarily-driven languages are not a prediction of just sonority/stress constraints; any positional markedness constraint relating segmental information to stress yields such predictions.

To see this, consider that in the above demonstration I did not make use of any property inherent only to sonority. The needed components are quite general:

(28) Components of a stress system driven by segmental feature X
(a) A positional markedness constraint C between features and stress
(b) A stress operation which repairs violations of C
(c) A segmental operation O which repairs violations of C
(d) A constraint which distinguishes among targets of O

For the cases described above, with roundness-driven stress created through sonority/stress constraints, this schema is instantiated in the following way:

(29) Example components of a stress system driven by roundness
(a) *UNSTRESSED[a, e, o]
(b) The footing operations of Pruitt (2010)
(c) Reduction
(d) IDENT(round)

With this schema in hand, we can readily generate novel pathological languages in an OT framework. For example, if we consider a language in which stress is generally left-aligning and trochaic, and in which stressed syllables are made heavy by germination, we can create an instance of the pattern.

(30) STRESSTOWEIGHT (SWP) (Gouskova 2003: 90 and references cited within): Assign a violation for every stressed nonheavy syllable.

(31) *VOICEDGEMINATE (*VOIGEM):
Assign a violation for every geminate consonant that is voiced.

(32) Voicing-driven stress

	pigeto	SWP	IDENT (voice)	*VOIGEM	ALLFTL	IDENT (other)
a.	(píge)to	1 W			L	L
b.	(pígge)to			1 W	L	1
c.	(píkke)to		1 W		L	1
d. →	pi(gétto)				1	1

Stressed syllables must be made heavy given the high ranking of SWP. Vowel lengthening must be ruled out by faithfulness, leaving gemination. Geminating the second consonant results in either a marked voiced geminate (32b) or devoicing (32c). The optimization avoids both by shifting stress.

As shown, the language described changes the location of stress based on the voicing of a following onset. In particular, stress is shifted if a following voiced onset can be avoided with a voiceless one. Such a prediction is likely pathological – that is, unattested and undesirable – and is an instance of the schema above:

(33) Example components of a stress system driven by voicing
  (a) SWP
  (b) The footing operations of Pruitt (2010)
  (c) Gemination
  (d) *VoiGem

This predicted language is avoided in Harmonic Serialism, even with the current definition of SWP. The local optimum here is a form that keeps the overall stress pattern of the language and (for example) geminates in the corresponding location. The (undesirable) global optimum is a form in which the overall stress pattern is broken and a different consonant is geminated. These candidates are too far apart to occur within the same tableau in a Harmonic Serialist derivation. HS can therefore not generate this pathological language.

Here we did not need to make use of any reference to unadjoined or adjoined syllables. This is because the position of interest was 'stressed'. In the cases I have considered here, there is only one type of position associated with 'stress', while *unstressed* position is more diverse. Thus, while gradualness alone is enough to do away with the pathological predictions of feature/stress positional markedness constraints on stressed position, we must restrict our constraints to foot dependents (and allow adjunction) to remove similar pathologies for unstressed position. This requirement is a useful one to bear in mind when developing other sorts of constraints that distinguish unstressed positions.

The approach outlined in this paper removes one class of pathological predictions for feature/stress interactions. However, not all such pathologies are excised. In particular, the focus of this paper is on interfaces that are typologically attested bidirectionally. For these, a particular segmental property is observed both to affect stress placement (e.g. sonority-driven stress) and to be affected by stress placement (e.g. reduction). However, some interfaces are unidirectional; aspiration can be motivated by stress placement, but underlying aspiration does not drive the placement of stress. These latter cases are problematic in that OT approaches predict bidirectionality in all cases, as shown by Blumenfeld (2006). HS does not avoid such pathologies, leaving this area open to future research.

5.5 Conclusion

In this paper, I have shown a variety of unattested, predicted interactions between features and stress in parallel Optimality Theory. I have demonstrated that these undesired predictions are greatly reduced if the serial Harmonic Serialist framework is used instead. Harmonic Serialism accounts for the observed interactions between segmental features and stress, but strictly limits them by its gradualness requirement. HS avoids a pathology in which any arbitrary feature can drive the assignment of stress. My approach to the pathology is aided by viewing the distinction between effects on 'stray syllables' and 'foot dependents' in terms of adjunction to recursive feet, rather than their stated names.

This paper contributes to a view of Harmonic Serialism as restricting typological predictions compared to Optimality Theory. Gradualness and an understanding of GEN allow for typological explanations resting on the distinction between local optima and global optima. However, this level of restriction is not overly severe – HS can preserve the good typological coverage of OT while not overpredicting.

References

Blumenfeld, Lev (2006) *Constraints on Phonological Interactions*. Doctoral dissertation, Stanford University, Stanford, CA.

Caballero, Gabriela (2008) *Choguita Raramuri (Tarahumara) Phonology and Morphology*. Doctoral dissertation, University of California, Berkeley.

de Lacy, Paul (2002) *The Formal Expression of Markedness*. Doctoral dissertation, University of Massachusetts Amherst. Amherst, MA: GLSA Publications.

Elfner, Emily (this volume) Stress-epenthesis interactions in Harmonic Serialism.

Gouskova, Maria (2003) *Deriving Economy: Syncope in Optimality Theory*. Doctoral dissertation, University of Massachusetts Amherst, Amherst, MA. Available on Rutgers Optimality Archive, ROA-610.

Hayes, Bruce (1980) *A Metrical Theory of Stress Rules*. Doctoral dissertation, MIT, Cambridge, MA. Published by Garland Press, New York, 1985.

Jensen, John (2000) Against ambisyllabicity. *Phonology* 17(2): 187–235.

Jesney, Karen (2008) Positional Faithfulness, non-locality, and the Harmonic Serialism solution. In Suzi Lima, Kevin Mullin, and Brian Smith (eds), *Proceedings of the 39th Meeting of the North East Linguistics Society (NELS 39)*. Amherst, MA: GLSA Publications.

Kager, René (1989) *A Metrical Theory of Stress and Destressing in English and Dutch*. Dordrecht: Foris.

Kager, René (2012). Stress in windows: language typology and factorial typology. *Lingua* 122: 1454–1493.

Kager, René and Violeta Martínez-Paricio (2014). The internally layered foot in Dutch word prosody. Paper presented at the Aspects of Germanic Prosody Workshop, University of Oxford.

Liberman, Mark and Alan Prince (1977) On stress and linguistic rhythm. *Linguistic Inquiry* 8: 249–336.

Martínez-Paricio, Violeta (2012) Superfeet as recursion. In Nathan Arnett and Ryan Bennett (eds), *Proceedings of the 30th West Coast Conference on Formal Linguistics (WCCFL 30)*, 259–269. Somerville, MA: Cascadilla Proceedings Project.

Martínez-Paricio, Violeta (2014) *An Exploration of Minimal and Maximal Metrical Feet.* Doctoral Disertation, University of Tromsø.

McCarthy, John J. (1982) Prosodic structure and expletive infixation. *Language* 58: 574–590.

McCarthy, John J. (2000) Harmonic serialism and parallelism. In Masako Hirotani (ed.), *Proceedings of the 30th Meeting of the North East Linguistics Society (NELS 30)*, 501–524. Amherst, MA: GLSA Publications.

McCarthy, John J. (2008a) The gradual path to cluster simplification. *Phonology* 25: 271–319.

McCarthy, John J. (2008b) The serial interaction of stress and syncope. *Natural Language & Linguistic Theory* 26: 499–546.

McCarthy, John J. and Alan Prince (1993a) Generalized Alignment. In Geert Booij and Jaap van Marle (eds), *Yearbook of Morphology*, 79–153. Dordrecht: Kluwer. Available on Rutgers Optimality Archive, ROA-7.

McCarthy, John J. (this volume) The theory and practice of Harmonic Serialism.

McCarthy, John J., Joe Pater, and Kathryn Pruitt (this volume) Cross-level interactions in Harmonic Serialism.

Moore-Cantwell, Claire (this volume) Contexts for epenthesis in Harmonic Serialism.

Prince, Alan (1990) Quantitative consequences of rhythmic organization. In M. Ziolkowski, M. Noske and K. Deaton (eds), *Parasession on the Syllable in Phonetics and Phonology*, 355–398. Chicago: Chicago Linguistic Society.

Prince, Alan and Paul Smolensky (1993/2004) *Optimality Theory: Constraint Interaction in Generative Grammar.* Malden, MA, and Oxford, UK: Blackwell.

Pruitt, Kathryn (2010) Serialism and locality in constraint-based metrical parsing. *Phonology* 27: 481–526.

Staubs, Robert (2013) Pathologies of feature-driven stress. In Lena Fainleib, Nicholas LaCara, and Yangsook Park (eds), *Proceedings of the 41st Meeting of the North East Linguistic Society (NELS 41).* Amherst, MA: GLSA Publications.

Staubs, Robert, Michael Becker, Christopher Potts, Patrick Pratt, John J. McCarthy and Joe Pater (2010) OT-Help 2.0 [computer program]. Computer program. Available at http://web.linguist.umass.edu/~OTHelp/.

van Oostendorp, Marc (1995) *Vowel Quality and Phonological Projection.* Doctoral dissertation, Katolieke Universiteit Brabant.

Chapter 6

Positional constraints in Optimality Theory and Harmonic Grammar[1]

Karen Jesney

6.1 Introduction

The ability of weighted constraints to model certain patterns that elude ranked constraints has sometimes been taken to be typologically problematic (e.g., Prince and Smolensky 1993/2004, Legendre, Sorace and Smolensky 2006). This paper argues that the greater power of weighted constraints can actually be advantageous, permitting a smaller set of more general constraints to be employed in the analysis of attested languages. With limitations on the set of constraints, the predicted typology is restricted beyond what is possible in Optimality Theory where more, and more specific, constraints are necessary. These principled revisions to Con mean that the languages generated by Harmonic Grammar need not be a superset of those generated by OT (see also Pater this volume, Potts *et al.* 2010).

The particular focus in this paper is upon constraints that govern patterns of featural licensing – i.e., patterns where a given marked feature surfaces only in specific phonological contexts. In OT these types of positional asymmetries have been captured using both positional

1 Many thanks to Joe Pater and John McCarthy for their guidance and encouragement throughout the many incarnations of this work, and to Michael Becker for his detailed commentary. Thanks also to Diana Apoussidou, Paul Boersma, Louis Goldstein, John Kingston, Chris Potts, Donca Steriade, Shelley Velleman, Rachel Walker, Matt Wolf, and the audiences at UMass, Harvard, USC, CLS 45, and MUMM 3.

faithfulness constraints and positional markedness constraints, and in many contexts there is little to discriminate between the two types of analysis. As this paper demonstrates, however, differences between the two constraint sets arise when featural licensing either requires a conjunction of contexts – i.e., being in context A *and* context B – or is possible in multiple (or disjoined) contexts – i.e., being in context A *or* context B. In OT conjunctive licensing patterns can only be modeled with positional markedness constraints, while disjunctive licensing patterns can only be modeled with positional faithfulness constraints. In HG, on the other hand, either set of constraints can account for both types of patterns. As a result, only positional markedness constraints are required in HG in order to capture a range of empirical phenomena that require both types of constraints in OT. I argue that this is a positive outcome, allowing the pathologies associated with positional faithfulness constraints to be avoided (*cf.* Jesney 2011), producing a more restrictive overall typology, and reducing duplication and redundancy within the theory.

The rest of this paper is structured as follows. Section 6.2 discusses the patterns of conjunctive and disjunctive licensing that are generated by the different constraint sets in the two theories, demonstrating that, given weighted constraints, positional markedness constraints are sufficient to the task. Two natural language case studies are presented: Kɔnni vowel place licensing and English [h] licensing. Section 3 considers the resulting typology and demonstrates how the use of a smaller constraint set in HG narrows the set of predicted languages relative to what is expected in OT. Section 6.4 addresses additional phenomena that have been argued to require analysis using positional faithfulness. Section 6.5 contrasts the typological effects of the HG approach with local constraint conjunction and other alternatives, and section 6.6 concludes.

6.2 Positional constraints and patterns of licensing

6.2.1 Conjunctive vs. disjunctive licensing

Two main types of constraints have been used to capture positional asymmetries in Optimality-Theoretic phonology: positional markedness constraints (Itô, Mester and Padgett 1995, Walker 2001a, 2005, 2011, Zoll 1996, 1998; see also Goldsmith 1989, 1990, Itô 1986, Steriade 1995), and positional faithfulness constraints (Beckman 1997, 1998,

Casali 1996, Lombardi 1999).[2] Each of these constraint types references a structure and context. As illustration, we can consider the case of the feature [+voice] on obstruents in interaction with the licensing position of syllable onset. The resulting constraints are given in (1).

(1) Example: Two types of constraints for capturing a positional asymmetry
  a. VoicedObstruent⇒Onset: Assign a violation mark to any [+voice] obstruent that is not associated with syllable onset position.
  b. Ident[±voice]/Onset: Assign a violation mark to any output segment in onset position that differs in specification for [±voice] relative to its input correspondent.

As the following OT tableaux show, if these constraints dominate the relevant conflicting general faithfulness (2) or general markedness constraints (3), [+voice] obstruents are restricted to onset position. Given an input like /bæd/, the onset voiced /b/ maps faithfully while the coda /d/ devoices.

(2) The marked feature maps faithfully only in onset when positional markedness dominates faithfulness

	/bæd/	VoicedObs⇒Onset	Ident[±voice]	*VoicedObs
a.	bæd	1		2
b. →	bæt		1	1
c.	pæt		2	

(3) The marked feature maps faithfully only in onset when positional faithfulness dominates markedness

	/bæd/	Ident[±voice]/Onset	*VoicedObs	Ident[±voice]
a.	bæd		2	
b. →	bæt		1	1
c.	pæt	1		2

2 The type of positional markedness constraints used in this paper are commonly known as 'positional licensing' constraints. Simple contextual markedness constraints – e.g., *VoicedObstruentCoda – are also widely used in modeling positional asymmetries. For all of the cases discussed in Section 6.2 of this paper, both types of markedness constraint have precisely the same properties. In OT they can be used to model conjunctive licensing patterns, but not disjunctive licensing patterns, whereas in HG they are able to model disjunctive licensing patterns as well. I use the more general term 'positional markedness' here in order to highlight the key role played by the markedness vs. faithfulness distinction in generating the different patterns.

With this small set of constraints and this input, ranked and weighted constraints make the same typological predictions. Voiced obstruents can be admitted in all positions, only in onset position, or in no position at all. Candidates that show devoicing in onset only are simply harmonically bounded and cannot be selected as optimal in either OT or HG.

In general the differences between positional markedness constraints and positional faithfulness constraints are relatively subtle and primarily situated in their ability to model patterns of directional spreading and the licensing of derived structure. There are, however, also differences in the types of stable positional restrictions that these types of constraints can model in OT. These differences become evident when multiple contextual constraints that reference the same marked feature are included in the system. In particular, OT positional markedness constraints can only model conjunctive patterns of licensing and positional faithfulness constraints can only model disjunctive patterns of licensing (see also Jesney 2012).

To illustrate this we will add a second licensing context – word-initial syllables – for [+voice] obstruents. The two possible ways of encoding this restriction are given in (4).[3]

(4) Example: Two types of constraints for capturing a second positional asymmetry
   a. VOICEDOBSTRUENT⇒INITIAL: Assign a violation mark to any [+voice] obstruent that is not associated with a word-initial syllable.
   b. IDENT[±voice]/INITIAL: Assign a violation mark to any output segment in a word-initial syllable that differs in specification for [±voice] relative to its input correspondent.

The following sections provide natural language examples of how multiple constraints like these interact. The table in (5) previews the conclusions. The first column lists languages composed of members of a set of four possible forms: [da], where the [+voice] obstruent is simultaneously in onset and the initial syllable, [sad], where the [+voice] obstruent is in the initial syllable but not in onset, [so.da] where the [+voice] obstruent is in onset but not in the initial syllable, and [so.nad], where the [+voice] obstruent is in neither onset nor the initial syllable. The following columns indicate whether each of these languages can be modeled

3 The selection of 'initial syllable' as the second licensing environment here is purely for ease of comparison with the natural language examples presented in Sections 6.2.2, 6.2.3 and 6.4.1; it is not intended as an empirical claim regarding the relative markedness of obstruent voicing in this position.

in OT and HG systems where positional restrictions are encoded using positional markedness constraints or positional faithfulness constraints. In each case the positional constraints are supplemented by the general markedness constraint *VOICEDOBSTRUENT and the general faithfulness constraints IDENT[±voice].

(5) Possible patterns of licensing in OT and HG with different types of positional constraints

		OT		HG	
		positional markedness (1a, 4a)	positional faithfulness (1b, 4b)	positional markedness (1a, 4a)	positional faithfulness (1b, 4b)
a.	{da, sad, so.da, so.nad} voiced obstruents everywhere	✓	✓	✓	✓
b.	{} voiced obstruents nowhere	✓	✓	✓	✓
c.	{da, sad} voiced obstruents in σ1	✓	✓	✓	✓
d.	{da, so.da} voiced obstruents in onset	✓	✓	✓	✓
e.	{da, sad, so.da} voiced obstruents in σ1 *or* onset	✗	✓	✓	✓
f.	{da} voiced obstruents in σ1 *and* onset	✓	✗	✓	✓

The most important observation here concerns the distinction between the types of patterns that can be modeled using positional markedness vs. positional faithfulness constraints. Either set of constraints can be used to model patterns (5a–d). The two patterns (5a) and (5b) are non-positional, relying simply upon the relative ranking or weighting of the general markedness and general faithfulness constraints. The patterns (5c) and (5d) are basic positional patterns of the type illustrated in (2) and (3). These too are easily accounted for in either system. The patterns in (5e) and (5f), on the other hand, reveal differences between the ranked and weighted constraint systems.

The language in (5e) shows a pattern of disjunctive licensing: [+voice] obstruents appear faithfully in either of the licensing environments – i.e., if they are in the initial syllable *or* if they are in onset. Only when neither of these conditions is met, as with the form *[so.nad], are voiced obstruents disallowed. This type of disjunctive licensing pattern can be modeled in OT with positional faithfulness constraints but not with positional markedness constraints. The language in (5f) shows a pattern of conjunctive licensing: [+voice] obstruents appear just when they are simultaneously in both licensing environments – i.e., when they are in the initial syllable *and* the onset. If both of these conditions are not met, as with the forms *[sad], *[so.da] and *[so.nad], voiced obstruents cannot appear. This type of pattern can be modeled in OT with positional markedness constraints, but not with positional faithfulness constraints.

As indicated in (5), both positional markedness and positional faithfulness constraints can model patterns of conjunctive *and* disjunctive licensing in Harmonic Grammar. This ability is a consequence of cumulative constraint interaction, as will be illustrated in the following sections. I argue that this greater power of constraints in HG is a considerable benefit, allowing the set of positional constraints employed by the grammar to be reduced, thereby creating a more elegant system that is nonetheless consistent with the types of patterns observed across languages.

We turn now to two case studies. The first case study – Kɔnni vowel place licensing – is used to demonstrate how, in OT, patterns of disjunctive licensing elude positional markedness constraints and patterns of conjunctive licensing elude positional faithfulness constraints. Both types of pattern can be straightforwardly modeled using only positional markedness constraints in Harmonic Grammar. The second case study – licensing of English [h] – extends this illustration, further considering how conjunctions and disjunctions of contexts interact.

6.2.2 *Kɔnni vowel place licensing*

Kɔnni is a Gur language spoken in the Northern Region of Ghana; all data and generalizations described here come from Cahill (2007). The vowel inventory is summarized in (6).[4]

4 Transcriptions directly reflect those provided by Cahill (2007), except for the following substitutions based on Cahill's (2007: 97) description of the consonant inventory: r→[ɾ], ch→[ʧ], j→[ʤ], y→[j].

(6) Vowel inventory of Kɔnni (Cahill 2007: 175)

[+ATR] vowels		[−ATR] vowels	
i ii	u uu	ɪ ɪɪ	ʊ ʊʊ
e ee~ie	o oo~uo	ɛ ɛɛ~ɪa	ɔ ɔɔ~ʊa
		a aa	

Short and long vowels contrast in Kɔnni, with long mid vowels typically realized as diphthongs. Within a non-complex word, all vowels agree for the feature [±ATR], and when a suffix is added, it adopts the [±ATR] value of the root vowels. We will abstract away from this vowel harmony pattern, and focus instead upon the disjunctive licensing restrictions.

There are two distinct environments in which the full range of contrastive vowel features are admitted in Kɔnni. The first licensing environment is the morpheme-initial syllable – a strong position discussed at length by Beckman (1998). As the verbs in (7) attest, all vowel qualities are represented in this context. In non-initial syllables, however, only high vowels – specifically [i] and [ɪ] – are found.[5]

(7) All vowel qualities are admitted in the initial syllable (Cahill 2007: 90)

[batɪ]	'to trap'	[kpegiri]	'to snap'
[jɛsɪ]	'to carve'	[sʊrɪsɪ]	'to rub'
[bɔbɪ]	'to tie'	[garɪsɪ]	'to pass'
[fogi]	'to take apart'	[kuri]	'to pound'

This morpheme-initial licensing condition can be formalized using the constraint in (8).[6]

(8) NonHigh⇒Initial: Assign a violation mark to any [−high] vowel that is not associated with a morpheme initial syllable.

In OT this markedness constraint must dominate the conflicting IdentHeight constraint in order for its effects to be observed – similar to the pattern seen in (2). The tableau in (9) reiterates this point, using

5 The second vowel of trisyllabic verbs can also be one of the high rounded vowels [u] or [ʊ] just in the case that the first vowel is [u] or [ʊ] and the intervening consonant is [g] (Cahill 2007: 90).

6 The addition of constraints identifying backness and vowel length as marked structures would allow the restrictions on these features to be captured as well. Importantly, these restrictions *do* appear to be distinct from those on vowel height, as the limitations on the vowel inventory in suffixes affect height and length, but not backness. These root vs. suffix effects are discussed briefly later in this section.

the hypothetical input /bɔbɛ/. With the ranking NonHigh⇒Initial ≫ IdentHeight ≫ *NonHigh, the first vowel maps faithfully to [ɔ] while the second vowel maps unfaithfully, losing its distinctive [−high] feature.

(9) Vowel height features map faithfully just when associated with the initial syllable

	/bɔbɛ/	NonHigh⇒Initial	IdentHeight	*NonHigh
a.	bɔbɛ	1		2
b. →	bɔbɪ		1	1
c.	bɪbɪ		2	

The output is the licit Kɔnni form [bɔbɪ] 'to tie'. Ranking NonHigh⇒Initial above faithfulness is crucial to this result; were the opposite ranking to hold, vowel height features would map faithfully in all positions.

The second licensing environment is nouns. As the data in (10) illustrate, a range of vowel qualities is admitted in all positions in nouns; being in the morpheme-initial syllable is not required.[7]

(10) All vowel qualities are admitted in nouns (Cahill 2007: 94–95, 99)

[gbárɪ́áŋ]	'earthworm'	[bítíéŋ]	'chin'
[dàjʊ̀ʊ́ŋ]	'rat'	[ʧàkùttù]	'blacksmith'
[jàltʊ́]	'hunter'	[bʊ́llɔ̀gɪ́ŋ]	'bag'
[gàɲɪàrà]	'weaver bird'	[ŋmárɪ́ʧɪ́áˈppʊ́rɪ́ŋ]	'speckled dove'

Like morpheme-initial privilege, noun-specific privilege is cross-linguistically well attested. Smith (2001) documents numerous examples from the phonological systems of unrelated languages. Child language provides further evidence; nouns are typically acquired before verbs (see Gentner and Boroditsky 2001 for a review), and, as in adult phonology, have been observed to license a greater range of contrasts (Adam and Bat-El 2008, Jesney and Tessier 2009). Notably, Kɔnni also shows noun-privilege effects in its tonal system. While nouns can contrast based solely on tone, the tonal patterns associated with verbs in Kɔnni are predictable based on inflectional properties of the sentence (Cahill 2007: 386). The constraint in (11) formalizes the noun-category licensing condition for non-high vowels.

7 While Cahill (2007) does not directly address the distribution of vowel place features in adjectives, from the data provided it appears that they pattern with nouns – e.g., [kúó-bíˈké] 'the small farm', *cf.* [kʊ́áŋ] (Cahill 2007: 187).

(11) NonHigh⇒Noun: Assign a violation mark to any [–high] vowel that is not associated with a noun.

In order for the effects of the constraint in (11) to be observed in OT, it must dominate IdentHeight. As the tableaux in (12) and (13) show, however, when both NonHigh⇒Initial and NonHigh⇒Noun dominate IdentHeight, the desired result is not obtained. Rather than [–high] vowels being licensed if they are in the initial syllable (of any lexical category) *or* in a noun (regardless of position), non-high vowels map faithfully only if they are simultaneously in an initial syllable *and* a noun. With this ranking, only nouns have distinctive vowel height, and then only in the initial syllable.

(12) Problem: Vowel height features in nouns map faithfully just when they are associated with the initial syllable

/gàɲɪàrà$_{Noun}$/	NonHigh⇒Initial	NonHigh⇒Noun	IdentHeight
a. ← gàɲɪàrà	2		
b. → gàɲɪìrì			2
c. gìɲɪìrì			3

(13) Problem: Vowel height features in verbs do not map faithfully in any position

/bɔbɪ$_{Verb}$/	NonHigh⇒Initial	NonHigh⇒Noun	IdentHeight
a. ← bɔbɪ		1	
b. → bɪbɪ			1

High-ranking NonHigh⇒Initial assigns violation marks to any non-initial [–high] vowel feature, regardless of whether it is found in a noun. This rules out the attested Kɔnni output in (12). Similarly, high-ranking NonHigh⇒Noun assigns violation marks to any instance of [–high] in a verb, ruling out attested outputs with this feature in the initial syllable (13).

This effect is a general one in OT. Whenever two positional markedness constraints that target overlapping sets of features both dominate a conflicting faithfulness constraint, the features will appear only in contexts that meet *both* of the licensing conditions. If further positional markedness constraints are added, the features will appear only in contexts that meet these licensing conditions as well. Positional markedness constraints in OT can only model conjunctive patterns of licensing – i.e., patterns where a feature is licensed just if it is simultaneously in context 1 (initial syllable) and context 2 (a noun), etc. It is not possible to express the type of disjunctive pattern seen in Kɔnni using these constraints in OT.

The reason for this is straightforward. As a general rule, markedness constraints act to restrict the set of contexts in which a given structure may appear. To the extent that markedness constraints dominate conflicting faithfulness constraints, the restrictions imposed by these constraints are enforced in the output (assuming that no conflicting markedness constraints outrank them). When multiple markedness constraints targeting overlapping structures are high ranking, the preference is for all of these restrictions to be enforced if possible. In the case of positional markedness constraints, the result is a general non-licensing of the structure in question unless all of the licensing conditions are simultaneously met.

In Harmonic Grammar, on the other hand, the missing disjunctive pattern can be readily modeled through the cumulative interaction of this set of constraints. The necessary weighting conditions are summarized in (14). For the Kɔnni pattern, the weight of IdentHeight must be greater than the independent weights of NonHigh⇒Initial (14a) and NonHigh⇒Noun (14b), even as the *summed* weight of the two positional markedness constraints is greater than the weight of IdentHeight (14c).

(14) HG weighting conditions for the Kɔnni disjunctive licensing pattern
w(IdentHeight) > *w*(NonHigh⇒Initial)
w(IdentHeight) > *w*(NonHigh⇒Noun)
w(NonHigh⇒Initial) + *w*(NonHigh⇒Noun) > *w*(IdentHeight)

The tableaux below illustrate the HG analysis, using a set of weights consistent with the weighting conditions in (14). Weights are given immediately below the constraint names; the total Harmony scores of candidates are given on the far right. These weights are partially arbitrary; any set of weights consistent with the conditions in (14) would yield the Kɔnni pattern. See Pater (this volume) for more on weighting conditions in HG.

(15) Non-high vowels in nouns map faithfully even when they are not associated with the initial syllable

/gàɲɪàɾà$_{Noun}$/	IdentHeight 3	NonHigh⇒Initial 2	NonHigh⇒Noun 2	
a. → gàɲɪàɾà		−2		−4
b. gàɲɪ̀ɾì	−2			−6
c. gìɲɪ̀ɾì	−3			−9

(16) Non-high vowels in the initial syllable map faithfully even when they are associated with a verb

/bɔbɪ$_{Verb}$/	IDENTHEIGHT 3	NONHIGH⇒INITIAL 2	NONHIGH⇒NOUN 2	
a. → bɔbɪ			−1	−2
b. bɪbɪ	−1			−3

In (15) the weight of IDENTHEIGHT is greater than the weight of NONHIGH⇒INITIAL, and so vowels in non-initial position, like the target /ɪa/ and /a/ in /gàɲɪàɾà/, map faithfully. The input here is a noun, and so all of the vowels satisfy the constraint NONHIGH⇒NOUN. This is the result attested in Kɔnni. In (16) the input is a verb, and so faithful mapping of the non-high vowel in the initial syllable leads to violation of NONHIGH⇒NOUN. Because the weight of IDENTHEIGHT is greater than the weight of NONHIGH⇒NOUN, however, the faithful form is selected.

The key case here is (17), where the input is the hypothetical form /ɲɪgasɪ/ – a verb that includes an illicit non-high vowel in the second syllable. The faithful candidate (17a) incurs violations of both of the positional markedness constraints, giving it a lower Harmony score than the unfaithful candidate (17b). The winning form – [ɲɪgɪsɪ] – is an attested verb in Kɔnni, meaning 'to flash' (Cahill 2007: 90).

(17) Non-high vowels map unfaithfully when they are not associated with the initial syllable or a noun

/nɪgasɪ$_{Verb}$/	IDENTHEIGHT 3	NONHIGH⇒INITIAL 2	NONHIGH⇒NOUN 2	
a. nɪgasɪ		−1	−1	−4
b. → nɪgɪsɪ	−1			−3

Given the importance of disjunctive licensing in Kɔnni and the fact that this is precisely the type of pattern that poses difficulties for positional markedness constraints in OT, one might wonder whether a positional faithfulness solution would be better suited to the data. Indeed, as previewed in Section 6.2.1, positional faithfulness constraints can model precisely this type of scenario. For Kɔnni, the required positional faithfulness constraints are the two in (18), along with the conflicting general markedness constraint *NONHIGH.

(18) Positional faithfulness constraints to account for the Kɔnni pattern
   a. IDENTHEIGHT/INITIAL: Assign a violation mark to any instance of an output segment in the morpheme-initial syllable that differs in vowel height features relative to its input correspondent.

b. IDENTHEIGHT/NOUN: Assign a violation mark to any instance of an output segment associated with a noun that differs in vowel height features relative to its input correspondent.

Ranking these constraints above *NONHIGH in OT allows the disjunctive licensing pattern to be captured.

Simply using positional faithfulness constraints is not an adequate solution, however, because positional faithfulness constraints cannot capture conjunctive licensing conditions that interact with disjunctive conditions like those described above. Within Kɔnni, this issue is highlighted by the role that root membership plays in licensing vowel features. While roots abide by the generalizations already described, the inventory of vowels in Kɔnni suffixes is restricted, so that only [i, u, e, o] appear with [+ATR] stems, and only [ɪ, ʊ, a] appear with [−ATR] stems (Cahill 2007: 88). Long vowels and the [−ATR] mid vowels [ɛ] and [ɔ] are not admitted in suffixes. This is true regardless of the lexical category of the word as a whole and regardless of the fact that the vowel may be in the initial syllable of the morpheme spelled out by the suffix. In effect, the full range of vowel features is admitted only when the relevant vowel is in a morpheme initial syllable *and* in a root or when it is in a noun *and* in a root.

The conjunctive aspect pattern of this pattern – i.e., the fact that being in a root plus some other strong environment is required – eludes positional faithfulness constraints in OT. If a root-specific positional faithfulness constraint dominates *NONHIGH, the set of contexts where vowel place features are freely admitted will extend to *all* positions in *all* roots – including non-initial syllables of verbs. High-ranking positional faithfulness constraints admit contrasts for the relevant features in *every* instance where those features are associated with the context identified by these constraints – a too liberal result in the Kɔnni case.[8]

8 Simply adding a general inventory constraint like *MID would be ineffective in saving the OT positional faithfulness analysis. In order to preserve mid vowels in nouns and in the initial syllable of verbs, IDENTHEIGHT/NOUN and IDENTHEIGHT/INITIAL would both need to dominate *MID. Given these rankings, however, mid vowels would incorrectly surface faithfully in affixes if these vowels were also in the initial syllable of the morpheme or in a noun. A more effective positional faithfulness approach would require inclusion of constraints that specify more detailed contexts, such as root-initial position vs. morpheme-initial position vs. in a noun root. These constraints effectively encode a conjunction of contexts within the definition of the constraints themselves and can therefore be used in OT to capture patterns like that found in Kɔnni. This approach requires a significant expansion of the constraint set; see Section 6.5 for related discussion.

This additional component of the pattern poses no difficulty for an HG analysis using positional markedness constraints; the key constraint is MID⇒ROOT, defined in (19).

(19) MID⇒ROOT: Assign a violation mark to any instance of a [–high, –low] vowel that is not associated with a root.

The vowel qualities targeted by this markedness constraint are a proper subset of those targeted by the other positional markedness constraints. Mid vowels are [–high], and therefore, depending upon their position within the form, may also violate NONHIGH⇒INITIAL and/or NONHIGH⇒NOUN.[9] The necessary weighting conditions for the full Kɔnni pattern are summarized in (20).

(20) HG weighting conditions for the Kɔnni disjunctive + conjunctive licensing pattern
  a. Vowel height features, including [–high, –low], are freely admitted when in a root and a noun (even non-initially) – *cf.* (14a)
  w(IDENTHEIGHT) > *w*(NONHIGH⇒INITIAL)
  b. Vowel height features, including [–high, –low], are freely admitted when in a root and an initial syllable (even of a verb) – *cf.* (14b)
  w(IDENTHEIGHT) > *w*(NONHIGH⇒NOUN)
  c. All [–high] vowels are disallowed if they are not in a noun or a morpheme-initial syllable (even if they are in a root) – *cf.* (14c)
  w(NONHIGH⇒NOUN) + *w*(NONHIGH⇒INITIAL) > *w*(IDENTHEIGHT)
  d. [–high, –low] vowels are disallowed in suffixes (even if the form as a whole is a noun, and the vowel is morpheme initial)
  w(MID⇒ROOT) > *w*(IDENTHEIGHT)

The Kɔnni pattern of vowel place licensing has both disjunctive and conjunctive aspects. Non-high vowels – and particularly mid vowels – are admitted if they are in a morpheme-initial syllable and in a root, or if they are in a noun and a root. In OT, positional markedness constraints can capture only the conjunctive aspects of this pattern, and positional faithfulness constraints can capture only the disjunctive aspects of this pattern. Neither set of constraints can capture both. If constraints are weighted as in HG, on the other hand, cumulative constraint interaction

9 I assume that the admissibility of the vowels [e] and [o] in suffixes is due to the fact that there are no low [+ATR] vowels in the inventory – i.e., [a] is unpaired. In order to harmonize and preserve a [–high] specification, vowels in suffixes added to [+ATR] words are therefore necessarily mid. This can be accommodated if the weight of the constraint enforcing agreement is greater than the weight of the constraint MID⇒ROOT.

allows positional markedness constraints to capture the conjunctive *and* disjunctive aspects of the pattern. The Konni distribution can be effectively modeled in HG using only a limited set of simple constraints.

6.2.3 *English [h] licensing*

The distribution of English [h] provides a second natural-language illustration of combined disjunctive and conjunctive licensing. While the licensing environments involved in the Konni case are arguably due to psycholinguistic prominence, contextual privilege driven by phonetic cues can be equally implicated in disjunctive and conjunctive licensing patterns. It is the constraint formalism and the consequent violation profiles that produce the HG vs. OT differences discussed in the previous sections. The substantive source of the licensing is immaterial to the range of patterns that can be modeled with these constraints.

As has been discussed extensively in the literature, the distribution of English [h] is restricted, with the segment appearing only when certain conditions are met (see, among others, Borowsky 1984, 1986, Davis and Cho 2003, Hammond 1999). The most commonly-recognized environments for English [h] are onset of a word-initial syllable (21) and onset of a stressed syllable (22).

(21) [h] is licensed if it is in onset *and* a word-initial syllable

[hə.ˈbɪ.ʧu.l̩]	'habitual'
[hə.ˈloʊ]	'hello'
[ˈhɛ.ðɚ]	'heather'

(22) [h] is licensed if it is in onset *and* a stressed syllable

[vi.ˈhɪ.kju.lɚ]	'vehicular'
[pɹo.ˈhɪ.bət]	'prohibit'
[ˈæl.kə.ˌhɑl]	'alcohol'

If one of these conjoined conditions is not met, [h] cannot appear. There are no words with [h] in the coda of a word-initial syllable, even when that syllable is stressed (23). Likewise, there are no words with [h] appearing in the onset of an unstressed non-initial syllable (24).

(23) [h] is not licensed in the coda of word-initial syllables

[ˈbɹɑ.mɪn] ~ *[ˈbɹɑh.mɪn]	'brahmin'
[ˌtʰɛ.ˈɹɑn] ~ *[ˌtʰɛh.ˈɹɑn]	'Tehran'

(24) [h] is not licensed in the onset of unstressed non-initial syllables

[ˈvi.ə.kl̩] ~ *[ˈvi.hə.kl̩]	'vehicle'	*cf.* 'vehicular'
[ˌpɹo.ə.ˈbɪ.ʃn̩] ~ *[ˌpɹo.hə.ˈbɪ.ʃn̩]	'prohibition'	*cf.* 'prohibit'

These environments closely mirror those where aspiration is found. Voiceless stops are aspirated in word-initial position, and in the initial position of stressed syllables (Davis and Cho 2003, Jensen 2000, Kreidler 1989, Selkirk 1982). Aspiration is absent from the coda of word-initial syllables, even when that syllable is stressed, and from most unstressed non-initial syllables – e.g., *[ˈætʰ.ləs] 'atlas', *[ˈɹæ.pʰəd] 'rapid'. Aspiration can, however, appear on unstressed non-initial syllables when the preceding syllable is also stressless – e.g., [ˌæ.bɹə.kʰə.ˈdæ.bɹə] 'abracadabra', [ˌmɛ.də.tʰə.ˈɹeɪ.ni.ən] 'Mediterranean' (Jensen 2000, Pater 2000).

This final context is arguably a less robust licensor of [h] than of aspiration. Davis and Cho (2003: 609) note that words with [h] in this environment are 'exceedingly rare' and identify only one example: [ˌtʰɛ.ɹə.hə.ˈmɑ.ɹə] 'Tarahumara'. Further distinguishing the distribution of [h] from the distribution of aspiration is the predictability and necessity of aspiration on voiceless stops in the contexts where it is licensed. The presence of [h] is unpredictable; it is not, for example, epenthesized in cases where a syllable would otherwise be onsetless. Nonetheless, it seems clear that the two distributions are connected. Articulatory evidence supports this; the extent of the glottal gesture in the various licensing contexts is similar for [h] and for aspiration (Goldstein 1992). Given these (incomplete) parallels, it seems reasonable to suppose that the phonetic enhancement observed with aspiration has found a categorical instantiation with English [h] licensing. For simplicity we will focus in the theoretical discussion on the distribution of [h] as represented in (21) through (24).

As with the Kɔnni pattern, the distribution of English [h] can be understood as resulting from a requirement that multiple licensing conditions be satisfied in order for the segment to appear. The relevant constraints are given in (25).

(25) Positional markedness constraints for English [h] licensing

a. H⇒ONSET: Assign a violation mark to any segment [h] that is not associated with a syllable onset.

b. H⇒INITIAL: Assign a violation mark to any segment [h] that is not associated with a morpheme-initial syllable.

c. H⇒STRESS: Assign a violation mark to any segment [h] that is not associated with a stressed syllable.

The claim here is that these are distinct licensing conditions that interact, and that the full pattern cannot be reduced to an onset-licensing requirement. This stands in contrast with stress-based resyllabification accounts, like that of Selkirk (1982), which attribute the absence of [h] in words like *vehicle* to a preference in English for intervocalic consonants to be syllabified as codas when the immediately preceding vowel is stressed (for related experimental findings involving other consonants, see, among others, Ishikawa 2002, Redford and Randall 2005, Turk 1994). The syllabification for the form *vehicle* under a resyllabification approach would be [ˈvih.ə.kl̩], with subsequent deletion of coda [h]. There are two reasons to question the adequacy of this account. First, alternative pronunciations of loanwords like *Tehran* and *Bahrain*, epenthesize a vowel immediately following target /h/; this allows the segment to be syllabified as an onset and preserved even though the preceding syllable retains its stress – i.e., [ˌtʰɛ.hə.ˈɹɑn] *Tehran*, [ˌbɑ.hə.ˈɹeɪn] *Bahrain*. While preservation of the [h] in these cases is unexpected, given that the epenthetic vowel is unstressed, it is clear that stress does not force coda syllabification. This effectively undermines the claim that target /h/ is deleted in words like *vehicle* due to stress-based resyllabification; minimally, resyllabification is not mandatory for [h], and therefore cannot be the sole factor governing the segment's distribution.[10] Second, articulatory evidence from work by Goldstein (1992) indicates that production of [h] varies across the onset contexts in which it appears. More extreme laryngeal gestures occur when the associated syllable is stressed than when it is unstressed, and, in the case of unstressed syllables, when the syllable is initial than when it is non-initial. In effect, onset syllabification is crucial in licensing [h], but stress and word-initial position are also key factors in determining the precise realization of the segment. To the extent that these factors are represented within the grammar, reducing the account to one relying on syllabification seems unwarranted.

As one would expect, OT achieves limited success with the constraints in (25). When the three positional markedness constraints dominate faithfulness, [h] is licensed just at the intersection of the privileged contexts – i.e., in the onset of stressed morpheme-initial syllables. In contrast, Harmonic Grammar is able to capture both the conjunctive and disjunctive aspects of the pattern with little difficulty. To ensure that [h] is admitted in the onset of initial syllables and the onset of stressed syllables, the weighting conditions in (26a) and (26b) are required. Supplementing these with the conditions in (26c) and (26d) ensures that [h] is banned outside of the attested contexts.

10 Thanks to John Kingston for pointing out the relevance of these data.

(26) HG weighting conditions for the English [h] disjunctive + conjunctive licensing pattern
a. [h] appears in word-initial onsets (even when the syllable is unstressed)
w(Max-C) > *w*(H⇒Stress)
b. [h] appears in the onset of stressed syllables (even when they are non-initial)
w(Max-C) > *w*(H⇒Initial)
c. [h] is barred from coda (even of initial syllables and stressed syllables)
w(H⇒Onset) > *w*(Max-C)
d. [h] is barred from the onset of unstressed non-initial syllables
w(H⇒Initial) + *w*(H⇒Stress) > *w*(Max-C)

The weights shown in the tableaux below meet these conditions, ensuring that [h] is restricted to the attested environments.

(27) [h] is admitted in the onset of an unstressed initial syllable

/hə.ˈloʊ/	H⇒Onset 4	Max-C 3	H⇒Initial 2	H⇒Stress 2	
a. → hə.ˈloʊ				−1	−2
b. ə.ˈloʊ		−1			−3

(28) [h] is admitted in the onset of a stressed non-initial syllable

/vi.ˈhɪ.kju.lɚ/	H⇒Onset 4	Max-C 3	H⇒Initial 2	H⇒Stress 2	
a. → vi.ˈhɪ.kju.lɚ			−1		−2
b. vi.ˈɪ.kju.lɚ		−1			−3

(29) [h] is barred from coda

/ˈtʰɛh.ɹɑn/	H⇒Onset 4	Max-C 3	H⇒Initial 2	H⇒Stress 2	
a. ˈtʰɛh.ɹɑn	−1				−4
b. → ˈtʰɛ.ɹɑn		−1			−3

(30) [h] is barred from the onset of an unstressed non-initial syllable

/ˈvi.hə.kl̩/	H⇒Onset 4	Max-C 3	H⇒Initial 2	H⇒Stress 2	
a. ˈvi.hə.kl̩			−1	−1	−4
b. → ˈvi.ə.kl̩		−1			−3

The problems for positional faithfulness in OT seen with the Kɔnni system in Section 6.2.2 extend to the current case. Positional faithfulness can capture the disjunctive aspects of the pattern – i.e., the fact that either stress *or* initial-syllable positioning is required for licensing – but not the fact that [h] must also be associated with an onset in each case. In OT positional faithfulness and positional licensing constraints make opposing predictions. As a result, neither set of constraints alone can capture complex patterns like Kɔnni vowel feature licensing or English [h] licensing. In HG, on the other hand, a single set of constraints can capture both conjunctive and disjunctive licensing effects. The simplicity of the resulting analyses is a distinct advantage of the HG approach.

6.3 Typological considerations

6.3.1 Typology size and static distributions

Alone, both positional markedness constraints and positional faithfulness constraints undergenerate in OT. Given that each of these sets of constraints can capture some types of positional asymmetries, however, it is reasonable to consider the consequences of a CON that includes both. To examine this, we will continue with the constraints defined while discussing the distribution of English [h].

The OT and HG typologies produced with only positional markedness constraints, only positional faithfulness constraints, and both positional markedness and positional faithfulness constraints are compared in (31). A total of six models are represented; the OT models are given on the left and the HG models are given on the right. All models include the general constraints MAX-C and *H. Additionally, models with positional markedness constraints included H⇒ONSET, H⇒INITIAL and H⇒STRESS, and models with positional faithfulness constraints include MAX-C/ONSET, MAX-C/INITIAL and MAX-C/STRESS.[11] Models with both positional markedness and positional faithfulness constraints include all six positional constraints. Typology computations were conducted using OT-Help (Staubs *et al.* 2010).

11 In parallel theories like the versions of OT and HG presented here, privileged contexts for positional faithfulness constraints are typically defined with respect to the output. Because violations of MAX are incurred precisely when there is no output correspondent of the element in question, however, a definition that ties the privileged position to the output is untenable in these cases (McCarthy 2005). Avoidance of this problem is a further reason to prefer an HG analysis that requires only positional licensing. For an alternative that avoids this problem within the context of Harmonic Serialism (Prince and Smolensky 1993/2004: 94–95, McCarthy 2006, 2008ab, 2009, Kimper 2011, Pruitt 2010), see Jesney (2011).

A total of eight inputs are considered. Two outputs – faithful and unfaithful mapping of target /h/ – are possible for each form, giving 256 potential combinations of optima. In fact, however, maximally 20 languages are produced with the constraints used here. All 20 are generated by HG with only positional markedness, HG with only positional faithfulness, and HG with both positional markedness and positional faithfulness. In other words, the full range of conjunctive and disjunctive patterns can be modeled in HG with any of the three constraint sets.

The picture in OT is rather more complex, but unsurprising given the discussion thus far. Positional markedness constraints in OT do not allow any type of disjunctive licensing pattern to be modeled, and so languages (31i) through (31s) are ruled out. Positional faithfulness constraints in OT do not allow any type of conjunctive licensing pattern to be modeled, and so languages (31e) through (31h) and languages (31m) through (31s) are ruled out. The English pattern is represented by (31r), and cannot be generated in OT with only positional markedness or positional faithfulness constraints. With both sets of positional constraints in OT, 19 of the languages – i.e., all except (31s) – are admitted.

(31) Possible patterns of licensing in OT and HG with different sets of positional constraints

		OT			HG		
		pos. lic. only	pos. faith only	pos. lic. and pos. faith	pos. lic. only	pos. faith only	pos. lic. and pos. faith
a.	[h] in all contexts	✓	✓	✓	✓	✓	✓
b.	[h] only if onset	✓	✓	✓	✓	✓	✓
c.	[h] only if stressed	✓	✓	✓	✓	✓	✓
d.	[h] only if initial	✓	✓	✓	✓	✓	✓
e.	[h] only if stressed and onset	✓	✗	✓	✓	✓	✓
f.	[h] only if stressed and initial	✓	✗	✓	✓	✓	✓
g.	[h] only if onset and initial	✓	✗	✓	✓	✓	✓
h.	[h] only if stressed and onset and initial	✓	✗	✓	✓	✓	✓
i.	[h] if stressed or if onset	✗	✓	✓	✓	✓	✓
j.	[h] if stressed or if initial	✗	✓	✓	✓	✓	✓
k.	[h] if onset or if initial	✗	✓	✓	✓	✓	✓
l.	[h] if onset or if initial or if stressed	✗	✓	✓	✓	✓	✓

m. [h] if stressed or if onset and initial	✗	✗	✓	✓	✓	✓
n. [h] if initial or if stressed and onset	✗	✗	✓	✓	✓	✓
o. [h] if onset or if stressed and initial	✗	✗	✓	✓	✓	✓
p. [h] if stressed and initial or if stressed and onset	✗	✗	✓	✓	✓	✓
q. [h] if initial and stressed or if initial and onset	✗	✗	✓	✓	✓	✓
r. [h] if onset and stressed or if onset and initial	✗	✗	✓	✓	✓	✓
s. [h] if onset and stressed or if onset and initial or if stressed and initial	✗	✗	✗	✓	✓	✓
t. [h] in no contexts	✓	✓	✓	✓	✓	✓

Two primary observations emerge from this comparison. First, all three of the HG models generate precisely the same set of languages. Unlike in OT, there is no typological benefit to be gained in HG from the inclusion of both positional faithfulness and positional markedness constraints. Second, the HG typologies include only one language that is not generated by OT with the two sets of constraints. Typological expansion due to cumulative constraint interaction is quite limited. An HG model with a single set of positional constraints is thus a viable solution, allowing conjunctive and disjunctive licensing patterns to be effectively modeled while reducing redundancy in Con.

6.3.2 Restricting the constraint set, restricting the typology

The possibility in HG of restricting the constraint set to just one type of contextual constraint – specifically positional markedness constraints – offers additional typological benefits that become evident when a wider range of phenomena are considered. We will focus in this section on the licensing of derived marked structures, which, as Zoll (1998) argues, show a preference for association with strong positions. This type of patterning requires analysis using positional markedness constraints; positional faithfulness constraints make precisely the wrong predictions (Zoll 1998; see also Itô and Mester 2003).

As an initial example, we will consider the Sanskrit morpheme denoting authorship, which, in addition to suffixation with [-a] (or [-iːya] in certain cases), induces lengthening and, when non-vacuous, introduction of the [a] vowel quality in the initial syllable of the base (Kiparsky 1972). This type of pattern is particularly interesting because the two exponents of the morpheme are not aligned with one another; instead, the vowel length and quality change are attracted to the strong initial (licensing) position. For illustrative purposes, we will focus upon the lengthening effect, which can be attributed to docking of a floating mora introduced by the morpheme.

(32) Sanskrit authorship morpheme triggers lengthening and lowering of the initial syllable (Kiparsky 1972: 202)[12]

a. Base form	b. Derived form
jinendra	jainendra
bida	baida
apiśali	a:piśala

Tableaux illustrating the necessity of positional markedness constraints for modeling such patterns are given in (33) and (34). (The results in (33) and (34) are shown using HG tableaux, but the patterns rely on simple harmonic bounding and therefore extend straightforwardly to OT.) The first tableau, (33), shows a successful analysis using the positional markedness constraint V:⇒INITIAL, which penalizes long vowels that appear outside the strong word-initial syllable. The optimal candidate here has lengthening of the initial vowel, as in the attested forms. The other candidates, all of which introduce the marked structure into a weak environment, are harmonically bounded. With the ordering of the constraints in (33), long vowels are admitted as faithful mappings in any position, but *derived* long vowels are limited to the position where they are best licensed.[13]

(33) Positional markedness leads to docking in the strong position

	/apiśali + a + μ/	IDENTVLENGTH 3	V:⇒INITIAL 2	
a. →	a:piśala	−1		−3
b.	api:śala	−1	−1	−5
c.	apiśala:	−1	−1	−5

12 Long vowels are indicated here using the symbol [:]. In all other respects the transcription follows Kiparsky (1972).

13 I assume throughout this section that the derived structures must be realized in the output. This is equivalent in HG to saying that the weight of a constraint forcing docking of morphologically-introduced floating features – e.g., MORPHREAL (Kurisu 2001) – is greater than the summed weight of the conflicting constraints.

The tableau in (34) shows that positional faithfulness constraints fail to achieve the same success with the Sanskrit data. The positional faithfulness constraint IDENTVLENGTH/INITIAL penalizes initial vowels that differ in length relative to their input correspondents. As a result, the floating mora actively avoids associating with the initial syllable; the candidate that violates IDENTVLENGTH/INITIAL is harmonically bounded. Including a general IDENTVLENGTH constraint does not alter these results. With positional faithfulness constraints, any candidate where the floating mora docks in the strong position is harmonically bounded by a candidate where the floating mora docks elsewhere.

(34) Positional faithfulness leads to avoidance of the strong position

/apiśali + a + μ/		IDENTVLENGTH/INITIAL 3	*V: 2	
a. ←	a:piśala	−1	−1	−5
b. →	api:śala		−1	−2
c. →	apiśala:		−1	−2

The preferences reflected in these tableaux persist as long as there are no additional constraints that favor association of the derived marked structure with some other position. For example, it would be possible to model a suffix's preference for lengthening an immediately preceding vowel – even if that vowel is in a relatively weak position – by including a constraint that is violated just in the case that the derived long vowel is not aligned with the suffix. Such a constraint would break the harmonic bounding pattern in (33), allowing a different candidate to prove optimal. Similarly, constraints disfavoring long vowels of a particular quality could block lengthening of the initial vowel in (33). What is crucial here is that positional markedness constraints alone allow marked structures to preferentially associate with strong positions, as is observed with the Sanskrit pattern. Positional faithfulness constraints cannot express this preference; indeed, all else being equal, positional faithfulness constraints *block* association of derived marked structures with strong licensing positions.[14]

14 Michael Becker suggests that effects of this type might be captured through positional faithfulness constraints of the form MAXFLOAT/INITIAL – 'Assign a violation mark to any floating element that does not have correspondent in the initial syllable of the output' – on analogy with the non-positional MAXFLOAT constraints of Wolf (2005). While this approach achieves the desired effect, it does so precisely because, unlike the positional IDENT constraints discussed

Zoll (1998), building on Kager's (1996) analysis of Guugu Yimidhirr, provides a more complex case that illustrates how static licensing-based inventory restrictions can interact with morpheme-specific alignment considerations. As the data in (35) illustrate, Guugu Yimidhirr limits long vowels to the first two syllables of the stem.

(35) Guugu Yimidhirr: long vowels are admitted in the first two syllables of the stem (Kager 1996: 66)

[wa:.d̪a]	'crow'	[da:.ɾa:l.ŋan]	'kangaroo'
[gam.bu:.gu]	'head'		

When vowel length is derived, this restriction still holds. The suffix [-ŋu], for instance, has the effect of lengthening the immediately preceding vowel (36a,b). Lengthening only occurs, however, if the preceding vowel is one of the first two vowels of the stem; no lengthening occurs if the targeted vowel falls outside of the licensing domain (36c).

(36) Guugu Yimidhirr: Lengthening induced by the suffix [-ŋu] is restricted to the first two syllables of the stem (Kager 1996: 93–94)

a.	/mayi/	[ma.yi:.ŋu]	'food-PURP'
b.	/ŋalgal/	[ŋal.ga:l.ŋu]	'smoke-PURP'
c.	/bad̪ibay/	[ba.d̪i.bay.ŋu]	'bone-PURP'
		*[ba.d̪i.ba:y.ŋu]	

This type of pattern is readily modeled with positional markedness constraints. The key constraint here is V:⇒HEADPWD, which limits long vowels to the two-syllable stem-initial licensing domain that Kager (1996) identifies as the head PWd. Weighting this constraint above IDENTVLENGTH ensures that long vowels map faithfully just if they appear within this two-syllable domain. V:⇒HEADPWD also conflicts with ALIGN-*ŋu*, a morpheme-specific constraint that requires the suffix [-ŋu] to immediately follow a heavy syllable. In cases where the stem is only two syllables long, the weighting of ALIGN-*ŋu* above IDENTVLENGTH allows the stem-final vowel to lengthen. In cases where

in this paper, it does not demand faithfulness between an input segment and its output correspondent. In effect, this constraint prefers *un*faithfulness in the initial syllable by attracting the floating element to that position; it is, essentially, a morpheme alignment constraint. To the extent that MAXFLOAT/INITIAL achieves the same result as the positional markedness constraints discussed in this section, it does so through morpheme-specific properties rather than general licensing preferences.

the stem is longer than two syllables, however, the added weight of violating V:⇒HeadPwd blocks lengthening. This same effect is achieved in OT with the ranking V:⇒HeadPwd ≫ Align-*ŋu* ≫ IdentVLength. (See Kager 1996 and Zoll 1998 for further discussion of the constraints involved.)

With positional markedness constraints, derived marked structures are, by default, attracted to positions that are the best licensers. Derived marked structures *can* appear in other positions, such as immediately preceding an associated suffix, but additional morpheme-specific constraints are necessary for this to occur. The Guugu Yimidhirr case illustrates both of these properties. Derived long vowels are restricted to a position where they are independently licensed, as captured with the V:⇒HeadPwd constraint. The selection of the second syllable as the locus for vowel lengthening, rather than the first, however, is a function of the morpheme-specific constraint Align-*ŋu*.

When the constraint sets required for patterns of conjunctive and disjunctive licensing are integrated with the type of alternations discussed in this section, rather dramatic typological differences are revealed. To illustrate, we will consider the marked feature [+voice], and two environments for its licensing – onset, and the word-initial syllable. The constraint set for the HG system is limited to VoicedObs⇒Onset and VoicedObs⇒Initial, along with general *VoicedObs and Ident[±voice] constraints. The OT system adds the positional faithfulness constraints Ident[±voice]/Onset and Ident[±voice]/Initial. Two inputs are considered: the simple input /bad.na.bad/, and the input /kat.na.kat [+voice]/, which includes a morphologically-introduced, [+voice] feature. The resulting typologies are summarized in (37).

(37) Predicted sets of possible outputs in HG with only positional markedness constraints vs. OT with positional markedness and positional faithfulness constraints

/kat.na.kat, [+voi]/→ /bad.na.bad/↓	**g**at.na.kat	ka**d**.na.kat	kat.na.**g**at	kat.na.ka**d**
a. **b**a**d**.na.**b**a**d**	HG and OT	OT	OT	OT
b. pat.na.pat	HG and OT	OT	OT	OT
c. **b**a**d**.na.pat	HG and OT	OT	OT	OT
d. **b**at.na.**b**at	HG and OT	OT	OT	OT
e. **b**a**d**.na.**b**at	HG	OT	OT	OT
f. **b**at.na.pat	HG and OT	OT	OT	

Six languages are generated by the HG system, each corresponding to one of the six configurations sketched in (5). As a faithful mapping, [+voice] may be realized in all contexts (37a), in no context (37b), if it is in the initial syllable (37c), if it is in onset (37d), if it is in onset *or* the initial syllable (37e), and if it is in onset *and* the initial syllable (37f). As a derived instance of the marked structure, the floating [+voice] feature consistently docks on the best licensor – i.e., the onset of the initial syllable – giving [gat.na.kat]. This is to be expected given the preferences of positional markedness constraints; derived marked structures preferentially target the context that allows all positional markedness constraints to be satisfied – i.e., the conjoined context. All other options are harmonically bounded. More generally, the HG system predicts that derived instances of a marked structure are limited to positions where underived instances of that structure are also licensed. The only exception comes from cases where the marked structure is not normally admitted in any position – e.g., cases where input /bad.na.bad/ maps to [pat.na.pat]. Licensing of derived marked structures is governed by the same basic conditions as licensing of non-derived marked structures.

The OT typology is considerably larger, comprising 22 languages. When both positional faithfulness and positional markedness constraints are at play, the set of available sites for docking of the floating [+voice] feature expands substantially. Depending upon the ranking of the constraints, the feature can associate with an initial onset, an initial coda, a non-initial onset, or a non-initial coda. These possibilities cross-classify almost exhaustively with the six patterns of faithful mapping. A number of the resulting languages are distinctly odd, imposing entirely different conditions on the licensing of floating and non-floating [+voice] features even in the absence of morpheme-specific alignment constraints. The OT-generated language with output forms [bad.na.pat] and [kat.na.gat], for example, generally licenses faithful realization of [+voice] just in the initial syllable, but prefers that derived [+voice] features target the onset of a non-initial syllable. Similarly, the language with outputs [bad.na.bat] and [kat.na.kad] normally admits voiced segments if they are in onset position or in the initial syllable, but explicitly avoids docking [+voice] features in these positions. With both sets of positional constraints, there is no necessary relation in the OT system between the licensing of derived and non-derived marked structures.

To summarize: The greater power of positional markedness constraints in HG allows fewer constraints to be included in Con, leading to a more restricted typology that excludes various dubious patterns (see also Potts *et al.* 2010). Far from expanding the typology of possible patterns in an unbounded fashion, HG's weighted mode of constraint interaction sometimes allows attested patterns to be modeled using fewer, more basic constraints, thereby constraining typological expansion in novel and important ways.

6.4 Other positional effects

Section 6.2 argued that positional markedness constraints in HG are able to capture a range of patterns that would require the addition of positional faithfulness constraints in OT. Section 6.3 extended the argument, demonstrating that restricting the set of contextual constraints to positional markedness has the benefit of tying the realization of derived marked structures to more general conditions on licensing, consistent with patterns observed in natural language. The possibility of limiting the set of contextual constraints to positional markedness does raise other issues, however, given that positional faithfulness constraints have uses beyond simply allowing disjunctive patterns of licensing to be captured in OT. This section turns to these issues.

Beckman (1998) identifies the three criteria in (38) as diagnostic of positional privilege, and argues that all three can be captured using positional faithfulness constraints.

(38) Diagnostics of positional privilege (Beckman 1998: 2)
  a. Positional maintenance of contrasts that are neutralized elsewhere
  b. Positional triggering of phonological processes
  c. Positional resistance to processes that apply elsewhere

The first of these, (38a), was argued in Section 6.2 to be effectively modeled by positional markedness constraints in HG. The diagnostics in (38b) and (38c) are both often connected with directionality in assimilation processes, which presents notable challenges for positional markedness constraints in parallel theories, regardless of the mode of constraint interaction (e.g., Beckman 1998, Walker 2001b, 2011, Zoll 1998; *cf.* McCarthy 2008a on direction of assimilation in Harmonic Serialism). The diagnostic in (38c) also has instantiations in the domain of alternations and blocking, however; these are of particular interest because the use of weighted constraints alters the landscape of potential analyses. This section thus first considers HG-OT differences in resistance to phonological processes, and then turns to the persistent challenge of directionality.

6.4.1 *Alternations and blocking*

In scenarios like those described in Section 6.3.2, a marked structure is present in the input but not associated with a segment in a licensing position. The task of the grammar, then, is to determine the optimal association of the marked structure. As we have seen, positional markedness constraints predict that derived marked structures will, all else being equal, preferentially target strong licensing positions, giving rise to alternations just in strong contexts. The alternations in Section 6.3.2 were morphologically driven, but purely phonological alternations can show similar behavior. In metaphony processes, for example, marked vowel features in weak positions trigger changes to the specifications of vowels in strong positions (Walker 2005, 2006, 2011). Metathesis processes in child language show similar effects, with marked segments shifting to strong positions where they are better licensed (Gerlach 2009, Marshall and van der Lely 2009, Velleman 1996). These types of alternations targeting strong positions are precisely what is predicted by positional markedness constraints, and stand in contrast with the type of blocking in strong positions that is predicted by positional faithfulness constraints.

This said, there are also established cases where strong positions resist alternations – indeed, this is the third diagnostic in (38) (see Becker, Nevins and Levine 2012 for additional evidence). In Optimality Theory, such patterns simply cannot be modeled using positional markedness constraints. In Harmonic Grammar, however, these same constraints can block alternations in strong positions in cases where the alternation would lead to markedness *reduction* – i.e., removal of a marked structure targeted by the positional markedness constraint. As an initial illustration, we will consider the constraint AGREEVOICE, which is violated when a sequence of two obstruents differ in [±voice] specification. This constraint interacts with the general faithfulness constraint IDENT[±voice] and the positional markedness constraint VOICEDOBS⇒INITIAL, leading to a restricted pattern of voicing assimilation that is blocked just in the strong initial syllable.

In the first tableau (39), the base form is monosyllabic, placing the final segment – voiced /d/ – in the privileged initial syllable. Adding the suffix /-pa/ leads to a violation of AGREEVOICE if both segments retain their input voicing specifications. As we would expect, if the weight of IDENT[±voice] is greater than the weight of AGREEVOICE, this voicing disagreement is tolerated, alternation is blocked, and the voicing specification of /d/ is preserved in the output. This contrasts with the second tableau (40), where the input /d/ is found in a non-initial syllable and its faithful mapping therefore incurs a violation of VOICEDOBS⇒INITIAL. Cumulative interaction is crucial here. Voiced segments are generally allowed in non-initial syllables based on the *w*(IDENT[±voice]) > *w*(VOICEDOBS⇒INITIAL) weighting condition, but when the summed weight of VOICEDOBS⇒INITIAL and AGREEVOICE is greater than the weight of the IDENT[±voice], assimilation is triggered in the weak context. The voicing alternation here is restricted to weak non-initial environments; segments in the strong initial syllable resist the markedness-reducing change.

(39) Voicing alternations are blocked in the initial syllable

/tad + pa/	IDENT[±voice] 4	AGREEVOICE 3	VOICEDOBS⇒INIT 2	
a. → tad.pa		−1		−3
b. tat.pa	−1			−4

(40) Voicing alternations are triggered in non-initial syllables

/tanad + pa/	IDENT[±voice] 4	AGREEVOICE 3	VOICEDOBS⇒INIT 2	
a. ta.nad.pa		−1	−1	−5
b. → ta.nat.pa	−1			−4

Given these constraints, this type of pattern, where (39a) and (40b) are co-optima, cannot be modeled in OT. With ranked constraints, a voicing alternation triggered by AGREEVOICE will either apply in all contexts or in no context, depending upon the relative ranking of IDENT[±voice] and AGREEVOICE. The ranking of the positional markedness constraint can determine whether voicing is ever allowed in a given context, but it cannot serve to selectively block alternations. It is only in HG that positional markedness constraints are able to block (markedness-reducing) alternations in strong positions.[15]

15 The properties of AGREEVOICE, and specifically its ability to trigger alternations that are markedness reducing per the positional markedness constraint VOICEDOBS⇒INITIAL, are important to this example. Markedness constraints favoring alternations that increase markedness show the opposite effect; they

The ability of positional markedness constraints to model this type of blocking in HG but not in OT should come as no surprise. As we have seen throughout this paper, in OT positional markedness constraints cannot capture patterns of disjunctive licensing. The constraint AGREEVOICE is a contextual markedness constraint, and, as such, acts much like the positional markedness constraints we have already seen. In effect, AGREEVOICE licenses [+voice] obstruents in any context where they are not immediately preceded or followed by a [−voice] obstruent. The pattern sketched above is thus precisely the type of disjunctive pattern that poses a challenge for positional markedness constraints in OT. A [+voice] obstruent is licensed if it is in the initial syllable *or* if it is not immediately adjacent to a [−voice] obstruent.

A natural-language example of this type of patterning comes from the licensing of independent place features on sonorant segments in the dialect of Tamil described by Christdas (1988; see also Beckman 1998). As the data in (41) show, sonorants appear in onset position with freely-specified place of articulation. In addition, coronal sonorants that disagree in Place with the following segment freely appear in the coda of the initial syllable; the overall pattern here is one of disjunctive licensing.[16]

(41) Independent place specifications on sonorants are permitted if they are in onset or in the initial syllable (Christdas 1988: 230–234)

/maarkaɻiy/	[maaɾ.xɜ.ɻɪ]	a month
/t̪eyʋam/	[t̪ey.ʋã]	'god'
/munʂiy/	[mun.ʂɪ]	'teacher'
/naɳpan/	[n̪aɳ.bã]	'friend'
/aaɾʋam/	[aaɾ.ʋã]	'eagerness'
/keeɭʋiy/	[keeɭ.ʋi]	'question'

can be selectively blocked just in *weak* positions by positional markedness constraints. The application of intervocalic voicing, for example, can be limited to stems if the weight of VOICEDOBS⇒STEM is sufficiently high. This is the effect seen in Section 6.3, where derived marked structures are limited to strong positions. In the case of markedness-*increasing* alternations, the predictions of positional markedness constraints are the same in OT and HG.

16 Coronal place includes dental, alveolar, retroflex, and palatal articulations in Tamil. Among these, retroflex place is generally barred from word-initial position, suggesting that post-vocalic position is an additional source of licensing required in this case (see Steriade 2001 for related discussion). I abstract away from this here; the pattern can be accommodated through the addition of constraints referring specifically to this subplace. For consistency, examples in this paper adopt the transcription used in Beckman (1998).

Outside of the initial syllable, the pattern is more complex. In word-final coda position, glides and nasals delete (with nasalization of the preceding vowel in the case of nasals), while liquids freely appear.

(42) Independent place specifications on coda liquids are permitted in word-final syllables (Christdas 1988: 159–161)

/kappal/	[kappɜl]	'ship'
/poɾuɭ/	[poɾuɭ]	'thing'
/tayiɾ/	[tayiɾ]	'yoghurt'
/pukaɻ/	[poxɜɻ]	'praise'

Despite the admissibility of independent place on liquids in forms like those in (42), a repair is consistently triggered when affixation places a non-homorganic consonant next to a sonorant in a non-initial syllable. For laterals, assimilation is the preferred repair when the following segment is coronal (43), and epenthesis is the preferred repair when the following segment is non-coronal (44).

(43) Assimilation is triggered when the suffix begins with a non-homorganic coronal (Christdas 1988: 319)

/ʋayal + t̪aan/	[ʋayɜl̪d̪ãã]	'field.EMPH'
/kappal + t̪aan/	[kappɜl̪d̪ãã]	'ship.EMPH'

(44) Epenthesis is triggered when the suffix begins with a non-coronal (Christdas 1988: 324, 328)

/kaatukaɭ + kk/	[kaadɯxɜɭɯkkɯ]	'ear.PL.DAT'
/ʋayal + kaɭ/	[ʋayɜlɯxɜ]	'field.PL'

Within HG, these data can be analyzed using the markedness constraints PLACE⇒INITIAL, which licenses independent place specifications on sonorants in the initial syllable, AGREEPLACE, which requires that adjacent consonants have the same place of articulation, and LATERAL⇒CORONAL, which penalizes lateral segments that are not also specified as coronal. These constraints conflict with IDENTPLACE and DEP-V. If the weight of each of the faithfulness constraints exceeds the weight of AGREEPLACE, sonorants (including coronal laterals) will freely appear with independent place features in the coda of the initial syllable, even when they are not homorganic with the following consonant (45). If the weight of faithfulness also exceeds the weight of PLACE⇒INITIAL, coronal laterals will freely appear in word final position bearing their own (coronal) place features (46).

(45) Independent place features freely appear on coda laterals in the initial syllable – even if the following segment is not homorganic

/keeḷʋiy/	LAT⇒COR 6	DEP-V 5	IDENTPLACE 4	AGREE 3	PLACE⇒INIT 3	
a. → keeḷ.ʋi				–1		–3
b. kee.ḷɯ.ʋi		–1				–5

(46) Independent place features freely appear on laterals when no consonant follows – even in a non-initial syllable

/kappal/	LAT⇒COR 6	DEP-V 5	IDENTPLACE 4	AGREE 3	PLACE⇒INIT 3	
a. → kap.pɜl					–1	–3
b. kap.pɜ.lɯ		–1				–5

In non-final codas where the following segment is not homorganic, both PLACE⇒INITIAL and AGREEPLACE are violated; this triggers a repair. Assimilation applies if the affix segment is coronal (47), and epenthesis applies if the affix segment is non-coronal (48).

(47) Laterals in non-initial syllables undergo assimilation when followed by a non-homorganic coronal consonant

/ʋayal+t̪aan/	LAT⇒COR 6	DEP-V 5	IDENT PLACE 4	AGREE 3	PLACE⇒INIT 3	
a. ʋa.yɜl.d̪ãã				–1	–1	–6
b. → ʋay.ɜl̪.d̪ãã			–1			–4
c. ʋa.yɜ.lɯ.d̪ãã		–1				–5

(48) Laterals in non-initial syllables trigger epenthesis when followed by a non-coronal consonant

/kaatukaḷ+kk/	LAT⇒COR 6	DEP-V 5	IDENT PLACE 4	AGREE 3	PLACE⇒ INIT 3	
a. kaa.dɯ.xɜḷk.kɯ				–1	–1	–6
b. kaa.dɯ.xɜʟk.kɯ	–1		–1			–10
c. → kaa.dɯ.xɜ.ḷɯk.kɯ		–1				–5

The alternations triggered by AGREEPLACE rely on cumulative interaction of this constraint with PLACE⇒INITIAL. Alternations are blocked when they do not allow violations of *both* constraints to be avoided. An OT ranking with the same ordering as seen above cannot achieve this type of selective blocking.

The particular usefulness of Harmonic Grammar here stems from the fact that multiple conditions interact in the licensing of Place. Lateral segments can bear independent place specifications when they are coronal and in an onset *or* when they are coronal and in the initial syllable *or* when they are coronal and not followed by a place-disagreeing segment. This final condition can be even further refined based on mappings like /ʋayal+**kaɭ**/→[ʋayɜluɯ**xɜ**], which show that being in a root is also crucial to the admissibility of coda laterals with independent place in non-initial syllables.[17] The full set of contexts is summarized in (49).

(49) Conditions for the licensing of independent place on lateral segments
  a. Coronal and Onset *or*
  b. Coronal and Initial syllable *or*
  c. Coronal and Not followed by a place-disagreeing segment and Root

Alternations are blocked in contexts that are sufficiently strong – i.e., in cases where one of the sets of conjunctions in (49) is satisfied.

This ability of positional markedness constraints in HG to block alternations extends to all situations where the alternation is markedness reducing. A constraint like PLACE⇒INITIAL is satisfied by segments that do not bear independent place features, effectively preferring segments that are either placeless or share their place specification. Similarly, a positional markedness constraint like VOICEDOBS⇒ONSET is satisfied by obstruents that are not specified as [+voice], effectively identifying [−voice] specification as unmarked. In each of these cases the markedness distinction is only relevant outside of the strong licensing context, as it is only there that violations of the positional markedness constraint can be incurred. This encoding of (un)markedness is key to the blocking effects seen here. Alternations that decrease markedness in weak positions can lead to removal of violations of the positional markedness constraints. If a markedness-reducing alternation is to be blocked in some context, then, it will be blocked in the strong position where no violation of the positional markedness constraint can be removed, and therefore no additional benefit can be derived.

17 Christdas (1988: 326–330) identifies the nominalizing suffix /-al/ as an exception to this generalization – e.g., /moot + al/ → [moodɜl] 'shove + N'. Lateral-final suffixes that *are* subject to deletion include /-kaɭ/ 'plural', /-aamal/ 'negative gerundive', /-aal/ 'conditional', /-aaɭ/ '3rd person feminine', and /-aaŋkaɭ/ '3rd person plural'.

While the cases seen above involve removal of a marked structure in the immediate environment of an affix, there is no inherent locality-based restriction on this type of selective blocking. For example, if an unmarked [−voice] feature is introduced in the course of morphological derivation, positional markedness predicts that this feature will preferentially target weak positions where a violation of the contextual constraint will be removed. Strong positions will be targeted only if overt realization of the unmarked feature is required by the grammar, and segments in weak positions cannot host the feature. This basic preference for the docking of unmarked features holds of both OT and HG. In HG, however, it is also possible to model patterns where actual blocking occurs, so that if the only host for the unmarked feature is a marked segment in the strong position, association will be blocked and the unmarked feature will receive no overt reflex. In OT it is also expected that unmarked features will target weak positions, but there can be no positional blocking. Cumulative constraint interaction is key to the ability of positional markedness to generate this type of pattern in HG; no OT ranking of these constraints will lead to selection of the same set of optima. Tableaux illustrating the HG blocking pattern are given below.

(50) [+voice] obstruents are generally allowed in all positions

/bad/	IDENT[±voice] 4	MORPHREAL 3	VOICEDOBS⇒ONSET 2	
a. → bad			−1	−2
b. bat	−1			−4

(51) Floating [−voice] features target obstruents in coda where possible

/gad, [−voice]/	IDENT[±voice] 4	MORPHREAL 3	VOICEDOBS⇒ONSET 2	
a. gad		−1	−1	−5
b. → gat	−1			−4

(52) Docking of unmarked features onto the strong onset position is blocked

/ban, [−voice]/	IDENT[±voice] 4	MORPHREAL 3	VOICEDOBS⇒ONSET 2	
a. → ban		−1		−3
b. pan	−1			−4

The predictions made by Harmonic Grammar with positional markedness constraints appear to be correct in at least some cases, as markedness-reducing alternations are blocked in strong positions. The Tamil case discussed above is one example. Another case comes from Turkish, where coda devoicing is blocked with greater probability in monosyllables than in longer words – i.e., voicing persists when the final consonant is found in the strong initial syllable (Becker, Ketrez and Nevins 2011). In Brazilian Portuguese, coalescence of stem-final [w] with the initial segment of the plural marker [-is] to yield surface [j] is blocked with greater frequency in monosyllables than in longer words; this pattern is also conceivably of the type predicted here (Becker, Clemens and Nevins 2011).

Other cases are more difficult to interpret. Giavazzi (2010), for example, reports that the Italian process of palatalization is blocked for segments immediately following a stressed vowel – e.g., /anˈtik-i/→[an.ˈti.ki] – but is required in other contexts – e.g., /ˈkomik-i/→[ˈko.mi.ʧi], *[ˈko.mi.ki]. This pattern *can* be analyzed using constraints that license dorsal place feature in post-tonic position, but it is unclear that this type of constraint is motivated by more general markedness considerations. Similarly, the blocking of [+nasal] spreading onto stressed vowels in Guaraní (Beckman 1998, Gregores and Suárez 1967) *can* be analyzed using positional markedness constraints that prefer preservation of oral place features in stressed contexts, but this is at odds with other markedness criteria that deem nasal vowels to be more marked than oral vowels.[18] Positional faithfulness does not standardly encode the type of markedness polarity that positional markedness constraints do, and therefore avoids this type of conflict between language-specific patterns and general markedness principles. Ultimately, investigation of a broader array of patterns should shed light on these issues and reveal to what extent the markedness-increasing vs. markedness-reducing nature of alternations plays a role in strong positions' resistance to change.

6.4.2 *Directionality and additional challenges*

Beyond the types of blocking effects discussed in the previous section, positional faithfulness constraints have been widely used to model

18 One possibility in both of these cases would be to define the spreading constraints so that they are violated by a failure to spread coronality / nasality onto weak targets – i.e., segments in unstressed contexts – but not by a failure to spread onto strong targets. This would avoid the need for positional faithfulness, while maintaining the pattern of encoding contexts on markedness rather than faithfulness constraints. See Section 6.4.2 for related discussion.

directionality in assimilation processes where elements in strong positions determine the feature value that is propagated onto elements in weak positions (e.g., Baković 2000, Beckman 1997, 1998, Lombardi 1999). The basic approach in these cases is simple. Given a sufficiently high ranking/weighting, a constraint like IDENT[±voice]/ONSET ensures that the input [±voice] specifications of segments in onset position are preserved; if adjacent segments are required to agree in this feature, then, it is necessarily the segments in the weak (coda) position that alter their input specifications. Positional markedness constraints do not have this same effect. As (53) shows, a constraint like VOICEDOBS⇒ONSET is equally satisfied by a [+voice] feature that originates in onset position and spreads to coda, and a [+voice] feature that originates in coda and spreads to onset. IDENT[±voice]/ONSET can distinguish between directions of spreading, but VOICEDOBSONSET cannot.

(53) Positional faithfulness distinguishes between directions of spreading; positional markedness does not

	VOICEDOBS⇒ONSET	IDENT[±voice]/ONSET
a. at.ba [−voi][+voi] → ad.ba [+voi]		
b. ad.pa [+voi][−voi] → ad.ba [+voi]		1

The licensing of marked features in weak positions through association with strong licensers, as in (53b), is precisely what occurs in cases of metaphony (Walker 2005, 2006, 2011), and so is arguably a positive prediction of positional markedness constraints. The problem lies in the fact these constraints cannot capture cases where features systematically spread from strong to weak positions regardless of markedness, as is the case in many assimilatory processes.

More generally, it seems that in order for constraints to capture some form of directionality, the preferred *source* of the spreading features must be distinguished in the constraint definition – either through input-output correspondence or through elements of the output representation. Positional faithfulness constraints are not alone in having this property. For instance, SPANHEAD constraints (McCarthy 2004), INITIAL(F) and FINAL(F) constraints (McCarthy 2009), and targeted constraints (Wilson 2000, 2003) all assign violations differently depending upon the source of the propagating features. It is not clear that positional faithfulness constraints enjoy a particular advantage over these other approaches in capturing directional phenomena. McCarthy (2004) identifies nasal harmony phenomena as particularly problematic for positional faithfulness, based on the potentially unbounded nature of the processes and their ignorance of the prosodic positions of either the trigger or target segments. Nasal spreading in Johore Malay (Onn 1976), for example, is initiated by underlying [nasal] segments regardless of their position in the word. This type of position-insensitive pattern is quite pervasive; indeed, Walker (1998) identifies the sonority of target segments as the key factor in blocking nasal spreading processes. Neither positional faithfulness constraints nor simple positional markedness constraints predict this type of directional spreading that ignores the prosodic position of the trigger and the targets.

Additional difficulties for positional faithfulness constraints come from a series of pathologies generated through interaction of positional faithfulness with constraints governing prosodic structure. While a full discussion of these issues is beyond the scope of this paper, the basic issue is straightforward: If the constraints that determine the default prosodic structure of a language are sufficiently low-ranking/low-weighted, input featural contrasts can be opaquely displaced onto the output prosodic structure. For example, with a sufficiently low ranking/weighting of the constraint ONSET, positional faithfulness constraints predict a system where input [−voice] intervocalic obstruents are syllabified in unmarked onset position, giving the mapping /pata/→[pa.ta], even as input [+voice] intervocalic obstruents are syllabified in marked coda position where they can devoice without incurring a violation of the positional faithfulness constraint, giving the mapping /pada/→[pat.a] (Beckman 1998: fn.37; see also Jesney 2011, Wilson 2000). As Wilson (2000, 2003) points out, similar opaque subversions of the expected prosodification can be generated through the interaction positional faithfulness constraints with certain constraints enforcing assimilation. Incorporating directionality into the spreading constraints themselves, rather than relying upon positional faithfulness, allows these types of unattested patterns to be avoided.

6.4.3 *Summary*

While the diagnostics of positional privilege identified by Beckman (1998) appear to be correct for a range of cases, it is not clear that positional faithfulness is the most effective, or only, means of accounting for the relevant phenomena. Contrary to the positional faithfulness prediction that strong (but not weak) positions block alternations that apply elsewhere (38c), there are well-documented cases where strong positions actually *attract* derived marked structures (see Sections 6.3.2 and 6.4.1). Similarly, despite the positional faithfulness association of strong positions with trigger behavior (38c) and weak positions with target behavior (38b), patterns in assimilation do not universally reflect this type of dichotomy (see Section 6.4.2). Such facts suggest that positional faithfulness is, at best, an inadequate account of positional privilege. Taken together with the fact that cumulative constraint interaction allows a greater range of the attested patterns to be modeled using only positional markedness constraints, it would seem that positional faithfulness constraints are truly unnecessary in Harmonic Grammar.

6.5 Simple vs. complex constraints

Certain aspects of the cumulative interaction seen in HG can be imitated in Optimality Theory through the use of complex, or highly-specified, constraints. One possible way of extending the range of analyses while retaining ranked constraints, then, is to allow for either positional constraints with highly-specified contexts or local conjunction of constraints (Smolensky 1993 *et seq.*). This section considers these possibilities in turn, and compares the basic results with those obtained through cumulative constraint interaction in HG.

To illustrate, we will use the simple system laid out in (5) where the marked feature is [+voice] and the two strong licensing environments are onset and the initial syllable. In the analyses sketched so far, the contextual constraints have each referred to only one of these licensing contexts. The contexts have been simple, and patterns where multiple contexts are relevant have been generated through the interaction of these simple constraints. One means of extending the OT typology is to specify the contexts of the constraints in greater detail. Thus, instead of only IDENT[±voice]/ONSET and IDENT[±voice]/INITIAL, CON might include a constraint IDENT[±voice]/INITIALONSET, which is violated just when a segment in the onset of an initial syllable differs in specification for [±voice] relative to its input correspondent. This effectively conjoins contexts within the very definition of the constraint, allowing positional faithfulness constraints in OT to model patterns where marked features are admitted only when they are simultaneously in the first context *and* the second context. In (54), for example, [+voice] maps faithfully just when it is in the onset and the initial syllable. This type of conjunctive licensing pattern eludes positional faithfulness constraints in OT when only constraints with simple contexts are included.

(54) Complex contexts allow positional faithfulness constraints to model patterns of conjunctive licensing in OT (locus of potential faithfulness violation is underlined)

	/bad.na.bad/	IDENT[±voice]/INITIALONSET	*VOICEDOBS
a.	<u>b</u>ad.na.bad		4
b.	<u>p</u>at.na.pat	1	
c.	<u>b</u>ad.na.pat		2
d.	<u>b</u>at.na.bat		2
e.	<u>b</u>ad.na.bat		3
f. →	<u>b</u>at.na.pat		1

The same approach cannot be used to extend the typology produced with positional markedness constraints in OT. A constraint like VOICEDOBS⇒INITIALONSET assigns violation marks to any instance of a voiced obstruent that is not in the onset of the initial syllable – precisely the same result as is obtained by simply ranking both VOICEDOBS⇒INITIAL and VOICEDOBS⇒ONSET above IDENT[±voice]. This type of highly-specified context does not allow the missing disjunctive pattern to be modeled with positional markedness constraints in OT.

Local constraint conjunction (Smolensky 1993, *et seq.*) is the second possibility for extending the OT typology. Under this approach, two (or more) constraints are conjoined to form a new constraint that is violated just when both of its conjuncts are violated within a specified domain. This conjoined constraint can then be ranked differently than either of its conjuncts, allowing apparently cumulative patterns to be modeled. As defined in (55), the conjunction of IDENT[±voice]/INITIAL and IDENT[±voice]/ONSET has precisely the same effect as the highly specified constraint IDENT[±voice]/INITIALONSET in (54). In other words, the conjunction of positional faithfulness constraints admits patterns whereby marked features are limited to positions of double privilege.

(55) [IDENT[±voice]/INITIAL & IDENT[±voice]/ONSET]$_{\text{Segment}}$: Assign a violation mark to any output segment that is in both onset and the initial syllable, and that differs in specification for [±voice] relative to its input correspondent.

Furthermore, because local constraint conjunction is logical disjunction when viewed from the perspective of constraint satisfaction – i.e., the constraint is satisfied when either conjunct 1 or conjunct 2 is satisfied – this approach also allows the missing disjunctive pattern to be modeled with positional markedness constraints. Ranking the constraint in (56) above Ident[±voice] predicts a pattern whereby voiced obstruents are admitted except when they are in neither the initial syllable nor a syllable onset.

(56) [VoicedObs⇒Initial & VoicedObs⇒Onset]$_{\text{Segment}}$: Assign a violation mark to any output voiced obstruent segment that is not associated with the initial syllable and that is not in onset position.

(57) Local constraint conjunction allows positional markedness constraints to model patterns of disjunctive licensing in OT (locus of potential markedness violation is underlined)

	/bad.na.bad/	[VoicedObs⇒Initial & VoicedObs⇒Onset]$_{\text{Segment}}$	Ident[±voice]
a.	bad.na.ba**d**	1	
b.	pat.na.pat		4
c.	bad.na.pat		2
d.	bat.na.bat		2
e. →	bad.na.bat		1
f.	bat.na.pat		3

As Pater (2009, this volume) discusses, while local constraint conjunction in OT and cumulative constraint interaction in HG can capture many of the same effects, local constraint conjunction is more powerful in many cases. The primary reason for this lies in the need to specify a domain for conjoined constraints. In (55) and (56) the specified domain is as small as possible – i.e., the segment – and it overlaps precisely with the locus of the conflicting constraint. In situations like this, the conjoined constraint simply replicates the effect of cumulative interaction in HG. Indeed, when these conjoined constraints are added to Con in HG, there is no additional effect on the typology.

This said, there is no inherent restriction on the size of the domain that can be specified for locally-conjoined constraints. Indeed, the only true restriction is that it be possible to simultaneously violate both conjuncts within *some* domain (Moreton and Smolensky 2002). As a result it would be possible to define a constraint like [IDENT[±voice]/INITIAL & IDENT[±voice]/ONSET]$_{PWd}$, that is equally violated by mappings like /bad.na.bad/→ [pad.na.bad], where the two violations are localized on the same initial segment, and mappings like /bad.na.bad/→[bat.na.pad], where the two violations simply occur in the same PWd. In effect, local constraint conjunction does not carry the same inherent locality restriction as HG cumulative constraint interaction (*cf.* Łubowicz 2005). When larger domains are defined for conjoined constraints, the typology is necessarily extended beyond what is generated in HG with cumulative constraint interaction alone.

The key point here is simply this: cumulative constraint interaction in Harmonic Grammar makes local constraint conjunction unnecessary for cases like those discussed in this paper. Patterns of both conjunctive and disjunctive licensing can be modeled using simple positional markedness constraints. For a range of cases, positional faithfulness constraints are not necessary, nor are locally-conjoined constraints or constraints with highly-specified contexts. Cumulative constraint interaction opens the door to analyses that are not possible in a theory that relies upon ranked constraints.

6.6 Conclusion

This paper has argued that the availability of cumulative constraint interaction in Harmonic Grammar provides for alternative analyses requiring fewer and more basic constraints. While the set of patterns that can be generated in HG with a given set of constraints is typically a superset of those that can be generated in OT with the same set of constraints, cumulativity can allow for analyses that require fewer constraints. This, in turn, can act to restrict the HG typology.

There remain, of course, many issues to be considered in determining the best set of constraints to employ in either theory. The examples in this paper have demonstrated that exclusively markedness-based accounts of both conjunctive and disjunctive licensing patterns are possible in HG. Other differences between the types of patterns that can be modeled using positional faithfulness and positional markedness constraints will benefit from further research. The patterns of alternation blocking discussed in Section 6.4 are of particular interest, as they suggest an enhanced role for markedness polarity in determining whether a strong position will attract or resist alternations. These are predictions particular to HG, and provide a further illustration of how altering the mode of constraint interaction can change the set of patterns that a given set of constraints can be used to model. The subset-superset relationship between the typologies generated in the two theories holds only insofar as the constraint set is held constant. When cumulativity allows a given set of constraints to model a greater range of patterns, fewer constraints may be required in HG, and the typology limited in consequence.

References

Adam, Galit and Bat-El, Outi (2008) Morphological knowledge without morphological structure: morphology–prosody interface in the first Hebrew verbs. In S. Arnom-Lotem, G. Danon and S. D. Rothstein (eds), *Generative Approaches to Hebrew Linguistics,* 197–222. Amsterdam: John Benjamins.

Baković, Eric (2000) *Harmony, Dominance, and Control.* PhD dissertation. Rutgers University.

Becker, Michael, Clemens, Lauren E. and Nevins, Andrew (2011) A richer model is not always more accurate: the case of French and Portuguese plurals. Ms. [www.phonologist.org/initialsyllfaith/becker_clemens_nevins_french_portuguese.pdf]

Becker, Michael, Ketrez, Nihan and Nevins, Andrew (2011) The surfeit of the stimulus: analytic biases filter lexical statistics in Turkish laryngeal alternations. *Language* 87 (1): 84–125.

Becker, Michael, Nevins, Andrew and Levine, Jonathan (2012) Asymmetries in generalizing alternations to and from initial syllables. *Language* 88 (2): 231–268.

Beckman, Jill (1997) Positional faithfulness, positional neutralization, and Shona vowel harmony. *Phonology* 14 (1): 1–46.

Beckman, Jill (1998) *Positional Faithfulness.* PhD dissertation. University of Massachusetts Amherst.

Borowsky, Toni (1984) On resyllabification in English. *Proceedings of the West Coast Conference on Formal Linguistics 3 Santa Cruz*: 1–15.

Borowsky, Toni (1986) *Topics in the Lexical Phonology of English.* PhD dissertation. University of Massachusetts Amherst.

Cahill, Michael C. (2007) *Aspects of the Morphology and Phonology of Konni.* Dallas, TX: SIL International and University of Texas at Arlington Publications in Linguistics.

Casali, Roderic (1996) *Resolving Hiatus.* PhD dissertation. UCLA.

Christdas, Prathima (1988) *The Phonology and Morphology of Tamil.* PhD dissertation. Cornell University.

Davis, Stuart and Cho, Mi-Hui (2003) The distribution of aspirated stops and /h/ in American English and Korean: an alignment approach with typological implications. *Linguistics* 41: 607–652.

Gentner, Dedre and Boroditsky, Lera (2001) Individuation, relativity and early word learning. In Melissa Bowerman and Stephen Levinson (eds), *Language Acquisition and Conceptual Development,* 215–256. Cambridge: Cambridge University Press.

Gerlach, S. (2009) Consonant harmony and non-local metathesis in phonological development. Paper presented at *The 45th Annual Meeting of the Chicago Linguistic Society.* Chicago, IL: University of Chicago.

Giavazzi, Maria (2010) *The Phonetics of Metrical Prominence and its Consequences for Segmental Phonology.* PhD dissertation. MIT.

Goldsmith, John (1989) Licensing, inalterability, and harmonic rule application. In *Proceedings of the 25th Annual Meeting of the Chicago Linguistic Society,* 145–156.

Goldsmith, John (1990) *Autosegmental and Metrical Phonology.* Cambridge, MA: Blackwell.

Goldstein, Louis (1992) Comments on chapters 3 and 4. In G. Docherty and D. R. Ladd (eds), *Papers in Laboratory Phonology II: Gesture, Segment, Prosody,* 120–124. Cambridge, UK: Cambridge University Press.

Gregores, Emma and Suárez, Jorge A. (1967) *A Description of Colloquial Guaraní.* The Hague: Mouton.

Hammond, Michael (1999) *The Phonology of English: A Prosodic Optimality–Theoretic Approach.* Oxford: Oxford University Press

Ishikawa, Keiichi (2002) Syllabification of intervocalic consonants by English and Japanese speakers. *Language and Speech* 45 (4): 355–385.

Itô, Junko (1986) *Syllable Theory in Prosodic Phonology.* PhD Dissertation. University of Massachusetts Amherst.

Itô, Junko and Mester, Armin (2003) *Japanese Morphophonemics: Markedness and Word Structure.* Cambridge, MA: MIT Press.

Itô, Junko, Mester, Armin and Padgett, Jaye (1995) Licensing and underspecification in Optimality Theory. *Linguistic Inquiry* 26: 571–614.

Jensen, John (2000) Against ambisyllabicity. *Phonology* 17: 187–235.

Jesney, Karen (2011) Positional faithfulness, non-locality, and the Harmonic Serialism solution. In Suzi Lima, Kevin Mullin and Brian Smith (eds), *Proceedings of the 39th Meeting of the North East Linguistic Society,* 403–416. Amherst, MA: GLSA. [ROA-1018].

Jesney, Karen (2012) Licensing in multiple contexts: An argument for Harmonic Grammar. In Ryan Bochnak, Nassira Nicola, Peet Klecha, Jasmin Urban, Alice Lemieux and Christina Weaver (eds), *Proceedings from the Annual Meeting of the Chicago Linguistic Society 45,* 287–301. Chicago, IL: Chicago Linguistic Society.

Jesney, Karen and Tessier, Anne-Michelle (2009) Gradual learning and faithfulness: Consequences of ranked vs. weighted constraints. In Anisa Schardl, Martin Walkow and Muhammad Abdurrahman (eds), *Proceedings of the 38th Meeting of the North East Linguistic Society,* 449–462. Amherst, MA: GLSA. [ROA-967].

Kager, René (1996) Stem disyllabicity in Guugu Yimidhirr. In Marina Nespor and Norval Smith (eds), *Dam Phonology: HIL Phonology Papers II,* 59–101. Den Haag: Holland Institute of Generative Linguistics.

Kimper, Wendell (2011) Locality and globality in phonological variation. *Natural Language and Linguistic Theory* 29 (2): 423–465.

Kiparsky, Paul (1972) Explanation in phonology. In Stanley Peters (ed.), *Goals of Linguistic Theory,* 189–227. Englewood Cliffs, NJ: Prentice–Hall.

Kreidler, Charles (1989) *The Pronunciation of English: A Course Book in Phonology.* Oxford: Blackwell.

Kurisu, Kazutaka (2001) *The Phonology of Morpheme Realization.* PhD dissertation. University of California Santa Cruz.

Legendre, Géraldine, Sorace, Antonella and Smolensky, Paul (2006) The Optimality Theory–Harmonic Grammar connection. In Paul Smolensky and Géraldine Legendre (eds), *The Harmonic Mind: From Neural Computation to Optimality-Theoretic Grammar, vol. 2: Linguistic and Philosophical Implications,* 339–402. Cambridge, MA: MIT Press.

Lombardi, Linda (1999) Positional faithfulness and voicing assimilation in Optimality Theory. *Natural Language and Linguistic Theory* 17: 267–302.

Łubowicz, Anna (2005) Locality of conjunction. In John Alderete, Chung-hye Han and Alexei Kochetov (eds), *Proceedings of the 24th West Coast Conference on Formal Linguistics,* 254–262. Somerville, MA: Cascadilla Press.

Marshall, Chloe R. and van der Lely, Heather K. J. (2009) Effects of word position and stress and onset cluster production: Evidence from typical development, specific language impairment, and dyslexia. *Language* 85: 39–57.

McCarthy, John J. (2004) Headed spans and autosegmental spreading. Ms., University of Massachusetts Amherst. [ROA–685].

McCarthy, John J. (2005) The length of stem-final vowels in Colloquial Arabic. In Mohammad T. Alhawary and Elabbas Benmamoun (eds), *Perspectives on Arabic Linguistics XVII–XVIII* 1–26. Amsterdam: John Benjamins.

McCarthy, John J. (2006) Restraint of analysis. In Eric Baković, John J. McCarthy and Junko Itô (eds), *Wondering at the Natural Fecundity of Things: Essays in Honor of Alan Prince* 195–219. [ROA–844.4].

McCarthy, John J. (2008a) The gradual path to cluster simplification. *Phonology* 25 (2): 271–319.

McCarthy, John J. (2008b) The serial interaction of stress and syncope. *Natural Language and Linguistic Theory* 26: 499–546.

McCarthy, John J. (2009) Harmony in Harmonic Serialism. Ms., University of Massachusetts Amherst. [ROA–1009].

McCarthy, John J. and Prince, Alan (1990) Prosodic morphology and templatic morphology. In Mushira Eid and John J. McCarthy (eds), *Perspectives in Arabic Linguistics II.* Amsterdam: John Benjamins.

Moreton, Elliott and Smolensky, Paul (2002) Typological consequences of local constraint conjunction. In Line Mikkelsen and Christopher Potts (eds), *Proceedings of the 21st West Coast Conference on Formal Linguistics* 305–319. Somerville, MA: Cascadilla Press.

Onn, Farid M. (1976) *Aspects of Malay Phonology and Morphology.* Urbana, IL: University of Illinois.

Pater, Joe (2000) Non–uniformity in English secondary stress: The role of ranked and lexically–specific constraints. *Phonology* 17: 237–274.

Pater, Joe (2009) Review of Smolensky and Legendre (2006). The Harmonic Mind. *Phonology* 26: 217–226.

Potts, Christopher, Pater, Joe, Jesney, Karen, Bhatt, Rajesh and Becker, Michael (2010) Harmonic Grammar with linear programming: from linear systems to linguistic typology. *Phonology* 27: 77–117.

Prince, Alan and Smolensky, Paul (1993/2004) *Optimality Theory: Constraint Interaction in Generative Grammar.* Technical Report, Rutgers University and University of Colorado at Boulder, 1993. Revised version published by Blackwell, 2004. [ROA–537].

Pruitt, Kathryn (2010) Serialism and locality in constraint–based metrical parsing. *Phonology* 27(3): 481–526.

Redford, Melissa A. and Randall, Patrick (2005) The role of juncture cues and phonological knowledge in English syllabification knowledge. *Journal of Phonetics* 33: 27–46.

Selkirk, Elisabeth O. (1982) The syllable. In Harry van der Hulst and Norval Smith (eds), *The Structure of Phonological Representations, Part II*, 337–383. Dordrecht: Foris.

Smith, Jennifer L. (2001) Lexical category and phonological contrast. In Robert Kirchner, Joe Pater and Wolf Wikely (eds), *PETL 6: Proceedings of the Workshop on the Lexicon in Phonetics and Phonology*, 61–72. Edmonton, AB: University of Alberta. [ROA–728].

Smolensky, Paul (1993) Harmony, markedness, and phonological activity. Paper presented at Rutgers Optimality Workshop 1. [ROA–87]

Staubs, Robert, Becker, Michael, Potts, Christopher, Pratt, Patrick, McCarthy, John J. and Pater, Joe (2010) *OT–Help 2.0.* [http://people.umass.edu/othelp/].

Steriade, Donca (1995) Underspecification and markedness. In John Goldsmith (ed.), *Handbook of Phonological Theory*, 114–174. Cambridge, MA: Blackwell.

Steriade, Donca (2001) Directional asymmetries in place assimilation. In Elisabeth Hume and Keith Johnson (eds), *The Role of Speech Perception in Phonology*, 219–250. San Diego, CA: Academic Press.

Turk, Alice (1994) Articulatory phonetic clues to syllable affiliation: gestural characteristics of bilabial stops. In Patricia A. Keating (ed.), *Papers in Laboratory Phonology III: Phonological Structure and Phonetic Form*, 107–135. Cambridge: Cambridge University Press.

Velleman, Shelley L. (1996) Metathesis highlights feature–by–position constraints. In Barbara Bernhardt, John Gilbert and David Ingram (eds), *Proceedings of the UBC International Conference on Phonological Acquisition*, 173–186. Somerville, MA: Cascadilla Press.

Walker, Rachel (1998) *Nasalization, Neutral Segments, and Opacity Effects.* PhD dissertation. University of California Santa Cruz. [ROA–405].

Walker, Rachel (2001a) Positional markedness in vowel harmony. In Caroline Féry, Antony Dubach Green and Ruben van de Vijver (eds), *Proceedings of HILP 5*, 212–232. University of Potsdam.

Walker, Rachel (2001b) Round licensing, harmony, and bisyllabic triggers in Altaic. *Natural Language and Linguistic Theory* 19: 827–878.

Walker, Rachel (2005) Weak triggers in vowel harmony. *Natural Language and Linguistic Theory* 23: 917–989.

Walker, Rachel (2006) Long distance metaphony: A generalized licensing proposal. Paper presented at *PhonologyFest Workshop.* Bloomington, IN: Indiana University.

Walker, Rachel (2011) *Vowel Patterns in Language.* Cambridge: Cambridge University Press.

Wilson, Colin (2000) *Targeted Constraints: An Approach to Contextual Neutralization in Optimality Theory.* PhD dissertation. Johns Hopkins University.

Wilson, Colin (2003) Analyzing unbounded spreading with constraints: Marks, targets, and derivations. Ms., UCLA.

Wolf, Matthew (2005) An autosegmental theory of quirky mutations. In John Alderete, Chung-hye Han and Alexei Kochetov (eds), *Proceedings of the 24th West Conference on Formal Linguistics,* 370–378. Somerville, MA: Cascadilla Press.

Zoll, Cheryl (1996) *Parsing Below the Segment in a Constraint Based Framework.* PhD dissertation. University of California Berkeley.

Zoll, Cheryl (1998) Positional asymmetries and licensing. Ms., MIT. [ROA–282].

Chapter 7

Positive constraints and finite goodness in Harmonic Serialism

Wendell Kimper

7.1 Introduction

In Optimality Theory (OT) (Prince and Smolensky, 1993/2004), constraints have traditionally been negatively rather than positively defined; that is, they assign *violations* rather than *rewards*. Because OT's GEN can perform multiple operations (and multiple instances of a single operation) simultaneously, negative constraints have been necessary. A positive constraint suffers from the *Infinite Goodness* problem (Prince, 2007); for any structure favored by a positive constraint, an infinite number of instances of that structure can be epenthesized, and there ceases to be an optimum.

In Harmonic Serialism (HS) (McCarthy, 2000, 2002, 2007), however, GEN is restricted to performing one operation at a time. This means that infinitely epenthetic candidates can no longer be entertained, and positive constraints are a viable possibility. In this paper, I discuss the properties of HS that render positive constraints feasible, as well as possible limitations on the types of constraints that may be viably positive.

Furthermore, I argue that positive constraints have certain advantages over their negatively-defined counterparts. In particular, negative constraints driving autosegmental spreading processes produce a number of pathological predictions; because they assign violations for unassimilated segments, they interact in unattested ways with processes that affect the number of segments in a word (Wilson, 2004, 2006). A positive spreading constraint, on the other hand, assigns rewards for assimilated segments; this, combined with the gradual harmonic improvement of HS, permits a superior account of harmony processes.

The paper is organized as follows. Section 7.2 discusses the infinite goodness problem and its resolution in HS, Section 7.3 shows how a positive constraint provides a better account of harmony processes than its negative counterparts, and 7.4 discusses potential limitations on the formulation of positive constraints in HS.

7.2 The Infinite Goodness Problem

The Infinite Goodness problem arises because positive and negative constraints are not simply the mirror image of each other – they assign marks differently. Compare, for example, positive and negative formulations of a constraint like ONSET.

(1) a. (–)ONSET: Assign –1 for every syllable that does not have an onset.
b. (+)ONSET: Assign +1 for every syllable that has an onset.

Both versions of the constraint demand that syllables have onsets; both prefer a candidate with an epenthetic onset (2a) over a faithful, onsetless candidate (2b). However, a positively formulated ONSET constraint will prefer a candidate with both an epenthetic onset and an epenthetic CV syllable (2c) over a candidate with a single epenthetic onset (2b), while the negative version of the constraint exerts no such preference.[1]

(2) Positive and negative ONSET

	ap	(–)ONSET	(+)ONSET
a.	.ap.	–1	
b.	.ʔap.		+1
c.	.ʔap.ʔə.		+2

In a language where ONSET dominates DEP, the negative version of the constraint will compel the insertion of an onset into the syllable in (2). A candidate like (2c) will be (correctly) harmonically bounded

1 Throughout this paper, I use integers instead of violation marks; positive integers represent rewards, and negative integers represent penalties. The optimum is, as always, the candidate with the highest value, though with the addition of reward-assigning constraints, 'highest' is no longer interchangeable with 'closest to zero'.

– it doesn't perform any better than its competitor on Onset, and gratuitously violates faithfulness. However, if a positive version of ONSET dominates faithfulness, a candidate like (2c) will be chosen over a candidate like (2b).

The problem with this is immediately apparent: if (+)ONSET favors a candidate with an additional epenthetic syllable over a candidate with mere onset epenthesis, it will favor a candidate with two epenthetic syllables over a candidate with only one epenthetic syllable, a candidate with three epenthetic syllables over a candidate with two, and so on. Because GEN can provide an infinite set of candidates with increasingly many epenthetic syllables (3):

> there is no candidate that has the maximal value ... and were the constraint asked to produce the class of forms that do maximally well on it, no output would be defined. If such a constraint is admitted, the theory ceases to exist. (Prince, 2007).

(3) The Infinite Goodness Problem

	ap	(–)ONSET	(+)ONSET
a.	.ap.	–1	
b.	.ʔap.		+1
c.	.ʔap.ta.		+2
d.	.ʔap.ta.ka.		+3
e.	.ʔap. + $\aleph_0$		+∞

With parallel OT's unrestrained GEN, positive constraints are untenable. Virtually any constraint satisfied by epenthesis necessarily encounters the infinite goodness problem – for any structure preferred by a positive constraint, infinitely many instances of that structure can be inserted, and optimality becomes impossible.[2] However, in Harmonic Serialism (HS), a derivational variant of OT with a number of independent typological advantages,[3] a sufficiently restricted GEN allows positive constraints to avoid the problem of infinite goodness.

In HS, GEN is restricted to producing candidates that each differ from the input by at most a single application of a single change (McCarthy,

2 One possible exception is constraints whose environment is co-extensive with the domain of evaluation; Joe Pater points out that e.g. a constraint requiring an utterance to begin with a consonant could avoid the problem if the domain of evaluation is restricted to single utterances.

3 See e.g. Elfner (2009); Kimper (2011); Jesney (2009); McCarthy (2008a, b, 2010b); Pruitt (2008); Wolf (2008) and other papers in this volume.

this volume).[4] This finite candidate set is evaluated by EVAL, just like in parallel versions of OT, but instead of exiting the derivation as the surface form, the optimum is sent back to GEN to serve as the input to a new step in the derivation. Once again, candidates differing from this new input by at most a single instance of a single change are generated and evaluated. The GEN→EVAL loop continues until the changes made by GEN no longer increase harmony – the derivation *converges* when the faithful candidate at the current step is chosen as optimal.

As a consequence, HS derivations must be *gradual* and *harmonically improving*. Because GEN is restricted, each step in the derivation can advance by at most a single change. Because each step involves choosing an optimum at EVAL, and the derivation converges if the faithful candidate is chosen, each step must improve performance with respect to the constraint hierarchy.

The gradual harmonic improvement involved in HS derivations means that it becomes possible for positive constraints to avoid the infinite goodness problem. Compare, for example, the candidates generated as a result of epenthesis by parallel OT's unrestrained GEN, which can perform an unbounded number of epenthesis operations, to those created by HS's restricted GEN, which can perform at most one instance of epenthesis (4).

(4) Possible Candidates in OT and HS

	OT	HS
Input	/ap/	.ap.
a.	.ap.	.ap.
b.	.ʔap.	.ʔap.
c.	.ʔap.ta.	–
d.	.ʔap.ta.ka.	–
e.	.ʔap.+$\aleph_0$	–

Candidate (4a), the faithful candidate, and candidate (4b), which has a single epenthetic segment, are possible candidates in both parallel OT and in HS. The problematic candidates (4c–e), on the other hand, are not possible members of this candidate set in HS. Because they involve

4 What constitutes a single change is a matter of ongoing research; see for example McCarthy (2010a) for a discussion of the kinds of evidence brought to bear on this question.

multiple applications of a single operation (epenthesis) and simultaneous application of multiple operations (epenthesis and core syllable creation), they cannot be produced by a restricted HS GEN. This result depends, of course, on the nature of the epenthesis operation – here I make the assumption that at most a single segment can be inserted at once, but any definition of epenthesis which is restricted to the insertion of a bounded number of units (including e.g. feature-by-feature insertion) will produce a similar result.

If (+)ONSET dominates DEP,[5] a HS derivation proceeds as follows. At the first step[6] in (5), epenthesizing an onset garners a reward on (+)ONSET and wins over the faithful candidate. At Step 2, epenthesizing does not improve performance on (+)ONSET, so the faithful candidate is chosen as optimal, and the derivation converges.

(5) Step 1

.ap.	(+)ONSET	DEP
a. .ap	W	L
b. → .ʔap.	+1	−1

Step 2: Convergence

.ʔap.	(+)ONSET	DEP
a. → .ʔap.	+1	
b. .ʔap.ʔ	+1	W−1

At any given step of the derivation, the candidate set produced by HS's restrained GEN is finite – this means that the potential for infinite goodness can't interfere with the ability to find an optimum at EVAL. Because there are not infinitely many candidates with increasing numbers of epenthesized onsets, there will in fact exist a candidate that garners the most rewards on a positive constraint. This means that at least some constraints in HS can be positive without running afoul of infinite goodness, but it does not necessarily mean that any constraint can be positive – for example, certain constraints run the risk of producing infinitely long derivations. See Section 7.4 for a discussion of the restrictions on the formulation of positive constraints.

5 For ease of exposition, faithfulness constraints will retain their traditional negative definitions throughout.

6 Following initial syllabification steps, not shown; for a concrete proposal regarding syllabification in HS, see Elfner (2009).

7.3 Wherefore positive constraints: Harmony

The previous section established that positive constraints are viable in HS; this section argues that, in at least some contexts, they are not only viable but also desirable. Autosegmental spreading has proven challenging to account for with a negative spreading constraint – Wilson (2004, 2006) identifies a number of pathological predictions made by common harmony constraints. In particular, commonly-used constraints make problematic predictions in systems where spreading is blocked. For example, in Malay, nasal harmony is blocked by obstruents (Onn, 1980; Walker, 2000):

(6) Nasal harmony in Malay
  a. Unbounded rightward spreading
    mĩnõm 'to drink'
    mãj̃ãn 'stalk (palm)'
    pəŋə̃ŋãh̃ãn 'central focus'
    mə̃nãw̃ãn 'to capture'
  b. Blocked by obstruents
    mãkan 'to eat'
    pəŋãw̃ãsan 'supervision'

One common approach to spreading is Agree, which assigns violations to a sequence of adjacent segments with different values for a given feature (Bakovic, 2000; Eisner, 1999; Lombardi, 1999, 2001; Pulleyblank, 2004). This constraint suffers from the *Sour Grapes* problem – if complete harmony is impossible, no spreading is predicited. Take, for example, an input like /pəŋawasan/ – a constraint like Agree(nas) will assign a violation for the [+nas][–nas] sequence [ŋa]. However, the attested output form [pəŋãw̃ãsan], where nasality has spread several segments to the right, performs no better – Agree(nas) will assign a violation to the [ãs] sequence. Harmony has not improved performance on Agree(nas), and has gratuitously violated faithfulness, so is predicted to be harmonically bounded. Nevertheless, languages like Malay, rather than giving up whenever total agreement is impossible, spread as far as is possible. See Wilson (2004); McCarthy (2004, 2010b) for further discussion of this pathology.

Another common approach to spreading, which does not suffer from the Sour Grapes problem, is to use gradient alignment constraints on feature domains (Archangeli and Pulleyblank, 2002; Cole and Kisseberth, 1994, 1995; Kirchner, 1993; Pulleyblank, 1996; Smolensky,

1993). Constraints like ALIGN assign violations to segments that intervene between the edge of the feature domain and the edge of the bounding domain.

However, in words where harmony is blocked, a sufficiently highly ranked ALIGN can block epenthesis. Consider Malay, where ALIGN dominates a constraint against word-final consonant clusters (which in turn dominates DEP). When harmony is blocked, an epenthesized segment increases the distance between the edge of the feature domain and the word edge – ALIGN will prefer a candidate with no epenthesis (7a) over a candidate with an epenthetic segment (7b).

(7) Epenthesis blocked by harmony

	nawakast	ALIGN-R(NAS,PWD)	*CC#
a. →	nãw̃ãkast	−4	−1
b.	nãw̃ãkasət	W−5	L

In Malay, final clusters are normally resolved by epenthesis (8a). However, in words where spreading is blocked, no epenthesis occurs (8b). More broadly, a constraint like ALIGN-R can produce a language where harmony does not occur at all (due to high-ranking faithfulness), but epenthesis is still blocked to the right of a nasal segment. These patterns are unattested, in Malay or elsewhere (Wilson, 2004, 2006).

(8) Epenthesis blocked by nasal segments

	ALIGN-R(NAS,PWD)	*CC#	DEP
a. tawakasət ~ tawakast		W	L
b. nãw̃ãkast ~ nãw̃ãkasət	W	L	W

ALIGN blocks epenthesis because it assigns a violation to each segment intervening between the edge of the feature domain and the edge of the word; epenthesizing a segment, then, adds to the number of violations assigned by ALIGN. Broadly speaking, the problem is that ALIGN is sensitive to the number of non-harmonized segments in a word; a number of the other pathologies identified by Wilson (2004) (including those related to deletion, allomorph selection, and reduplication) also arise because of this kind of segment counting.

McCarthy (2010b) notes that serial evaluation plays a role in resolving some of the problematic predictions of ALIGN and other harmony constraints (including, to some extent, the Sour Grapes problem). However, sensitivity to the number of non-harmonized segments in a word remains

– segment-counting pathologies[7] like harmony-dependent epenthesis blocking are still predicted with ALIGN in HS. The competitors at the nth step of a HS derivation in (9) are exactly the same as those in the OT derivation in (7).

(9) Epenthesis blocked by harmony

nãw̃ãkast	ALIGN-R(NAS,PWD)	*CC#	DEP
a. → nãw̃ãkast	−4	−1	
b. nãw̃ãkasət	W−5	L	W−1

Because a positively-defined spreading constraint assigns rewards for autosegmental associations rather than assigning violations for failure to assimilate, it will not be sensitive to the number of non-harmonized segments present. Unlike ALIGN, a positive spreading constraint (10) will not suffer from segment-counting pathologies.

(10) (+)SPREAD(F): Assign +1 for each segment linked to F as a dependent.

This definition relies on a basic assumption about autosegmental structure: that each feature domain has one and only one head. For the present proposal, it does not matter which segment serves as head; for further discussion of headed domains in autosegmental spreading, see Cole and Kisseberth (1994); McCarthy (2004); Smolensky (2006) and others.

In this case the relevant single change performed by GEN is the creation of a single autosegmental link (or, more generally, spreading the relevant feature to a single segment). A precise definition and illustration of the operation are given in (11).

(11) For a feature $\pm F_i$ and a segment S, create an autosegmental association between F_i and S.[8]

Harmony thus proceeds one segment a time, as in the derivation in (12). At the first step, the candidate with nasal spreading wins over the faithful candidate. This is again the case at the second and third steps; spreading will continue to iterate, segment by segment, until there are no more segments left unassociated or until spreading is blocked.

7 See McCarthy (2010b) for another HS-based approach to addressing this class of pathologies.

8 If binary features are assumed, this operation may need to include delinking of the opposing feature value.

The latter is true in (12); at the fourth step, spreading conflicts with (+)ObsOral. The faithful candidate is chosen as optimal, and the derivation converges.

(12) Gradual harmony blocked by an obstruent

Step 1

	nawakast	(+)ObsOral	(+)Spread(nas)	Ident(nas)
a.	nawakast	+2	W	L
b. →	nãwakast	+2	+1	−1

Step 2

	nãwakast	(+)ObsOral	(+)Spread(nas)	Ident(nas)
a.	nãwakast	+2	W+1	L
b. →	nãw̃akast	+2	+2	−1

Step 3

	nãw̃akast	(+)ObsOral	(+)Spread(nas)	Ident(nas)
a.	nãw̃akast	+2	W+2	L
b. →	nãw̃ãkast	+2	+3	−1

Step 4

	nãw̃ãkast	(+)ObsOral	(+)Spread(nas)	Ident(nas)
a. →	nãw̃ãkast	+2	+3	
b.	nãw̃ãk̃ast	W+1	L+4	W−1

Because it assigns rewards for harmony rather than violations for failure to harmonize, (+)Spread(nas) will not be sensitive to the number of segments to the right of the blocking obstruent in [nãw̃ãkast]. Consider a candidate with epenthesis to break up the final cluster – both the faithful candidate (13a) and the candidate with epenthesis (13b) perform equally well on (+)Spread(nas).

(13) Epenthesis is not blocked

Step 4

	nãw̃ãkast	(+)Spread(nas)	(+)CV
a.	nãw̃ãkast	+3	W+3
b. →	nãwakasət	+3	+4

A positive spreading constraint, therefore, cannot block epenthesis. Furthermore, a constraint like (+)Spread(nas) is prevented from *inducing* epenthesis by the restricted Gen in HS. As we saw for (+)ONSET above, a positive constraint like (+)SPREAD(F) in parallel OT would suffer from the infinite goodness problem. In this case, it would cause infinite harmony by epenthesis – inserting and harmonizing a segment would earn a reward, so infinitely many epenthesized segments would be preferred.

In HS, however, epenthesis and harmony cannot occur concurrently. (+)SPREAD(F), therefore, can't induce epenthesis on its own – the intermediate step necessary will not constitute an improvement. An epenthetic candidate (14b) performs no better on (+)SPREAD(F) than the faithful candidate (14a). Because there is no possible candidate at this step with both epenthesis and harmony, the faithful candidate is chosen and the derivation converges.

(14) Epenthesis is not blocked
Step 4

	nãw̃ãkast	(+)SPREAD(NAS)	(+)CV
a.	nãw̃ãkast	+3	W+3
b. →	nãwakasət	+3	+4

The solution presented here has centered around pathologies involving epenthesis, but is general to the set of pathologies involving the number of non-harmonized segments in a word. A positive spreading constraint exerts no preference against non-harmonized segments, and because a HS GEN is restricted to making gradual changes, a positive constraint cannot itself induce processes that increase the number of segments available to harmony.

This proposal is very much in the spirit of Wilson (2004), where Targeted Constraints are used to address the set of segment-counting pathologies in harmony. Targeted Constraints involve both a reward-assigning component and a penalty-assigning component, and there is considerable (though not exhaustive) overlap between the predictions of the Targeted Constraints approach and that sketched here.

A full comparison is beyond the scope of this paper; this section has presented a brief discussion of the merits of positive evaluation in analyzing autosegmental spreading, and has served in part as a proof of principle for positively defined constraints, but it is not intended to be taken as a fully developed theory. For a more detailed proposal, including a comparison between reward-only constraints and reward-plus-penalty approaches like Targeted Constraints, see Kimper (2011).

7.4 The scope of the solution

Because HS derivations are gradual and harmonically improving, they provide a possible way out of the infinite goodness problem suffered by positive constraints in parallel versions of OT. The previous section showed how a positive constraint – with the restricted GEN of HS – proves desirable in the realm of autosegmental spreading. However, it is important to note that the solution to infinite goodness presented by HS is not universal; that is, it is not true that any and all constraints can necessarily be positively defined, even when GEN is restricted to producing a limited candidate set.[9]

This is because a finite candidate set is not in and of itself sufficient to prevent infinite goodness – HS derivations only *converge* when the single changes produced by GEN fail to produce harmonic improvement. This means that in order to avoid infinite goodness, a derivation must be sufficiently gradual to prevent a positive constraint from inducing an infinite loop of instant satisfaction.

With a positive version of ONSET, the derivation in (5) converges at the second step – it's not possible to both epenthesize a segment and create the supporting structure that would make it an onset; epenthesis and syllable creation constitute separate steps. This is also true for the positive spreading constraint in the previous section; because it is not possible to simultaneously insert a segment and create an autosegmental link, any further satisfaction requires a two-step process, and the first of those steps (epenthesis) must be *independently* harmonically improving in order to take place.

This means that the degree to which infinite goodness can be avoided will depend crucially on the degree to which GEN is restricted. Consider, for example, a context-free markedness constraint like (+)OBSORAL, which assigns a reward to every oral obstruent. With an epenthesis operation capable of inserting a fully-specified segment, a constraint like this would induce infinitely long derivations – at each step, epenthesis of an oral obstruent would garner an additional reward, and (under the right ranking conditions) the derivation would fail to converge. On the

9 Flemming (2001) uses a positively defined contrast maximization constraint; because this constraint evaluates inventories rather than derivations, the distinction between parallel and serial versions of OT is not relevant, and the present solution to the infinite goodness problem does not extend to that case.

other hand, requiring epenthesis of root nodes and segmental features to occupy different derivational steps would render (+)OBSORAL harmless.

In general terms, a constraint may be viably positive iff the structure which earns the reward depends critically on some prerequisite condition whose creation requires a distinct derivational step. So (+)ONSET is viable so long as construction of a core syllable and epenthesis of a segment require distinct derivational steps; (+)SPREAD(F) is viable so long as insertion of a segment and creation of an autosegmental link require distinct derivational steps; (+)OBSORAL is viable so long as insertion of a segment and specification of that segment as oral require distinct derivational steps.

It is unlikely to be the case that *all* constraints have a viable positive formulation, given independently motivated definitions of gradualness in HS. But that there is gradualness at all in HS means at least *some* constraints may be viably positive, and Section 7.3 suggests why this might be a desirable outcome.

Even if conditions on GEN were such that all constraints could be positively defined, it is not clear that a CON with only positive constraints would be desirable. In particular, a positive constraint can reward satisfaction, but not varying degrees of failure to satisfy – this makes edge-based alignment phenomena difficult to capture, since the generalization that increasing distance from the desired edge incurs increasing degrees of violation cannot be expressed using only positive evaluation.

Additionally, while one of the benefits of positive constraints is that they can help to alleviate some instances of the *too many repairs* problem (as was the case in the spreading example in the previous section), one side effect is losing the ability to cohesively handle certain conspiracies. A constraint like *NC (Pater, 1999), for example, crucially relies on its negative definition to produce the attested range of effects. Capturing the same set of effects would require a range of distinct positive constraints, one for each attested repair. This loses the generalization that this set of alternations are all seeking to solve the same problem, and the constraints needed would likely introduce problems of their own. For example, post-nasal voicing is an attested repair for *NC violations; a positive constraint producing this result would need to assign rewards to sequences of a nasal followed by a voiced stop. However, such a constraint would be predicted to induce alternations of its own – it could not only induce post-nasal voicing, but also pre-voiced nasalization.

Given these challenges, it does not appear as though a CON consisting of all and only positive constraints is the way forward. While we saw in

Section 7.3 above that a positively-defined constraint produces superior results for autosegmental spreading, some generalizations, like *NC, are better suited to negative definition. For most constraints, though, a sufficiently restricted Gen in HS will mean that positive and negative versions have the same practical effect (as we saw with Onset in Section 7.2 above) – further research is needed to identify points of departure, and which of those constraints are better suited to positive definition and which are better suited to negative definition.

7.5 Conclusion

In this paper, I have shown that positively-defined constraints, which suffered from the Infinite Goodness problem in parallel versions of OT, are a viable option in HS. Because GEN is restricted to performing at most a single instance of a single operation, the candidate set does not contain candidates with infinitely many epenthesized instances of the structure preferred by a positive constraint. With a sufficiently restricted GEN, most kinds of OT constraint can be implemented positively without running afoul of infinite goodness. Some constraints remain better suited to negative definitions, so a CON which may admit both positive and negative constraints is proposed.

Furthermore, I have argued that there is at least one domain where positive constraints produce a better result than their negative counterparts; namely, autosegmental spreading. Negatively defined harmony constraints penalize unassimilated segments, and therefore are able to influence processes like epenthesis which increase the number of segments in a word. A positively defined harmony constraint, on the other hand, avoids a number of the pathological predictions of its negative predecessors – because it assigns rewards to assimilated segments, it is not sensitive to the number of segments which have failed to harmonize, and therefore exerts no undue segment-counting influence, and is appropriately limited to merely inducing assimilation.

References

Archangeli, Diana and Douglas Pulleyblank (2002) Kinande vowel harmony: Domains, grounded conditions and one-sided alignment. *Phonology* 19: 139–188.

Bakovic, Eric (2000) *Harmony, Dominance, and Control*. Doctoral dissertation, Rutgers University, New Brunswick, NJ. Available on Rutgers Optimality Archive, ROA-360.

Cole, Jennifer S. and Charles Kisseberth. (1994) An Optimal Domains theory of harmony. Unpublished paper, University of Illinois, Urbana. Available on Rutgers Optimality Archive, ROA-22.

Cole, Jennifer S. and Charles Kisseberth. (1995) Nasal harmony in Optimal Domains Theory. Unpublished paper, University of Illinois, Urbana. Available on Rutgers Optimality Archive, ROA-49.

Eisner, Jason. (1999) Doing OT in a straitjacket. Unpublished paper, Johns Hopkins University, Baltimore, MD. Available at http://www.cs.jhu.edu/~jason/papers/eisner.ucla99.handout.large.pdf.

Elfner, Emily (2009) Syllabification and stress-epenthesis interactions in Harmonic Serialism. Available on Rutgers Optimality Archive, ROA-1047.

Flemming, Edward (2001) Scalar and categorical phenomena in a unified model of phonetics and phonology. *Phonology* 18: 7–44.

Gouskova, Maria (2003) *Deriving Economy: Syncope in Optimality Theory*. Doctoral dissertation, University of Massachusetts Amherst, Amherst, MA. Available on Rutgers Optimality Archive, ROA-610.

Jesney, Karen (2011) Positional faithfulness, non-locality, and the Harmonic Serialism solution. In Suzi Lima, Kevin Mullin and Brian Smith (eds), *Proceedings of the 39th Annual Meeting of the North East Linguistic Society*, 239–440. Amherst, MA: GLSA Publications. Available on Rutgers Optimality Archive, ROA-1018.

Kimper, Wendell (2011) *Competing Triggers: Transparency and Opacity in Vowel Harmony*. Doctoral dissertation, University of Massachusetts Amherst, Amherst, MA.

Kimper, Wendell (2011) Locality and globality in phonological variation. *Natural Language & Linguistic Theory* 29: 423–465.

Kirchner, Robert. (1993) Turkish vowel harmony and disharmony: An Optimality Theoretic account. Unpublished paper, UCLA, Los Angeles. Available on Rutgers Optimality Archive, ROA-4.

Lombardi, Linda (1999) Positional faithfulness and voicing assimilation in Optimality Theory. *Natural Language & Linguistic Theory* 17: 267–302.

Lombardi, Linda (ed.) (2001) *Segmental Phonology in Optimality Theory: Constraints and Representations*. Cambridge; New York: Cambridge University Press.

McCarthy, John J. (2000) Harmonic serialism and parallelism. In Masako Hirotani (ed.), *Proceedings of the North East Linguistics Society* 30, 501–524. Amherst, MA: GLSA Publications. Available on Rutgers Optimality Archive, ROA-357.

McCarthy, John J. (2002) *A Thematic Guide to Optimality Theory*. Cambridge: Cambridge University Press.

McCarthy, John J. (2003) OT constraints are categorical. *Phonology* 20: 75–138.

McCarthy, John J. (2004) Headed spans and autosegmental spreading. Unpublished paper, University of Massachusetts Amherst, Amherst, MA. Available on Rutgers Optimality Archive, ROA-685.

McCarthy, John J. (2007) Restraint of analysis. In Sylvia Blaho, Patrik Bye and Martin Krämer (eds), *Freedom of Analysis*, 203–231. Berlin and New York: Mouton de Gruyter.

McCarthy, John J. (2008a) The gradual path to cluster simplification. *Phonology* 25: 271–319.

McCarthy, John J. (2008b) The serial interaction of stress and syncope. *Natural Language & Linguistic Theory* 26: 499–546.

McCarthy, John J. (2010a) Studying Gen. *Journal of the Phonetic Society of Japan* 13: 3–12.

McCarthy, John J. (2010b) Autosegmental spreading in Optimality Theory. In John Goldsmith, Elizabeth Hume and Leo Wetzels (eds), *Tones and Features*, 195–222. Berlin and New York: Mouton de Gruyter.

Onn, Farid M. (1980) *Aspects of Malay Phonology and Morphology: A Generative Approach*. Kuala Lumpur: Universiti Kebangsaan Malaysia.

Pater, Joe (1999) Austronesian nasal substitution and other NC effects. In René Kager, Harry van der Hulst and Wim Zonneveld (eds), *The Prosody-Morphology Interface*, 310–343. Cambridge: Cambridge University Press.

Prince, Alan (2007) The pursuit of theory. In Paul de Lacy (ed.), *The Cambridge Handbook of Phonology*, 33–60. Cambridge: Cambridge University Press.

Prince, Alan and Paul Smolensky (1993/2004) *Optimality Theory: Constraint Interaction in Generative Grammar*. Malden, MA, and Oxford: Blackwell.

Pruitt, Kathryn. (2008) Iterative foot optimization and locality in stress systems. Unpublished paper, University of Massachusetts Amherst, Amherst, MA. Available on Rutgers Optimality Archive, ROA-999.

Pulleyblank, Douglas (1996) Neutral vowels in Optimality Theory: A comparison of Yoruba and Wolof. *Canadian Journal of Linguistics* 41: 295–347.

Pulleyblank, Douglas (2004) Harmony drivers: No disagreement allowed. In Julie Larson and Mary Paster (eds), *Proceedings of the Twenty-eighth Annual Meeting of the Berkeley Linguistics Society*, 249–267. Berkeley, CA: Berkeley Linguistics Society.

Smolensky, Paul. (1993) Harmony, markedness, and phonological activity. Unpublished paper, University of Colorado. Available on Rutgers Optimality Archive, ROA-87.

Smolensky, Paul (2006) Optimality in phonology II: Harmonic completeness, local constraint conjunction, and feature domain markedness. In Paul Smolensky and Géraldine Legendre (eds), *The Harmonic Mind: From Neural Computation to Optimality-Theoretic Grammar, Volume 1: Cognitive Architecture*, 26–160. Cambridge, MA: MIT Press/Bradford Books.

Walker, Rachel (2000) Long-distance consonantal identity effects. In Roger Billerey and Brook Danielle Lillehaugen (eds), *Proceedings of the 19th West Coast Conference on Formal Linguistics*, 532–545. Somerville, MA: Cascadilla Press.

Wilson, Colin. (2004) Analyzing unbounded spreading with constraints: Marks, targets, and derivations. Unpublished paper, UCLA.

Wilson, Colin. (2006) Unbounded spreading is myopic. Unpublished paper, UCLA. Available at http://www.linguistics.ucla.edu/people/wilson/Myopia2006.pdf.

Wolf, Matthew (2008) *Optimal Interleaving: Serial Phonology-Morphology Interaction in a Constraint-Based Model*. Doctoral dissertation, University of Massachusetts Amherst, Amherst, MA. Available on Rutgers Optimality Archive, ROA-996.

Chapter 8

Contexts for epenthesis in Harmonic Serialism

Claire Moore-Cantwell

8.1 Introduction

A theory of phonological grammar should produce phonological patterns that actually exist in the world's languages, while at the same time failing to predict typologically unattested patterns. Models based on Optimality Theory (Prince and Smolensky 1993/2004) make typological predictions through the mechanism of *factorial typology*. Given a set of constraints in Con, every logically possible arrangement of those constraints represents a pattern that is predicted to be possible in human language. If no possible arrangement of those constraints can produce a particular pattern, a language with that pattern is predicted not to exist.

OT as proposed by Prince and Smolensky solves the problem of 'conspiracies' pointed out by Kisseberth (1970), by producing in a principled way many possible repairs for the same marked structure. This allows an account of within-language conspiracies of the type discussed by Kisseberth, and also makes typological predictions: when a constraint militating against a marked phonological structure interacts with two or more constraints preferring different specific repairs, the factorial typology predicts that different languages should repair the same marked structure in different ways (see Pater 1999 for a worked-through example).

This ability to produce multiple repairs overgenerates, however: it frequently predicts a repair for a particular type of marked structure that is not found cross-linguistically. Theoretical modifications of OT that have been proposed to solve this 'too-many-solutions' problem include the P-map (Steriade 2001/2008), Targeted Constraints (Wilson 2001), and Procedural Constraints (Blumenfeld 2006). It remains a challenge, however, to rule out particular repairs for the cases in which they are unattested, while at the same time allowing for attested multiple repairs for a single constraint (see e.g. McCarthy 2002 and Pater 2003).

Harmonic Serialism (Prince and Smolensky 1993/2004, McCarthy 2007, *et seq.*), uses constraint interaction together with a restricted GEN and a serial evaluation framework. While serialism does not provide a general solution to the too-many-solutions problem, it does constrain the set of possible repairs to a particular marked structure. Specifically, the only repairs which can resolve the marked structure are those which can do so in a *harmonically improving* fashion. A harmonically improving derivation is one in which a form's harmony under the constraint ranking increases at each derivational stage. Because of this requirement, there are some conceivable output forms which are 'globally' optimal, but which are not reachable. Examples of solutions to too-many-repairs problems that make use of this property include McCarthy (2008), Pizzo (2010), Jesney (2011) and Staubs (this volume).

In this paper, I propose a restriction on the set of operations contained in GEN in Harmonic Serialism (HS), which leads to the prediction that epenthesis can be used to repair syllable-structure and segmental markedness, but cannot repair metrical markedness (such as stress clashes and lapses). Epenthesis to repair these structures is predicted by parallel OT but is unattested cross-linguistically.

I propose that the operation of epenthesis in GEN is constrained in two closely related ways. First, it must be distinct from the operation of prosodic structure building. This means that segmental epenthesis and syllable or foot building cannot occur in the same step. Second, epenthesis must satisfy the prosodic markedness constraint EXHAUSTIVITY (Selkirk 1995). This constraint forbids 'level skipping', in the prosodic tree – violations would include a syllable adjoined directly to a prosodic word without an intervening foot, or a segment adjoined directly to a foot without also being parsed into a syllable. The proposal is that epenthesis obeys EXHAUSTIVITY in that segments may be epenthesized directly into an existing syllable (the next level up on the prosodic hierarchy) but not into an existing foot, prosodic word, prosodic phrase, etc.

In HS with these restrictions on GEN, epenthesis may be used to resolve violations of syllable-structure markedness constraints and segmental phonotactic markedness constraints, but it cannot be used to resolve violations of metrical markedness constraints, which depend on structure above the level of the syllable. Outputs that use epenthesis to resolve violations of metrical markedness constraints never surface as optimal because they are unreachable in the derivation. Although they are globally optimal under some constraint rankings, the required derivational paths are not harmonically improving under any constraint ranking.

The paper is structured as follows: Section 8.2 provides an overview of the types of markedness which epenthesis is observed to resolve cross-linguistically. Section 8.3 introduces and formally defines the proposed restrictions on GEN, and discusses their consequences. In particular, Sections 8.3.3 and 8.3.4 demonstrate that epenthesis cannot resolve a violation of any of *CLASH, FOOTBINARITY, *LAPSE, or NONFINALITY. Section 8.4 briefly discusses attested epenthesis-triggering environments and demonstrates that HS with the proposed restrictions on GEN accounts for these cases. Finally, Section 8.5 concludes.

8.2 Typology of epenthesis-triggering environments

8.2.1 What occurs

Broselow (1982) shows that epenthesis can be used to resolve three types of markedness: syllable-structure markedness, segmental phonotactic markedness, and word-subminimality.[1] An epenthetic vowel can resolve syllable-structure markedness by allowing a bad coda to be re-syllabified as an onset, for example. Broselow gives Swahili as an example of such a case.

(1) Obstruents forbidden in codas in Swahili:
tiket ~ tiketi̲ *ticket*
ratli ~ rati̲li *pound*

Epenthesis can also resolve phonotactic markedness which is defined purely in terms of segmental content, and which does not depend on higher-level structure. One of Broselow's examples is Winnebago epenthesis, which separates an obstruent and a sonorant – a sequence that is forbidden regardless of the syllable structure in which it occurs. Likewise, in Mohawk a consonant cannot be followed by a glottal stop,

1 Broselow's terminology differs somewhat from mine – she calls these 'syllabic', 'segmental' and 'prosodic' markedness, respectively.

and when it otherwise would in the surface form, an epenthetic vowel intervenes.

(2) Winnebago: obstruent-sonorant sequences forbidden (Dorsey's Law)
hoʃwaʒa → hoʃa̲waʒa *be sick, 2pl*

(3) Mohawk: Consonant+glottal stop clusters forbidden
o+nʌst+ʔ → onʌste̲ʔ *corn, nom.*

Another example of epenthesis for segmental markedness comes from English plural formation, where epenthesis breaks up sequences of sibilants.

(4) English plural:
bɹʌʃ+s → bɹʌʃə̲z

In none of these three cases can epenthesis be construed as resolving syllabic markedness. English and Winnebago both permit sibilant codas, and Mohawk permits obstruent coda clusters, so the epenthesis in (2–4) is apparently not to eliminate bad codas, contrary to Swahili.

The final situation where epenthesis is attested is in resolving word-subminimality. Hayes (1995) lists about thirty languages exhibiting what he calls the 'minimal word syndrome'. Broselow gives an example from Mohawk, where monosyllabic words are forbidden and a long vowel is epenthesized at the beginning of the word to resolve them.

(5) Mohawk: Monosyllabic words forbidden
k+ek+s → i̲k̲+ek+s *I eat*

All three of these types of pattern have standard analyses in Parallel OT using well-motivated constraints.

8.2.2 *What doesn't occur*

Blumenfeld (2006, ch. 4) notes a typological gap: epenthesis does not apparently occur to resolve metrical markedness such as stress clash or lapse or foot-subminimality. He examines the factorial typology produced in Parallel OT by constraints such as NONFINALITY, FOOTBINARITY, *CLASH, and *LAPSE interacting with DEP and other prosodic constraints such as PARSE and the WEIGHT-TO-STRESS PRINCIPLE. He shows possible rankings of these constraints which produce languages in which epenthesis is used to avoid violations of these constraints. I will summarize these findings here. Before beginning, though, I provide the notational system that I will use throughout the rest of this paper in (6). Note in particular that foot boundaries are indicated with square brackets. I will also be underlining epenthetic segments.

(6) Notational system:

Syllable boundaries	( )
Foot boundaries	[ ]
Prosodic Word boundaries	{ }
Unparsed segments	< >

Marked prosodic structures arise because of the pressure of conflicting high-ranked constraints that motivate foot building, such as PARSE, or the WEIGHT-TO-STRESS PRINCIPLE (Prince and Smolensky 1993/2004).

(7) WEIGHT-TO-STRESS PRINCIPLE (WSP): Assign a violation to every heavy syllable that is not stressed.[2]

In particular, the WSP can interact with other prosodic markedness constraints to produce epenthesis. I will illustrate this first with the constraint NONFINALITY.

(8) NONFINALITY: Assign a violation to every stressed syllable that is final in a prosodic word.

When the WSP and NONFINALITY are both ranked above DEP ('Assign a violation to every segment in the output which is not present in the input'), then a syllable will be epenthesized just in case it avoids a violation of both. The surface forms of such a language would follow a pattern like that in (9).

(9) Epenthesis repairs NONFINALITY violations (unattested)

a. /batki/ → {[(bát)(ki)]}

b. /batki + ta/ → {[(bát)(ki)](ta)}

c. /baduk/ → {[(ba)(dúk)](ʔə)} *{[(ba)(dúk)]}

d. /baduk + ta/ → {[(ba)(dúk)](ta)}

2 Constraint definitions have been modified from their sources to follow the prescription for constraint definition in McCarthy (2003). As far as I can tell this does not change the violation profiles.

In this language all heavy syllables (syllables with a coda) are stressed, and just in case a stressed heavy would be final in the prosodic word, a syllable is epenthesized. This language can be produced by ranking WSP and NONFINALITY over DEP.

(10) WSP, NONFINALITY ≫ DEP

/baduk/	WSP	NONFINALITY	DEP
a. → {[(ba)(dúk)](ʔə)}			2
b. {[(ba)(dúk)]}		1W	L
c. {[(bá)(duk)]}	1W		L

In this tableau, the WSP forces the heavy syllable to be stressed. This syllable is final in the prosodic word in candidate (b), violating NONFINALITY. However, in the winning candidate, (a), an extra syllable is epenthesized, intervening between the heavy syllable and the prosodic word edge.

When *CLASH and the WSP are both high ranked, epenthesis can likewise obtain.

(11) *CLASH: Assign a violation to every pair of adjacent stressed syllables

With both constraints ranked above DEP, all heavy syllables will be stressed, but an extra syllable will be epenthesized just in case two heavy syllables are next to each other.

(12) Epenthesis repairs stress clash (unattested)

a. /baduk/ → [(ba)(dúk)]
b. /baduk + ti/ → [(ba)(dúk)](ti)
c. /baduk + kit/ → [(ba)(dúk)][(ʔə)(kít)] * [(ba)(dúk)][(kít)]

(13) WSP, *CLASH ≫ DEP

/baduk + kit/	WSP	*CLASH	DEP
a. → [(ba)(dúk)][(ʔə)(kít)]			2
b. [(ba)(dúk)](kit)	1W		L
c. [(ba)(dúk)][(kít)]		1W	L

The case of *LAPSE looks very similar.

(14) *LAPSE: Assign a violation to every pair of adjacent unstressed syllables.

If the WSP and *Lapse both outrank Dep, then for some inputs, a stress lapse will be avoided through epenthesis of an extra syllable. This is demonstrated in the tableau in (15).

(15) WSP, *Lapse ≫ Dep

	/bakdupikib/	WSP	*Lapse	Dep
a. →	[(bák)(du)][(pí)(ʔə)][(kíb)]			2
b.	[(bak)(dú)][(pi)(kíb)]	1 W		L
c.	[(bák)(du)][(pi)(kíb)]		1 W	L

In this tableau, the WSP forces all the heavy syllables ([bak], [kib]) to be stressed, ruling out the candidate with alternating stress in b. Because of the placement of the heavy syllables, stressing both of them results in a stress lapse, as in c. Candidate a. avoids both violations by epenthesizing an extra syllable and building a new foot around it.

The language where WSP, *Lapse ≫ Dep is a language where all heavy syllables are stressed, and a syllable is epenthesized and a foot built with it just in case stressing all heavies would result in a stress lapse. Such a language would have surface alternations like those in (16).

(16) Epenthesis repairs stress lapse (unattested)

a.	/baduk/	→	[(ba)(dúk)]
b.	/baduk + pikit/	→	[(ba)(dúk)][(pi)(kít)]
c.	/batki/	→	[(bát)(ki)]
d.	/batki + pikit/	→	[(bát)(ki)][(pí)(ʔi)][(kít)] *[(bát)(ki)][(pi)(kít)]

Other conflicting constraints can interact with Nonfinality, *Clash, and *Lapse to force epenthesis, including faithfulness to underlying stress, constraints on foot shape, or even alignment constraints.

The final case of predicted epenthesis to resolve metrical markedness that I will discuss involves FootBinarity interacting with the structure-motivating constraint Parse-syll.

(17) FootBinarity (FtBin): Assign a violation to every foot that does not contain two syllables.

(18) Parse-syll: Assign a violation to every syllable that is not parsed into a foot.

Foot binarity can be enforced at a syllabic or at a moraic level of analysis. In this paper I will focus on binarity at the syllabic level only, since only that level would enforce epenthesis of an entirely new vowel. Binarity at the moraic level can be enforced by epenthesis of a mora without a new segment.

If FOOTBINARITY and PARSE-SYLL both outrank DEP, then a syllable would be epenthesized just in case the foot would otherwise be subminimal. This is a language in which every input syllable is footed, and epenthesis applies to make every foot binary. This would be a language in which all prosodic words are even-parity.

(19) Epenthesis repairs subminimal feet (unattested):

a. /bata/ → [(bá)(ta)]

b. /bataka/ → [(bá)(ta)][(ká)(ʔə)] *[(bá)(ta)][(ká)] *[(bá)(ta)](ka)

c. /bataka + pi/ → [(bá)(ta)][(ká)(pi)]

(20) PARSE-SYLL, FTBIN ≫ DEP

/bataka/	PARSE-SYLL	FTBIN	DEP
a. → [(bá)(ta)][(ká)(ʔə)]			2
b. [(bá)(ta)][(ká)]		1W	L
c. [(bá)(ta)](ka)	1W		L

In this tableau, candidate c. fails because it does not parse all the existing syllables into feet, and candidate b. fails because one of the feet is subminimal. Candidate a. solves both these problems, by epenthesizing an extra syllable.

McCarthy and Prince (1986), and McCarthy and Prince (1993), argue that FTBIN *can* be repaired by epenthesis, just in case there is only one foot in the prosodic word. Specifically, they argue that foot binarity plays a role in determining the size of a minimal word in languages with word minimality (such as Mohawk, (5)). They argue that the 'minimal word syndrome' (Hayes, 1995) is an emergent effect that arises out of the joint action of two contraints: a requirement that words have at least one foot, and a limit on foot size (of which FTBIN is an example). A language like Mohawk, in which words must be at least bisyllabic, would be analyzed with FTBIN and a constraint requiring prosodic words to have at least one foot.

(21) PROSODICWORDHEAD (PRWDHEAD): Assign a violation to every prosodic word that does not have a head foot.

PRWDHEAD forces all prosodic words to have a head foot. If a language does not allow degenerate (monosyllabic) feet, then the language cannot have words that are smaller than the smallest allowable foot. This can be modeled with PRWDHEAD, FTBIN ≫ DEP, as seen in (22), in which an actual word of Mohawk is derived.

(22) PRWDHEAD, FTBIN ≫ DEP

	/k+ek+s/	PRWDHEAD	FTBIN	DEP
a. →	{[(í̲:̲)(keks)]}			1
b.	{[(kéks)]}		1W	L
c.	{(keks)}	1W		L

In the winner, a., a new syllable is epenthesized and incorporated into the foot so that the prosodic word can have at least one foot, and that foot can be well-formed. This approach predicts that the smallest word a language allows will be the same size as the smallest foot a language allows. For example, if a language allows degenerate monosyllabic, monomoraic feet then it will also allow monosyllabic, monomoraic words. If the smallest feet a language allows are monosyllabic but bimoraic then its smallest words will also be monosyllabic but bimoraic.

More recent work (Piggott 1993, 2010, Garrett 1999) argues that this prediction is false. Piggott and Garrett demonstrate that many languages have minimal word requirements that are not the same as their minimal foot requirements. An example given by Piggott is the Arawan language Paumari, in which monomoraic syllables can be footed in odd-parity words (for example: (má)(sikò), *'ear'*), but the smallest prosodic words are bimoraic (koá, *'mouse'*, but *ko). Garrett further shows that some languages have minimal word sizes that are unattested as minimal foot sizes cross-linguistically. An example is 'CCV or CVC' in Yakima (Hargus and Beavert, 2006). Piggott argues for an independent constraint requiring word-minimality. In light of these arguments I will treat the case of epenthesis to avoid a violation of foot sub-minimality as a true gap that needs to be accounted for. The case of epenthesis to avoid subminimal words will be taken up again in Section 8.4.3.

In summary, while epenthetic material is observed to resolve violations of syllable-structure markedness, segmental phonotactic markedness, and word-subminimality, it is not observed to resolve violations of the metrical markedness constraints NONFINALITY, *CLASH, *LAPSE, and FOOTBINARITY. Parallel OT predicts that it should, however, in that each of these constraints can interact with some conflicting constraint that motivates foot building (WSP, PARSE-SYLL, etc.) to force epenthesis just in case it would avoid a violation of both constraints.

It should be noted that although no categorical epenthesis processes have been found which occur specifically to resolve violations of the metrical markedness constraints discussed here, Hall (2011) discusses variable processes of epenthesis which are conditioned by metrical factors. Examples include French schwa epenthesis (Coté 2000), Dutch schwa epenthesis (Booij, 1995), and epenthesis at Intonational-phrase edges in Galician (Martinez-Gill 1997). Hall makes the generalization that variable epenthetic processes can sometimes be affected by metrical considerations, but no categorical epenthesis process is conditioned by metrical markedness. I leave this apparent difference between categorical processes and variable ones for future work.

8.3 Eliminating epenthesis pathologies in HS

8.3.1 Restricting GEN

In Parallel OT, the set of operations in GEN and their possible interactions is unbounded, and it is the constraint set and ranking that do all of the work of choosing an output from an input. Operations in HS's GEN are much more restricted, and they cannot combine within a derivational stage. Because of this, the exact content of GEN plays a much greater role in determining the final outcome of the derivation. Thus, the study of GEN plays a central role in the development of HS. A theory of the content of GEN must be based on typological considerations. In this section, I will make a claim about GEN, proposing a specific definition of epenthesis with limited structure-building abilities. I will then demonstrate that this definition of epenthesis restricts the set of potential environments for epenthesis in ways that agree with the typological facts given in Section 8.2 above.

Specifically, in order to account for the typological generalization that epenthesis occurs to repair bad syllable structure, but not bad foot structure, I propose that epenthesis is a separate operation from the creation of new prosodic structure (syllable-building, for example), and further that epenthesis obeys EXHAUSTIVITY in that segments can only be epenthesized into existing syllables, not into any higher-level prosodic structure (feet, prosodic words, phrases).

I propose to define epenthesis as in (23).

(23) EPENTHESIS: Insert a segment at any linear position in the input string.

For Example:

Input:	batka
Candidates:	əbatka
	bəatka
	baətka
	batəka
	⋮

This operation can be used to resolve segmental markedness like the constraint against adjacent sibilants that affects the shape of the English plural:

(24) OCP-SIBILANT, ≫ DEP

	/bɹʌʃ +z/	OCP-SIBILANT	DEP
a. →	bɹʌʃəz		1
b.	bɹʌʃz	1W	L
c.	bəɹʌʃz	1W	1

But EPENTHESIS does not include parsing into prosodic structure. If a segment is epenthesized into a string that has some prosodic structure, it's left out of that structure.

Epenthesized segments can later be incorporated, gradually, into prosodic structure, as shown in (25).

(25) Epenthesis in action

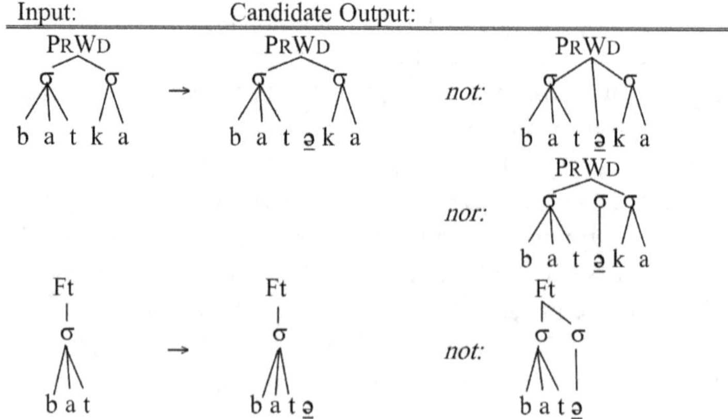

Epenthesized segments can later be incorporated, gradually, into prosodic structure, as shown in (26). Recall from (6) that angled brackets indicate an unparsed segment.

(26) Parsing into prosodic structure

Input		Epenthesis		Syllabification		Parsing I		Parsing II
{(bat)(ka)}	→	{(bat)<ə̲>(ka)}	→	{(bat)<(ə̲)>(ka)}	→	{(bat)(ə̲)(ka)}		
[(bát)]	→	[(bát)]ə̲	→	[(bát)](ə̲)	→	[(bát)(ə̲)]		
{[(bát)]}	→	{[(bát)]}ə̲	→	{[(bát)]}(ə̲)	→	{[(bát)](ə̲)}	→	{[(bat)(ə̲)]}

In order for these chains of forms to arise in a derivation, they must be harmonically improving. Each step must do a better job than the one before it of satisfying the constraint set. In the next section I will show that the step of epenthesis is not harmonically improving with respect to the relevant metrical markedness constraints and faithfulness constraints. This means that epenthesis can never occur in such a situation, even when the epenthetic segment would eventually, once prosodified, resolve a violation of a high ranked metrical markedness constraint.

8.3.2 Motivation

In this section, I will briefly provide some theory-internal motivation for the above restriction on GEN. I hope this discussion will lead towards a more general theory of GEN for prosodic structure building.

The Strict Layer Hypothesis (Selkirk 1981, 1984, Nespor and Vogel, 1986) imposes the restriction of EXHAUSTIVITY on prosodic structure.

(27) EXHAUSTIVITY (Selkirk 1995):
No C_i immediately dominates a constituent C_j, $j < i-1$,
e.g. 'No PWd immediately dominates a σ.'

EXHAUSTIVITY simply states that no levels of the prosodic hierarchy are 'skipped.' A syllable cannot be adjoined directly to a prosodic word, or a foot to a prosodic phrase. EXHAUSTIVITY is violated in surface forms in some languages, which show evidence especially that syllables can be adjoined directly to the prosodic word, and not footed (Selkirk 1995 and references therein). Therefore, Selkirk argues that it should be a violable constraint in an OT-type framework.

The restriction on GEN proposed above can be restated more generally, and in terms of EXHAUSTIVITY:

(28) EPENTHESIS RESPECTS EXHAUSTIVITY:
No epenthetic constituent C_j, may be immediately dominated by a C_i, where $j < i-1$,
e.g. 'No epenthetic σ may be immediately dominated by a PrWd.'

This generalization is not a violable constraint in CON, but rather is a statement about the content of GEN. It does not restrict the behavior of epenthetic segments *after* they are epenthesized, but only restricts the process of epenthesis. An epenthetic segment, for example, cannot be epenthesized into an existing prosodic word, but if it were epenthesized before prosodic structure building happened, then it could later be parsed into a prosodic word without first being syllabified. If it were possible to epenthesize whole feet, or whole syllables, then according to this generalization, it would be possible to epenthesize a whole syllable into an existing foot, but not into an existing prosodic word. Likewise, it would be possible to epenthesize a whole foot into an existing prosodic word, but not into an existing prosodic phrase.

I will argue in Section 8.3.3 that it might be desirable to allow a process of whole-syllable epenthesis into GEN.[3]

8.3.3 *Epenthesis to resolve metrical markedness is not harmonically improving*

In order for epenthesis to resolve a violation of *LAPSE, an input must go through a specific series of derivational stages. These are illustrated in (29). First, foot building operations must produce the bad structure, the stress lapse. Then, epenthesis must occur. Then, the epenthetic vowel must be syllabified and parsed into a foot. Only after this last step is the stress lapse removed.

(29) Required derivational stages

		⋮
	(foot-building operations)	[(bák)(du)](pi)[(kíb)(ti)]
1.	EPENTHESIS	[(bák)(du)](pi)ə̲[(kíb)(ti)]
2.	Syllabification	[(bák)(du)](pi)(ə̲)[(kíb)(ti)]
3.	Foot Building	[(bák)(du)][(pí)(ə̲)][(kíb)(ti)]

3 I am assuming that there is no operation which can epenthesize an empty prosodic category, such as a foot or a syllable. I know of no independent evidence that such operations are necessary, and they would allow a 'loophole' through which segmental epenthesis could be used after all to resolve prosodic markedness like subminimal feet.

Although syllabification and foot building would each be harmonically improving, epenthesis at the first stage is not. The candidate in (30b.), with epenthesis, is harmonically bounded since it does nothing to help the *LAPSE violation, and incurs a gratuitous violation of DEP.

(30) *LAPSE ≫ DEP

	[(bák)(du)](pi)[(kíb)(ti)]	*LAPSE	DEP
a. →	[(bák)(du)](pi)[(kíb)(ti)]	1	
b.	[(bák)(du)](pi)ə[(kíb)(ti)]	1	1W

Candidate b. is harmonically bounded by candidate a., and the derivation converges at this step, on the epenthesis-less candidate.

The form in (29, 3.), in which a segment has been epenthesized, syllabified, and parsed into a new foot, is more harmonic given the constraint set and the ranking of *LAPSE ≫ DEP, than the candidate with no epenthesis. However, because of the restriction on GEN, this form is not available as a candidate at the first stage, and because it is not in the candidate set it cannot be selected. This form will never make it into the candidate set, and thus will never be chosen as the winner.

In a similar way, epenthesis to resolve violations of NONFINALITY and FTBIN isn't harmonically improving. Example (31) shows the derivational stages required to reach a form that uses epenthesis to satisfy one of these constraints.

(31) Required derivational stages

		NONFINALITY	FTBIN
		⋮	⋮
	(parsing operations)	{[(bák)]}	[(bák)]
1.	EPENTHESIS	{[(bák)]}ə	[(bák)]ə
2.	Syllabification	{[(bák)]}(ə)	[(bák)](ə)
3.	Foot Building	{[(bák)](ə)}	[(bák)(ə)]

Epenthesis is not harmonically improving in either of these cases. The violation of NONFINALITY is not removed until after the epenthetic segment has been syllabified and parsed into a prosodic word. The violation of FTBIN is not removed until the segment has been syllabified and parsed into a foot. In both cases, epenthesis at step 1 is harmonically bounded, since it does not fix the violation of the markedness constraint, but does introduce a gratuitous violation of DEP.

8.3.4 *The special case of *CLASH*

In this section, I will discuss the constraint *CLASH, arguing that its definition should be grid based. Only under a grid-based definition does the constraint behave like *LAPSE, NONFINALITY, and FOOTBINARITY. Under the definition in (11), a violation of *CLASH can be resolved in a single step. This is demonstrated in (32).

(32) *CLASH ≫ DEP

	[(ba)(dúk)][(pít)(ki)]	*CLASH	DEP
a. →	[(ba)(dúk)]ə[(pít)(ki)]		1
b.	[(ba)(dúk)][(pít)(ki)]	1W	L

According to the definition in (11), *CLASH is violated whenever two stresses are adjacent, and in candidate a., epenthesis has applied, causing the stressed syllables to now be non-adjacent. Presumably later in the derivation the epenthesized syllable will get prosodified, but it does not need to get prosodified in order to separate the two stressed syllables, removing the violation of *CLASH.

Kager (1993) defines a stress clash simply as a pair of adjacent stressed syllables, and this is how the OT constraint has typically been defined. Earlier work on stress clash, however, provides a more detailed definition of what counts as 'adjacent' stresses for the purposes of clash assessment. Liberman and Prince (1977) and Prince (1983) define a stress clash in terms of the metrical grid. In a metrical grid, the syllable is the minimal unit of prominence, and each syllable gets one grid mark. Syllables get additional grid marks for additional levels of prominence, such as secondary word stress, primary word stress, phrasal stress, etc. This is illustrated in (33) with the name of the Dundee soccer team (clashes are circled):

(33)

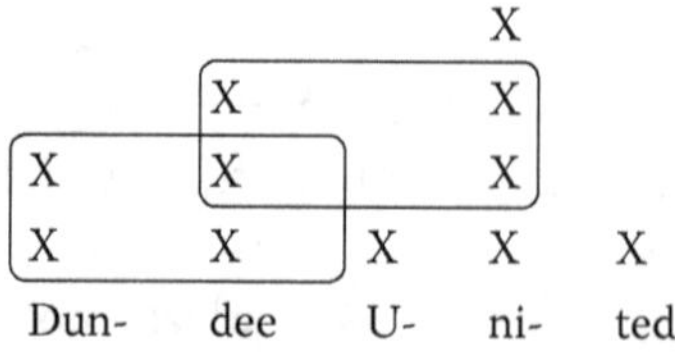

Prince (1983) defines a stress clash in terms of the metrical grid thus: 'if two entries are adjacent, with no intervening entry one level down, they will be said to "clash".' Two entries are adjacent if there is no intervening grid mark at their same level of prominence. Thus, the top two grid marks of the syllable 'dee' are adjacent to the middle two grid marks of 'ni-'. In (33), there are two stress clashes, 'Dun-' with 'dee', and 'dee' with 'ni-'. Prince justifies this definition of adjacency by showing that clashes like the second do trigger stress retraction. In fact, this example is frequently realized as in (34).

(34)

			X	
X			X	
X	X		X	
X	X	X	X	X
Dun-	dee	U-	ni-	ted

In (34), 'dee' and 'ni-' are no longer clashing. Although they are adjacent on the second level up of the grid, they are not adjacent on the next level down – they are separated by 'U-'.

Consider (35), in which material is epenthesized between two stressed syllables. In a., the epenthesized material is parsed into a syllable, and therefore projects a grid mark at the lowest level, which intervenes between the stresses making them no longer adjacent. In b., the epenthesized material is unsyllabified, and therefore doesn't project a grid mark, and thus doesn't resolve the stress clash.

(35)

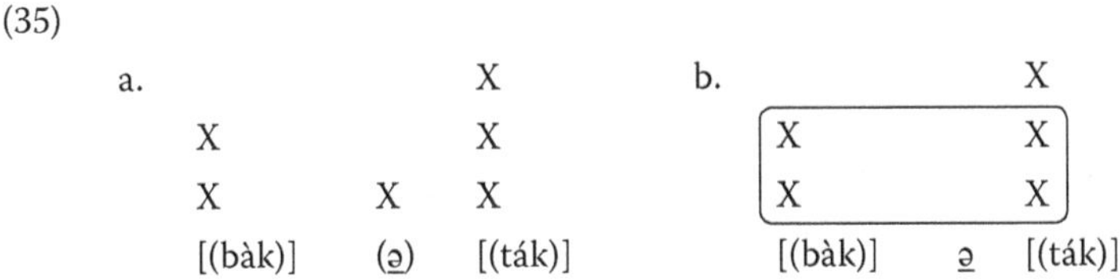

Thus, only epenthesized material that is fully syllabified can remove the stress clash. Clashes that involve higher levels of prominence would require higher levels of prosodic structure to intervene in order to resolve the stress clash. However, under the definition of epenthesis given in (23), and the definition of adjacency given above, epenthesis can never be used to resolve stress clashes, even at a low level of prominence. The derivational stages that would be required for epenthesis to resolve a stress clash are given in (36).

(36) Required derivational stages

	⋮
(foot-building operations)	[(bàk)][(ták)]
1. EPENTHESIS	[(bàk)] ə̲ [(ták)]
2. Syllabification	[(bàk)](ə̲)[(ták)]

By the definition of adjacency above, the stress clash is not resolved in the first step with epenthesis, and since it is not resolved, this step is not harmonically improving with respect to these constraints.

(37) *CLASH ≫ DEP

	[(ba)(dúk)][(pít)(ki)]	*CLASH	DEP
a. →	[(ba)(dúk)][(pít)(ki)]	1	
b.	[(ba)(dúk)] ə̲ [(pít)(ki)]	1	1W

Although the stress clash would be resolved in the second step after syllabification, that step cannot be reached, since the epenthetic candidate is not optimal in step 1. In fact, the derivation converges on [(ba)(dúk)][(pít)(ki)].

8.4 Attested epenthesis

I've argued for a restriction on GEN in HS preventing segments from being epenthesized into prosodic structures higher than the level of the syllable. I show that HS with this restriction is restrictive enough that it never produces segmental epenthesis as a repair for violations of the metrical markedness constraints *CLASH, *LAPSE, NONFINALITY, and FTBIN. In this section I will briefly demonstrate that segmental epenthesis can still be used in the attested cases, namely to repair bad syllable structure and segmental phonotactic markedness. I will argue that attested epenthesis to resolve word-subminimality is not segmental epenthesis, but rather morpheme epenthesis, and as such should be dealt with slightly differently.

8.4.1 For segmental markedness

The success of epenthesis in resolving a segmental markedness problem is not contingent on the existence of any prosodic structure. Because of this, an epenthetic segment can alleviate segmental markedness without being parsed, and so it can always do so in a single step. An example of epenthesis to resolve the adjacent-sibilants problem in the English plural is given in (24), repeated below. The epenthetic segment does not need to be parsed into any prosodic structure to separate the sibilants from each other, and thus satisfies OCP-sibilant in a single step.

(24) OCP-Sibilant, ≫ Dep

	/bɹʌʃ +z/	OCP-Sibilant	Dep
a. →	bɹʌʃəz		1
b.	bɹʌʃz	1W	L
c.	bəɹʌʃz	1W	1

8.4.2 *For syllable structure*

Elfner (2009, this volume) presents an analysis of stress-epenthesis interactions in HS, in which she uses a single operation of epenthesis of a segment plus adjunction of that segment to an existing syllable – that is, epenthesis into a syllable. As long as a segment can be epenthesized and simultaneously parsed into a syllable, epenthesis can be used to resolve syllable-level markedness. To illustrate, consider the case of Kwak'wala, where vowels are epenthesized to avoid syllable-initial clusters (Elfner, 2009, pp 16–19)

The Kwak'wala root /p'ɬa/, 'to blink', must first be syllabified:

(38) Syllabification of /p'ɬa/ : Step 1

	/p'ɬa/	SonNuc	*Complex	ParseSeg	Syll-Head	Dep-V
a. →	p'(ɬa_μ)			1		
b.	(p'$ɬ_\mu$)a	1W		1		
c.	p'ɬ(a_μ)			2W		
d.	(p')ɬa			2W	1W	

The first step is the creation of what Elfner calls a 'core syllable' – a syllable with an onset and a moraic nucleus. Parsing two segments into a core syllable is a single step. The constraint nicknamed SonNuc militates against the creation of core syllables whose nuclei are not sonorous enough, like that in candidate b. Candidate d. represents another type of syllable, the 'minor syllable', whose nucleus is not moraic. Because its nucleus is not moraic, this syllable doesn't violate SonNuc, but it does violate Syll-Head, which militates against mora-less syllables. In the second step, formation of a minor syllable with the initial consonant will be optimal:

(39) Syllabification of /p'ɬa/ : Step 2

p'(ɬa$_\mu$)	SonNuc	*Complex	ParseSeg	Syll-Head	Dep-V
a. → (p')(ɬa$_\mu$)				1	
b. (p'ɬa$_\mu$)		1W			
c. (p'$_\mu$)(ɬa$_\mu$)	1W				
d. p'(ɬa$_\mu$)			1W		

Once a minor syllable is formed, a vowel can be epenthesized to repair the Syll-Head violation.

(40) Syllabification of /p'ɬa/ : Step 3

(p')(ɬa$_\mu$)	SonNuc	*Complex	ParseSeg	Syll-Head	Dep-V
a. → (p'ə̠)(ɬa$_\mu$)					1
b. (p')(ɬa$_\mu$)				1W	L
c. p'(ɬa$_\mu$)			1W		L

The derivation converges here because the structure (p'ə̠)(ɬa$_\mu$) is maximally harmonic under the constraint ranking. Any further changes would only make it worse. In a similar way, epenthesis can be used to resolve complex codas, or to avoid codas altogether, through a series of stages.

(41) Required derivational stages

Input	Core syllabification	(Adjunction)	Minor syllable formation	Epenthesis
/p'ɬa/	p'(ɬa)		(p')(ɬa)	(p'ə̠)(ɬa)
/tapk/	(ta)pk	(tap)k	(tap)(k)	(tap)(kə̠)
/tap/	(ta)p		(ta)(p)	(ta)(pə̠)

Minor syllable formation occurs in each case to avoid a violation of syllable-structure markedness constraint. In the cases with clusters it is *Complex, and in the coda case it is *Coda. After the minor syllable is formed, epenthesis can apply into that syllable, resolving that syllable's markedness (it is headless).

8.4.3 For word-minimality

The last type of epenthesis discussed by Broselow (1982) is epenthesis for word minimality. Above, I claimed that epenthesis does not occur to remove a violation of FootBinarity. I also cited Garrett (1999) and Piggott (1993, 2010) as arguing that word minimality is unrelated to foot-minimality. Piggott argues for a constraint MINWD, whose exact definition is language specific, and which is enforced separately from FTBIN.

(42) MINIMALWORD (MINWD) (Piggott, 2010, p.10)
A Prosodic Word contains more that one syllable (or mora).

With the definition of epenthesis given above, epenthesis cannot be used to resolve a violation of this constraint either. This is because an epenthetic segment would have to be first syllabified, and then parsed into a syllable in order to satisfy the constraint.

(43) Required derivational stages

		⋮
	(prosodic-word building)	{(bak)}
1.	EPENTHESIS	{(bak)}ə̲
2.	Incorporation into the prosodic word	{(bak)ə̲}
3.	Syllabification	{(bak)(ə̲)}

Only at the final of these steps does epenthesis resolve a violation of MINWD. At the first step, epenthesis is not optimal, because the epenthetic segment does not remove any violations, and incurs a gratuitous violation of DEP.

This seems like a wrong prediction, since epenthesis can apparently be used to resolve word-minimality. However, I propose that the epenthesis used to resolve word-minimality is not segmental epenthesis, but rather morpheme epenthesis, and thus is not subject to the same constraints.

Wolf (2008) proposes morpheme-epenthesis as an operation in GEN which violates a faithfulness constraint DEP-M(FS), which militates against insertion of feature-structure that is not present in the input. He argues for this operation from cases like the augmentative [-pa] in Pitjantjatjara. In Pitjantjatjara, whenever a word ends with a consonant, the semantically vacuous [-pa] is added (/mankur/ → [mankurpa], but /mankur + tu/ → [mankurtu]). As an epenthetic sequence, [-pa] is problematic for two reasons. First, it contains a labial, which is otherwise unattested as an epenthetic segment cross-linguistically, and second, it has more material than is necessary to satisfy the requirement that words end in a vowel. Epenthesis of a vowel would be sufficient to solve the problem; why does the [p] show up as well? The answer that Wolf gives is that [-pa] is not an example of segmental epenthesis, but rather of morpheme epenthesis.

Sequences epenthesized to resolve word-minimality are often similarly typologically marked. For example, the epenthetic vowel that is used for word minimality in Mohawk (Broselow, 1982) is completely different from the epenthetic vowel used for other types of marked structures in the language: [e] is used to resolve stop-glottal sequences, while [i] is used to form a minimal word.

(44) Mohawk epenthesis:
o+nʌst+ʔ → onʌsteʔ *corn, nom.*
k+ek+s → iːkeks *I eat*

A similar discrepancy is observed in Swahili (Hinnebusch and Mirza, 1998).

(45) Swahili epenthesis:
ratli → ratili *pound*
la → kula *eat*

In Swahili, what is epenthesized for word minimality is not only different from what is epenthesized for coda avoidance, but also bears all the hallmarks of the Pitjantjatjara [-pa]: [k] is typologically unattested as an epenthetic segment, and it is superfluous. Swahili's disyllabic minimal word requirement would just as easily be filled with a vowel only. Furthermore, [ku-] has the same phonological shape as Swahili's infinitive morpheme. I propose that this epenthetic material *is* the infinitive morpheme being recruited for a purely phonological purpose. Another clear example of this is epenthetic *yi-* in Navaho, which Young and Morgan (1987) have argued is a semantically null morpheme, and which is recruited to repair minimal words.

Cases that look more like segmental epenthesis do exist – examples include Lardil (Piggott, 2010) and Choctaw (Lombardi and McCarthy, 1991). My proposal is that even in these cases, the material epenthesized for word-minimality is really a semantically vacuous morpheme. This makes two predictions. First, it should never, or only very rarely, be the case that what is epenthesized for word-minimality requirements matches what is epenthesized for other types of markedness in the same language. Second, the typology of epenthetic material used to repair word-minimality should look different from the typology of epenthetic vowels in general.[4] Both of these predictions remain to be tested.

4 Thanks to Paul de Lacy for pointing this out to me.

Providing a fully worked-out account for morpheme-epenthesis in HS is a job for future work. However, such an account, if it is to allow morpheme-epenthesis to repair word-subminimality, must allow morphemes to be epenthesized at least partially prosodified. For example, the Swahili epenthetic morpheme –ku- must be already syllabified before it is epenthesized. The ability to insert already-prosodified morphemes will be necessary independently to model prosodically conditioned allomorphy, in which the choice of allomorphs depends on how they will be prosodified into the existing string (for example, Kager, 1996).

If the epenthetic morpheme is inserted already syllabified, then according to the Exhaustivity condition in (28), it can be inserted into an existing foot. This means that after the formation of a subminimal prosodic word, the material inside the prosodic word must be footed first, after which a syllabified morpheme can be epenthesized into that foot, thereby eliminating the word minimality violation. In Swahili, these two derivational stages would go something like this:

(46) Word-minimality epenthesis in Swahili, step 1

	{(la)}	PARSE-SYLL	WORDMIN	DEP-M(FS)
a. →	{[(la)]}		1	
b.	{(la)}	1W	1	

(47) Word-minimality epenthesis in Swahili, step 2

	{(la)}	PARSE-SYLL	WORDMIN	DEP-M(FS)
a. →	{[(ku)(la)]}			1
b.	{[(la)]}		1W	L

8.5 Conclusion

In this paper, I've proposed a hypothesis about the content of Harmonic Serialism's GEN, namely that it contains an operation of epenthesis which is separate from subsequent operations of prosodic structure building, and which obeys absolutely the EXHAUSTIVITY constraint of Selkirk (1981, 1984), and Nespor and Vogel (1986). I've argued that under this hypothesis, epenthesis of a segment can be used to resolve syllable-structure markedness and segmental phonotactic markedness, but will never be used to resolve violations of *CLASH, *LAPSE, *NONFINALITY, and FTBIN. This is because epenthesis before prosodic structure building is harmonically bounded, since it cannot immediately remove the markedness violation, but incurs an additional faithfulness violation.

This paper is a contribution to the general effort to discover what kinds of too-many-repairs problems can be solved in a serial framework like Harmonic Serialism. As I have shown here, restricting GEN restricts the space of possible repairs to any markedness constraint. This paper also contributes to the research program of finding reasonable (typologically sensible) constraints on the space of possible operations in HS's GEN.

References

Blumenfeld, Lev (2006) *Constraints on Phonological Interactions.* Doctoral dissertation, Stanford University.

Booij, Geert (1995) *The Phonology of Dutch.* Oxford: Clarendon Press.

Broselow, Ellen (1982) On predicting the interaction of stress and epenthesis. *Glossa* 16: 115–132.

Côté, Marie-Hélène (2000) *Consonant Cluster Phonotactics: A perceptual approach.* Doctoral dissertation, MIT.

Elfner, Emily (2009) Syllabification and stress-epenthesis interactions in Harmonic Serialism. Ms, University of Massachusetts Amherst. Available on Rutgers Optimality Archive, ROA-1047.

Elfner, Emily (This volume) Stress-Epenthesis Interactions in Harmonic Serialism.

Edward Garrett (1999) Minimal words aren't minimal feet. In Matthew K. Gordon (ed.), *UCLA Working papers in Linguistics* 1, 68–105.

Hall, Nancy (2011) Vowel Epenthesis. In Marc van Oostendorp, Colin J. Ewen, Elizabeth Hume and Keren Rice (eds), *The Blackwell Companion to Phonology*, 1576–1596. Chichester: Wiley.

Hargus, Sharon and Virginia Beavert (2006) Word-initial clusters and minimality in Yakima Sahaptin. *Phonology* 23: 21–58.

Hayes, Bruce (1995) *Metrical Stress Theory: Principles and case studies.* Chicago, IL: The University of Chicago Press.

Hinnebusch, Thomas J. and Sarah Mirza (1998) *Kiswahili, msingi wa kusema kusoma na kuandika: Swahili, a foundation for speaking, reading, and writing.* Lanham, MD: University Press of America.

Jesney, Karen (2011) Positional faithfulness, non-locality, and the Harmonic Serialism solution. In Suzi Lima, Kevin Mullin and Brian Smith (eds), *Proceedings of the 39th Annual Meeting of the North East Linguistic Society*, 429–440. Amherst, MA: GLSA Publications. Available on Rutgers Optimality Archive, ROA-1018.

Kager, René (1993) Alternatives to the iambic-trochaic law. *Natural Language and Linguistic Theory*, 11: 381–432.

Kager, René (1996) On affix allomorphy and syllable counting. In Ursula Kleinhenz (ed.), *Interfaces in Phonology*, 155–171, [Studia Grammatica 41.] Berlin: Akademie Verlag.

Kisseberth, Charles W. (1970) On the functional unity of phonological rules. *Linguistic Inquiry* 1: 291–306.

Liberman, Mark and Alan Prince (1977) On stress and linguistic rhythm. *Linguistic Inquiry* 8: 249–336.

Lombardi, Linda and John McCarthy (1991) Circumscription in Choctaw morphology. *Phonology* 8: 37–72.

Martínez-Gil, Fernando (1997) Word-final epenthesis in Galician. In Fernando Martínez-Gil and Alfonso Morales-Front (eds), *Issues in the Phonology and Morphology of the Major Iberian Languages*, 270–340. Washington, DC: Georgetown University Press.

McCarthy, John J. (2007) Restraint of analysis. In Sylvia Blaho, Patrik Bye and Martin Krämer (eds), *Freedom of Analysis*, 203–231. Berlin and New York: Mouton de Gruyter. Also appears in Eric Bakovic, Junko Ito, and John J. McCarthy (eds) (2006) *Wondering at the Natural Fecundity of Things: Essays in Honor of Alan Prince*. Linguistics Research Center, UC Santa Cruz. Available at http://repositories.cdlib.org/lrc/prince/10.

McCarthy, John J. (2008) The gradual path to cluster simplification. *Phonology*, 25: 271–319.

McCarthy, John J. (2009) Studying GEN. *Journal of the Phonetic Society of Japan*, 13: 3–12.

McCarthy, John J. and Alan Prince (1986/1996) *Prosodic Morphology 1986*. New Brunswick, NJ: Rutgers University Center for Cognitive Science.

McCarthy, John J. and Alan Prince (1993) *Prosodic Morphology I: Constraint Interaction and Satisfaction*. Technical Report #3, Rutgers University Center for Cognitive Science.

Nespor, Marina and Irene Vogel (1986) *Prosodic Phonology*. Dordrecht: Foris.

Pater, Joe (1999) Austronesian Nasal Substitution and Other NC Effects. In René Kager, Harry van der Hulst, and Wim Zonneveld, (eds), *The Prosody Morphology Interface*, 310–343. Cambridge: Cambridge University Press.

Pater, Joe (2003) Balantak metathesis and theories of possible repair in Optimality Theory. Ms, University of Massachusetts Amherst. Available at http://people.umass.edu/pater/pater-balantak.pdf.

Piggott, Glyne (1993) Satisfying the minimal word. In *McGill working papers in linguistics*. Montreal: McGill University.

Piggott, Glyne (2010) Cyclic spell-out and the typology of word minimality. ms. McGill University, Montreal.

Pizzo, Presley (2010) Stress-feature interactions in harmonic serialism. Paper presented at RUMMIT (Rutgers-UMass-MIT phonology meeting).

Prince, Alan (1983) Relating to the grid. *Linguistic Inquiry* 14: 19–100.

Prince, Alan and Paul Smolensky (1993/2004) *Optimality Theory: Constraint Interaction in Generative Grammar*. Malden, MA, and Oxford: Blackwell.

Selkirk, Elisabeth (1995) The prosodic structure of function words. In Jill Beckman, Laura Walsh Dickey, and Suzanne Urbanczyk (eds), *Papers in Optimality Theory*. Amherst MA: GLSA.

Selkirk, Elisabeth, O. (1984) *Phonology and Syntax: The Relation Between Sound and Structure*. Cambridge, MA: MIT Press.

Selkirk, Elisabeth O. (1981) On the nature of phonological representation. In Terry Meyers, John Laver and John Anderson (eds) *The Cognitive Representation of Speech,* Amsterdam: North Holland.

Staubs, Robert (This volume) Serial Restrictions on Feature/Stress Interactions.

Steriade, Donca (2001/2008) The phonology of perceptibility effects: The P-map and its consequences for constraint organization. In Kristin Hanson and Sharon Inkelas (eds), *The Nature of the Word: Studies in Honor of Paul Kiparsky,* 151–180. Cambridge, MA: MIT Press. Originally circulated in 2001.

Wilson, Colin (2001) Consonant cluster neutralization and targeted constraints. *Phonology* 18: 147–197.

Wolf, Matthew (2008) *Optimal Interleaving: Serial Phonology-Morphology Interaction in a Constraint-Based Model.* Doctoral dissertation, University of Massachusetts Amherst.

Chapter 9
Stress-epenthesis interactions in Harmonic Serialism[1]

Emily Elfner

9.1 Introduction

In many languages, epenthetic vowels disturb an otherwise regular stress pattern. In Dakota (Shaw 1976, 1985), for example, stress regularly falls on the second syllable, except when the second syllable contains an epenthetic vowel. In these forms, stress falls instead on the first syllable. This can be seen by comparing the data in (1), which show the basic stress pattern, and (2), which show initial stress for a class of verb and noun stems that predictably end in the epenthetic vowel /a/:[2]

(1) Regular second-syllable stress in Dakota (Shaw 1985: 31)

a.	/č^{h}i-kte/	č^{h}i.kté	'I kill you'
b.	/ma-ya-kte/	ma.yá.kte	'You kill me'
c.	/wiča-ya-kte/	wi.čhá.ya.kte	'You kill them'

1 Many thanks to John McCarthy and Joe Pater, as well as the members of the UMass phonology group and an anonymous reviewer, for feedback on this work. This research has been partially supported by a doctoral fellowship from the Social Sciences and Humanities Research Council of Canada and by grant BCS-0813829 from the National Science Foundation to the University of Massachusetts, Amherst.

2 As Shaw (1985: 32, 116–120) notes, there is evidence for the underlying representations in (2) from phonotactic distribution (the final consonant is limited to voiceless unaspirated stops and affricates, and voiced fricatives), as well as morphological alternations.

(2) Initial-syllable stress (Shaw 1985: 32)
   a. /čap/ čá.pa̲ 'beaver'
   b. /šuk/ šú.ka̲ 'to stagger'
   c. /šič/ ší.ča̲ 'to be bad'

Contrasting data can be found in Swahili, where the placement of stress is not affected by epenthetic vowels, as in loanwords (Alderete 1995, Ashton 1944, Broselow 1982, Polomé 1967). This can be seen by comparing the data in (3), which show the regular penultimate stress pattern, and (4), which show an optional process of [i]-epenthesis following coda consonants in loan words:

(3) Penultimate stress in Swahili
   a. jíko 'kitchen'
      jikóni 'in the kitchen'
   b. nilimpíga 'I hit him'
      nitakupíga 'I should hit him'

(4) Penultimate stress in words with epenthetic vowels
   a. tíket ~ tikéti 'ticket'
   b. rátli ~ ratíli 'pound'

The pattern in languages like Dakota presents a problem for Classic OT (Prince and Smolensky 1993/2004) because it shows a type of opaque interaction that cannot be adequately described in a strictly parallel framework. In Classic OT, faithfulness constraints govern input-output relations and markedness constraints militate against marked output forms. In languages where epenthetic vowels differ from their non-epenthetic counterparts only in their lack of input correspondents, there is no simple markedness constraint that can derive the opaque pattern: in Dakota, stress is expected to fall on the second syllable, whether or not the vowel is epenthetic.

Within Classic OT, Alderete (1995, 1999) argues that such patterns arise from a dispreference for epenthetic vowels to be contained in the heads of prosodic constituents, and he proposes to augment the tools of Classic OT with the positional faithfulness constraint HEAD-DEP. However, while this constraint can be used to capture the basic patterns, as in the contrast between languages like Dakota and Swahili, it can be shown to lead to pathological typological predictions of a type that are also found with other positional faithfulness constraints, as discussed in Jesney (2011b) and in Section 9.2 of this paper, as well as to make incorrect predictions with respect to attested data.

In this paper, I propose that the interactions exhibited by stress and epenthesis are best captured using Harmonic Serialism (HS; among others, McCarthy 2000, 2002: 159–163, 2008a, b: et. seq., Prince and Smolensky 1993/2004: 94–95), a version of OT that combines constraint interaction with serial derivation. HS is like Classic OT, except for a restriction on GEN that requires that candidates differ from their input by no more than a single operation. Each step in the derivation is an OT evaluation, and the derivation continues to pass through GEN and EVAL until the output is identical to the input, at which point it converges. Unlike other serial implementations of OT such as Stratal OT (Bermúdez-Otero to appear, Kiparsky 2003, to appear) and OT-with-Candidate-Chains (OT-CC; McCarthy 2007), the order of operations in HS is controlled solely by the ranking and interaction of basic OT constraints.

In HS, stress assignment and epenthesis are separate operations in GEN whose order of application is controlled by the relative ranking of those markedness constraints responsible for driving epenthesis (syllable structure markedness constraints) and by a constraint punishing stressless words. Because epenthesis and stress are separate operations, they must be applied sequentially rather than concurrently, as they would in Classic OT. As in Classic OT, factorial typology predicts that languages will differ with respect to whether epenthesis or stress is prioritized by the constraint ranking. This prediction appears to hold. In languages like Dakota, stress markedness outranks syllable structure markedness, requiring stress to precede epenthesis. Stress ignores the epenthetic vowel because the vowel is not yet present, resulting in the opaque pattern. In languages with transparent interactions like Swahili, the constraints are ranked in the opposite order, with the result that epenthesis precedes stress. The stress pattern is not disturbed in this case because the epenthetic vowel is present at the point in the derivation where stress is assigned. The HS analysis shares many of the properties of other serial analyses of epenthesis including rule-based ones (Broselow 1982, Farwaneh 1995, Michelson 1983, 1988, 1989, Selkirk 1981), but it derives the serial interactions through constraint ranking rather than rules. Moreover, HS retains many of the advantages of Classic OT by deriving output forms through constraint interaction and by making clear typological predictions based on factorial typology.

This paper is organized as follows. Section 9.2 provides an overview of the difficulties associated with an account of stress-epenthesis interactions in Classic OT using HEAD-DEP. Section 9.3 presents the HS analysis of the basic contrast between languages with opaque stress-epenthesis interactions (Dakota) and transparent interactions (Swahili). Section 9.4 develops analyses of 'mixed' languages in HS, those with both transparent and opaque interactions, including Mohawk (9.4.1) and Levantine Arabic (9.4.2). Section 9.4.3 discusses the typological predictions of the HS analysis, and Section 9.4.4 provides a comparison between the HS account of stress-epenthesis interactions and that of other serial implementations of OT, including Stratal OT and OT-CC. Section 9.5 concludes the paper.

9.2 Parallelism versus Serialism in Stress-epenthesis Interactions

Alderete (1995, 1999) proposes to augment the tools of Classic OT with the constraint HEAD-DEP, a version of the anti-epenthesis faithfulness constraint DEP. HEAD-DEP is a positional faithfulness constraint violated by epenthetic vowels that are the heads of prosodic constituents. Alderete (1995: 8) defines this constraint as below using the formal characterization of correspondence given in McCarthy and Prince (1995, 1999):

(5) HEAD-DEP: Every segment contained in a prosodic head in S_2 has a correspondent in S_1. If β is contained in a prosodic head in S_2, then $\beta \in$ Range(R).

This constraint is violated by every epenthetic vowel that is contained within a prosodic head. This would include, for example, the head syllable of a foot or the head foot of the prosodic word.

When HEAD-DEP is added to the constraint set, Classic OT can account for the contrast between languages with an opaque stress-epenthesis interaction like Dakota and a transparent interaction like Swahili. In Dakota, HEAD-DEP is ranked above FT=IAMB, the constraint that militates against feet that are not right-headed, resulting in a non-iambic stress pattern when the second syllable contains an epenthetic vowel:[3]

3 The tableaux in this paper are in the comparative format (Prince 2002). Violations are represented by integers. *W*s and *L*s appear in the rows with losing candidates and compare the performance of the losing candidate to that of the winner. A *W* indicates that a constraint favors the winner over the loser, while an *L* indicates that a constraint favors the loser over the winner. A well-formed tableau has no *L*s to the left of a *W* in a given row; this indicates that the winner performs better than all competing candidates with respect to the constraint hierarchy.

(6) Opacity in Classic OT (Dakota)

/čap/	HEAD-DEP	FT=IAMB
→ (čá.pa)		1
(ča.pá)	1 W	L

In Swahili, HEAD-DEP is ranked below stress constraints such as ALIGNHDR, which calls for the head foot to be right-aligned. This ranking prevents stress from being assigned to the antepenultimate syllable when the penultimate syllable contains an epenthetic vowel:

(7) Transparency in Classic OT (Swahili)

/ratli/	ALIGNHDR	HEAD-DEP
→ ra(tí.li)		1
(rá.ti)li	1 W	L

It is clear from the above analysis that Classic OT with the constraint HEAD-DEP is capable of dealing with the basic contrast between transparent and opaque systems.

Adding the constraint HEAD-DEP to the constraint set in Classic OT is not without its problems. One problem with HEAD-DEP arises from consideration of factorial typology. HEAD-DEP is a positional faithfulness constraint which militates against epenthesis in prominent prosodic positions. In each of the cases examined by Alderete (1995, 1999) and others, opaque interactions arise because stress constraints are violated to accommodate the constraints driving epenthesis. In OT terms, this occurs because syllable structure markedness constraints dominate the constraints governing stress assignment. This can be seen in a more complete analysis of opaque stress assignment in Dakota, where the tableau is expanded to include the syllable structure markedness constraint NOCODA, which drives epenthesis, and a candidate where epenthesis does not occur. In order for candidate (a) to be optimal, both NOCODA and HEAD-DEP must dominate the stress constraint FT=IAMB:

(8) Opacity in Classic OT (Dakota)

/čap/	HEAD-DEP	NOCODA	FT=IAMB	DEP
→ (čá.pa)			1	1
(ča.pá)	1 W		L	1
(čáp)		1 W	L	L

However, if the ranking between NoCoda and Ft=iamb were to be reversed, the candidate without epenthesis would win:

(9) Predicted language (Dakota'): epenthesis blocked in prominent positions

/čap/	Head-Dep	Ft=iamb	NoCoda	Dep
→ (čáp)			1	
(čá.pa)		1 W	L	1 W
(ča.pá)	1 W		L	1 W

Under this ranking, NoCoda, as the lower-ranked constraint, is violated to avoid epenthesis in a prominent position, showing an alternative strategy to avoid violating Head-Dep. However, because Head-Dep is a positional-faithfulness constraint, epenthesis will only be blocked under this ranking when a vowel would be epenthesized into a prominent position. If a vowel were to be epenthesized into a syllable other than the second, epenthesis would not be blocked:

(10) Epenthesis not blocked in non-prominent positions (hypothetical)

/čapat/	Head-Dep	Ft=iamb	NoCoda	Dep
→ (čá.pa)ta				1
(ča.pát)			1 W	L
(čá.pa)ta		1 W		1
(čá.pat)		1 W	1 W	L

The Head-Dep analysis thus predicts the existence of a language where epenthesis is avoided in positions that would be prominent, but occurs freely in non-prominent positions. To the best of my knowledge, such languages are unattested. This is part of a more general typological problem with positional faithfulness constraints; see Jesney (2011b) for discussion.

Conversely, the Head-Dep account also appears to be overly restrictive in its typological predictions. Alderete (1995: 16) draws attention to a prediction made by the account, namely that extrametrical epenthetic vowels will never affect stress assignment. This prediction appears to be explicitly contradicted in data from Levantine and Iraqi Arabic (Abu-Salim 1982, Broselow 2008, Farwaneh 1995, Kiparsky to appear), where extrametrical epenthetic vowels do result in irregular stress patterns. In these dialects, final syllables are extrametrical, as can be evidenced by a regular pattern of antepenultimate stress when the penult is a light syllable (Broselow 2008: 140):

(11) Iraqi Arabic, Antepenultimate stress: light penult
  a. (ʃá.ri)ka 'company'
  b. (ʕáː.la)mi 'my world'
  c. mu(máθ.θi)la 'actress'

However, when a word underlyingly ends in certain final sequences of two consonants, stress falls on the light penultimate syllable rather than the antepenult, as expected from the pattern in (11):

(12) Iraqi Arabic, Opaque Pattern: Final CC clusters
  /kitab-t/ kitábi̲t 'I wrote' *kítabi̲t

This pattern presents a problem for HEAD-DEP because the epenthetic vowel is in the final syllable, which is extrametrical. The shift of stress rightward cannot be attributed to HEADDEP, because the transparent form, *[(kí.ta)bit], would not include the epenthetic vowel in the head of any prosodic constituent. HEAD-DEP predicts that epenthetic vowels can only affect stress assignment when they would be placed in prosodically prominent positions, and therefore cannot account for the data in (12) (Alderete 1995: 16).[4]

These data, as well as the arguments presented above relating to typological predictions, present serious challenges to the HEAD-DEP account of stress-epenthesis interactions, and suggest that alternatives should be considered. This paper develops an analysis of these and other patterns using Harmonic Serialism, a serial framework whose derivations are driven by the ranking and interaction of violable OT constraints. This framework provides a way of preserving many of the insights of Classic OT without the need to rely exclusively on surface patterns.

9.3 HS and Stress-epenthesis Interactions[5]

As in Classic OT, unfaithful operations in HS are motivated by the satisfaction of markedness constraints that outrank relevant faithfulness constraints. For instance, epenthesis occurs when a constraint on syllable structure markedness such as NOCODA or *COMPLEX outranks DEP.

4 Broselow (2008) proposes an alternative analysis of the Iraqi Arabic pattern that relies on morphological constraints desiring contrast and paradigm uniformity rather than HEAD-DEP. For reasons of space, I do not discuss this alternative in detail here.

5 For an introduction to HS, see McCarthy (this volume).

Similarly, feet are built when a constraint such as PROSODICWORDHEAD (PWDHD, which is violated by words without a head foot) outranks a constraint against the insertion of prominences that are not present in the input (DEPPROMINENCE).

In this paper, I follow the proposal in McCarthy (2007) that the notion of gradualness in HS is tied to faithfulness constraints, such that a single change in HS is equivalent to an unfaithful mapping. This can be defined as follows:

(13) Gradualness Requirement on GEN:
Candidates differ from their input only by the application of one phonological operation, where an operation incurs a violation of a basic faithfulness constraint.

For example, epenthesis and deletion, as well as segmental changes such as assimilation, are all distinct operations in GEN because they violate basic faithfulness constraints (i.e. DEP, MAX, IDENT). Prosodic structure is less clearly defined, and in this, I follow this line of reasoning and most closely the proposals in McCarthy (2008b, 2010) and Pruitt (2010), where stress, but not syllabification, constitutes a separate operation.[6] Syllabification, which is claimed not to be contrastive in any language (Blevins 1995, Clements 1986, Hayes 1989, McCarthy 2003), does not violate any faithfulness constraints, and therefore does not constitute a separate operation in an HS derivation, under the definition in (13). However, other prosodic constituents, like stress as defined by foot assignment, can be contrastive among languages, indicating that the building of foot structure differs from syllabification by being an unfaithful act, violating DEPPROMINENCE. In other words, syllables can be built or modified at any time as segments are deleted or epenthesized. Foot structure requires a separate operation to build and undo that cannot be applied concurrently with any other operation. Most importantly for the current proposal, stress and epenthesis cannot apply in the same step, as both are faithfulness-violating operations.

In languages with both stress and epenthesis, each of the relevant markedness constraints will outrank the corresponding faithfulness constraint. However, in Classic OT, with its parallel evaluation of stress

6 See Elfner (2009), Pater (2012), Moore-Cantwell (this volume) and Torres-Tamarit (this volume) for discussion of treating syllabification as a distinct operation.

and epenthesis, it is never possible to determine the relative ranking of undominated markedness constraints with respect to each other. For example, as discussed above, Swahili is a language with predictable penultimate stress (PWdHd » DepProm) and epenthesis to avoid codas (NoCoda » Dep). This language shows a transparent interaction between stress and epenthesis because penultimate epenthetic vowels may be stressed (e.g. /ratli/ > [ratíli] 'pound').

In Classic OT, both possible rankings of PWdHd and NoCoda will result in the same transparent interaction between stress and epenthesis attested in Swahili. As can be seen in the following tableau, the opaque form (candidate b) is harmonically bounded by the transparent form:

(14) Classic OT: PWdHd and NoCoda are equally satisfied by winning candidate

/ratli/	NoCoda	PWdHd	AlignHdR	DepProm	Dep
a. → ra(tí.li)				1	1
b. (rá.ti)li			1 W	1	1

In HS, however, the markedness constraints responsible for motivating stress and epenthesis conflict with each other, and as such, must be ranked. It is this ranking that will determine the order in which operations are applied. More specifically, if NoCoda outranks PWdHd, epenthesis will precede stress because the ranking indicates a preference for satisfying NoCoda before PWdHd. The opposite ranking would indicate instead a preference for stress to be assigned before epenthesis. Languages thus differ with respect to whether or not the epenthetic vowel is present at the moment of stress assignment. In transparent languages, epenthesis precedes stress and stress can fall on the epenthetic vowel. In opaque languages, stress is applied before epenthesis occurs and cannot fall on the epenthetic vowel because it is not yet present. The basic typology of transparent and opaque interactions in this way follows from one of the fundamental properties of OT: constraint permutation and factorial typology.

The HS analysis of this basic contrast can again be illustrated with Swahili and Dakota. In Swahili, a transparent language, epenthesis precedes stress, as captured by ranking NoCoda over PWdHd. The following two tableaux illustrate the derivation. In Step 1, epenthesis occurs to satisfy higher-ranked NoCoda:

(15) Swahili, Step 1: Epenthesis

	/ratli/	NoCoda	PWdHd	DepProm	Dep	*Operations*
a. →	ra.ti.li		1		1	*epenthesis*
b.	(rát.li)	1 W	L	1 W	L	*stress*
c.	rat.li	1 W	1		L	*no change*

High-ranked NoCoda eliminates all candidates that do not epenthesize a vowel, including candidate (b), which assigns stress to satisfy PWdHd. The gradualness requirement ensures that there is no candidate that can satisfy both NoCoda and PWdHd.

The output of Step 1, *ra.ti.li*,[7] then becomes the input to Step 2. Now that the input satisfies NoCoda, the derivation turns to the next-highest ranked constraint, PWdHd. The optimal candidate, then, is *ra(tí.li)*, which assigns stress transparently to the penultimate syllable and satisfies PWdHd. Candidates with alternative stress patterns, such as *(rá.ti)li*, are eliminated based on the ranking of constraints generally responsible for stress assignment in the language; in this case, for example, AlignHdR outranks AlignHdL.

(16) Swahili, Step 2: Stress

	ra.ti.li	NoCoda	PWdHd	DepProm	Dep	*Operations*
a. →	ra(tí.li)			1		*stress*
b.	ra.ti.li		1 W	L		*no change*

Finally, the derivation will converge in Step 3, because there is no operation that can be applied that will produce a candidate that is more harmonic with respect to the constraint hierarchy.

In contrast, the opaque pattern in languages such as Dakota can be derived by assuming the opposite ranking of syllable structure markedness and PWdHd. In Dakota, PWdHd outranks NoCoda. Because stress will be assigned before epenthesis occurs, stress falls on the initial syllable rather than the second syllable. Step 1 is shown in the following tableau:[8]

7 Here and elsewhere, I use italics to refer to intermediate forms when cited in the text of the paper. Slanted brackets / / are reserved for underlying forms, and square brackets [ ] for surface forms. For example: /pat/ > *pati* > [patʃi]. 'No change' candidates are those which are syllabified, but show no other change.

8 Another possible candidate for this intermediate step is that the final consonant is left unparsed rather than parsed as a coda consonant. For simplicity, I will not consider this candidate here. See Section 9.4 for further discussion of unparsed segments in other languages.

(17) Dakota, Step 1: Stress

	/čap/	PWdHd	NoCoda	DepProm	Dep	*Operations*
a. →	(čáp)		1	1		*stress*
b.	ča.pa	1 W	L	L	1 W	*epenthesis*
c.	čap	1 W	1	L		*no change*

The epenthesis candidate in the above tableau is initially eliminated by high-ranked PWdHd. Because PWdHd is satisfied in Step 1, the epenthesis candidate can now win in Step 2. Once epenthesis occurs, the stress pattern becomes opaque, because the surface form violates the stress constraints that are normally satisfied in the language, including the preference for iambic feet governed by the constraint Ft=iamb. However, provided that NoCoda (the epenthesis-driving constraint) outranks Ft=iamb, the opaque form *(čá.pa)* emerges as optimal, as shown in the following tableau:

(18) Dakota, Step 2: Epenthesis

	(čáp)	PWdHd	NoCoda	Ft=iamb	DepProm	Dep	*Operations*
a. →	(čá.pa)			1		1	*epenthesis*
b.	(čáp)		1 W	L		L	*no change*

In order for the form with opaque stress to surface, the derivation must converge on this form. This means that when the output of Step 2 is fed into the next step, the *no change* candidate must be the most harmonic. In other words, the transparent form *(ča.pá)* cannot be among the candidates considered in Step 3 because it would be more harmonic than opaque *(čá.pa)*, given the constraint hierarchy. This follows from the assumptions regarding Gen in HS: if stress assignment is itself a faithfulness-violating operation, any further changes to stress placement must also be faithfulness-violating operations. Under this hypothesis, stress shift requires the application of two operations, one to remove the foot, and another one to reassign it.[9] In the case of *(čá.pa)*, we would first have to demote the first syllable from head status by removing the foot (> *ča.pa*) and then by parsing an iambic foot, which assigns head status to the second syllable (> *(ča.pá)*). However, this chain of events would require an intermediate step where stress is removed from the word. Because stress assignment

9 Another possibility is that the head status of one of the vowels would be removed in one step, and that the reassignment of head status would occur in the following step. I do not distinguish between these two possibilities here, as they both require the application of two distinct operations.

itself was a harmony-improving operation, its removal cannot be harmonically improving. The intermediate candidate without word stress cannot win without decreasing the harmony of the derivation, and will not be chosen as the optimal candidate. In Dakota, stress will therefore remain stranded on the initial syllable, even though it violates the basic stress constraints in the language.

This is illustrated in the tableau for Step 3 of the above derivation. This step considers, alongside the faithful candidate *(čá.pa)*, the candidate *ča.pa* (which deletes the foot). The candidate *(ča.pá)* (which both deletes a foot and adds a foot), on the other hand, is not considered because it is not produced as a candidate at this step. As can be seen below, *ča.pa* incurs a violation of high-ranked PWDHD, and for this reason, loses to the *no change* candidate. The derivation instead converges on the opaque form *(čá.pa)*, even though it violates FT=IAMB. This is illustrated in the following tableau for Step 3:

(19) Dakota, Step 3: Convergence

(čá.pa)	PWDHD	NOCODA	FT=IAMB	DEP	***Operations***
a. → (čá.pa)			1		no change
b. ča.pa	1 W		L		stress deletion

The above discussion of Swahili and Dakota illustrate how HS can be used to account for opaque and transparent stress-epenthesis interactions. This account assumes only constraint ranking and the HS assumptions regarding the gradualness requirement on GEN. While the analysis shares with rule-based serialism the idea that the ordering of stress and epenthesis operations is responsible for deriving opacity in stress-epenthesis interactions, the HS analysis differs from the rule-based one because it assumes that ordering derives from the ranking and interaction of violable constraints, as in Classic OT.

9.4 Mixed Languages

In addition to this basic contrast, some languages show 'mixed' systems, where epenthetic vowels are stressed in some environments but not others. These languages are important because they show that epenthesis processes are independent from one another, and may interact differently with respect to stress assignment. In this section, I show how the complex array of data in two of these languages, Mohawk (Michelson 1983, 1988, 1989) and Levantine Arabic (Abu-Salim 1982, Farwaneh 1995) can be accounted for using HS.

9.4.1 Mohawk

9.4.1.1 CCC and CR epenthesis

Mohawk (Michelson 1983, 1988, 1989) has several epenthesis processes that interact in different ways with stress. The language is described as having three epenthetic vowels, [e], [i] and [a]. Of these, [e]-epenthesis is the only one that is clearly phonological in nature and arises in three contexts to break up certain consonant clusters. The other epenthetic vowels are less clearly motivated by syllable structure markedness constraints: [i]-epenthesis occurs only to fulfil a disyllabic word minimality requirement, while [a]-epenthesis patterns as a type of 'joiner' morpheme that surfaces to avoid consonant sequences when morphemes are concatenated. Because this paper is concerned with epenthesis as motivated by syllable structure, I will not discuss [i]-epenthesis or [a]-epenthesis, but leave a full analysis of Mohawk to future research. For more details on these other processes, see Michelson (1983, 1988, 1989).

In Mohawk, the epenthetic vowel [e] interacts non-uniformly with respect to stress assignment, in contrast with Swahili and Dakota: [e] can be stressed when it breaks up an underlying three-consonant cluster but not when it breaks up a consonant-sonorant sequence (CR, where C is an oral consonant and R is a sonorant from the set /n, r, w/). This can be seen in the following data: (20) shows the regular pattern of penultimate stress, (21) shows that epenthetic [e] is stressed when it is inserted to break up a CCC cluster and falls in penultimate position, and (22) shows that stress avoids epenthetic [e] when its origin is a CR cluster, and falls instead on the antepenultimate syllable:[10, 11, 12]

10 There is a regular process of stressed vowel lengthening, in which all stressed vowels in open syllables are long. As seen in (22), stressed vowel lengthening also does not occur when the stressed vowel precedes epenthetic [e] in a CR cluster. I will not discuss this process in this paper, though a complete analysis of Mohawk would need to take this process into account.

11 As can be seen in (21)d–e and (22)b–f, the final syllable is extrametrical in words which end in a sequence [eʔ]. This [e] is also epenthetic, and occurs to break up a final underlying consonant-/ʔ/ sequence. This epenthetic vowel also avoids stress, but I will not discuss it at present in order to keep the presentation relatively simple. See Michelson (1983, 1988, 1989) and Section 9.4.1.3 below for discussion and more examples.

12 Note that some lengthened stressed vowels are marked with a grave accent as in [watè̲:̲skute̲ʔ] 'a roast'. This preserves the notation in Michelson (1983, 1988, 1989), which indicates that the vowel is both stressed and has a falling tone resulting from coalescence with a following [ʔ] or [h].

(20) Regular penultimate stress (Michelson 1989: 44)

a.	k-ketskwas	[kkétskwas]	'I raise it'
b.	k-ata?kerahkw-ha?	[kata?keráhkwa?]	'I float'
c.	k-atirut-ha?	[katirútha?]	'I pull it'
d.	k-riht-ha?	[keríhtha?]	'I cook'
e.	wak-ashet-u	[wakashé:tu]	'I have counted it'
f.	k-akʌ?rokew-as	[kakʌ?roké:was]	'I'm dusting'

(21) CCC [e]-epenthesis ('Type II,' Michelson 1989: 42)

a.	wak-nyak-s	[waké̱nyaks]	'I get married'
b.	s-rho-s	[sé̱rhos]	'you coat it with something'
c.	ak-tshe-?	[aké̱tshe?]	'my container, jar'
d.	ʌ-k-the?t-?	[ʌké̱the?te̱?]	'I will pound'
e.	w-at-?skut-?	[watè̱:skute̱?]	'a roast'

(22) CR [e]-epenthesis ('Type I,' Michelson 1989: 40–41)

a.	w-akr-as	[wáke̱ras]	'it smells'
b.	ʌ-k-r-ʌ?	[ʌ́ke̱rʌ?]	'I will put it into a container'
c.	ʌ-k-rahkw-?	[ʌ́ke̱rahkwe̱?]	'I will use it as a container'
d.	wa?-k-riht-?	[wà:ke̱rihte̱?]	'I will cook it'
e.	t-ʌ-k-rik-?	[tʌ́ke̱rike̱?]	'I will put together side by side'
f.	wa?-t-k-atat-nak-?	[wa?tkatáte̱nake̱?]	'I scratched myself'

The rule-based serial analysis in Michelson (1983, 1988, 1989) accounts for this difference by assuming the existence of two epenthesis rules, where CCC epenthesis (Vowel Insertion I) precedes stress assignment and CR epenthesis (Vowel Insertion II) follows stress assignment. In HS, the two types of epenthesis are instead motivated by two different markedness constraints. More specifically, epenthesis occurs in CCC clusters in order to avoid the creation of a complex onset or coda (a violation of *Complex), while epenthesis in CR clusters occurs in order to avoid either a sequence of rising sonority across a syllable boundary (with syllabification C.R) or to avoid a complex onset (with syllabification .CR). Consonant sequences which show rising sonority across a syllable boundary violate the markedness constraint SyllableContact (Davis and Shin 1999, Gouskova 2004, Murray and Vennemann 1983, Rose 2000). I assume the following definition of the constraint; see the cited references for refinements and further discussion:

(23) SyllableContact (SyllCon): assign one violation mark for every sequence of two heterosyllabic consonants A.B, where A is less sonorous than B.

Because each type of epenthesis is motivated by a different markedness constraint, it is possible to rank PWdHd between the two constraints.

Because CCC epenthesis is transparent to stress assignment, *COMPLEX must outrank PWDHD so that the epenthetic vowel will be visible when stress is assigned in Step 2 of the derivation. The proposed derivation for CCC epenthesis is illustrated in the following tableaux:[13]

(24) a. Mohawk CCC cluster, Step 1: Epenthesis

	/wak-nyak-s/	*COMPLEX	PWDHD	*Operations*
a. →	wa.ken.yaks		1	*epenthesis*
b.	(wák.nyaks)	1 W	L	*stress*
c.	(wákn.yaks)	1 W	L	*stress*
d.	wak.nyaks	1 W	1	*no change*
e.	wakn.yaks	1 W	1	*no change*

b. Mohawk CCC cluster, Step 2: Stress

	wa.ken.yaks	*COMPLEX	PWDHD	*Operations*
a. →	wa(kén.yaks)			*stress*
b.	wa.ken.yaks		1 W	*no change*

As in Swahili, the derivation for /wak-nyak-s/ will converge on Step 3 with stress falling transparently on the epenthetic vowel.

The constraint motivating CR epenthesis, SYLLCON, is ranked below PWDHD. This means that stress will be assigned before epenthesis occurs to satisfy PWDHD. Because the epenthetic vowel is not yet present, stress falls on the first syllable of the word, which is then the penultimate syllable and the preferred syllable for main word stress:

(25) Mohawk CR cluster, Step 1: Stress

	/w-akr-as /	*COMP	PWDHD	SYLLCON	*Operations*
a. →	(wák.ras)			1	*stress*
b.	(wá.kras)	1 W		L	*stress*
c.	wa.ke.ras		1 W	L	*epenthesis*
d.	wak.ras		1 W	1	*no change*
e.	wa.kras	1 W	1 W	L	*no change*

Note that candidate (b) in the above tableau is eliminated because it violates high-ranking *COMPLEX, even though it avoids violating both PWDHD and SYLLCON. This is one way in which constraint ranking

13 For simplicity, I will not include the faithfulness constraints DEP and DEPPROM in the tableaux from this point on. In all of the languages that will be considered here, they are low-ranked.

is important in HS beyond merely inducing ordering: here, constraint ranking is also responsible for choosing among the candidates that apply stress, in a way familiar from Classic OT.

As in Dakota, epenthesis occurs in the next step, this time to satisfy SYLLCON. Stress assignment becomes opaque, resulting in a violation of the stress constraint ALIGNHDR, which is ranked below SYLLCON:[14]

(26) Mohawk CR cluster, Step 2: Epenthesis

	(wák.ras)	*COMP	PWDHD	SYLLCON	ALIGNHDR	*Operations*
a. →	(wá.ke)ras				1	*epenthesis*
b.	(wák.ras)			1 W	L	*no change*

Owing to the gradualness requirement on GEN, stress cannot be altered without violating higher-ranked constraints, and the derivation converges on the opaque form in Step 3:

(27) Mohawk CR cluster, Step 3: Convergence

	(wá.ke)ras	*COMP	PWDHD	SYLLCON	ALIGNHDR	*Operations*
a. →	(wá.ke)ras				1	*no change*
b.	wa.ke.ras		1 W		L	*stress deletion*

As in Swahili and Dakota, HS derives the two epenthesis patterns found in Mohawk by ranking the syllable structure constraints independently with respect to PWDHD.

9.4.1.2 Stressed [e] in an open syllable

The HS analysis derives the difference in behavior of epenthetic [e] from the markedness constraints driving epenthesis. In this way, the HS analysis differs fundamentally from the HEAD-DEP analysis argued for in Alderete (1995, 1999), which aims to account for the above patterns using a surface-based explanation. More specifically, Alderete argues that epenthetic [e] is stressed when it occurs in a closed syllable (CCC clusters) and unstressed when it occurs in an open syllable (CR clusters), making it possible to account for the two patterns by ranking HEAD-DEP between two markedness constraints: HEAD-DEP is outranked by WSP (the Weight-to-Stress principle), which is violated by unstressed heavy

14 I have arbitrarily chosen the candidate [(wá.ke)ras] as the winner; however, a parse where the foot is trisyllabic, as in [(wá.ke.ras)], is also possible. I know of no ranking argument in Mohawk that would distinguish between these two candidates. Note that cost-free resyllabification also applies with epenthesis in these forms.

syllables, but outranks ALIGNHDR. The ranking arguments can be seen in the following two tableaux:

(28) a. WSP » HEAD-DEP

	/wak-nyak-s/	WSP	HEAD-DEP
a. →	wa(ké̱n.yaks)		1
b.	(wá.ke̱n)yaks	1 W	L

b. HEAD-DEP » ALIGNHDR

	/w-akr-as/	HEAD-DEP	ALIGNHDR
a. →	(wá.ke̱)ras		1
b.	wa(ké̱.ras)	1 W	L

The following example, however, which shows a stressed epenthetic vowel in an open syllable, suggests that the generalization that stressed [e] is only found in a heavy syllable is not always true in Mohawk (Michelson 1988: 58):

(29) k-rʔok-s [ké̱roʔks] 'I chop with an axe'

In this form, epenthetic [e] is epenthesized to break up an underlying CCC cluster, but ends up in an open syllable in the output because epenthesis interacts with metathesis, which occurs to eliminate sequences of [Cʔ] in the output (Michelson 1988: 56–58).[15] The epenthetic vowel in this case ends up being stressed, even though it does not surface in a closed syllable, and provides a counterexample to the predictions of the HEAD-DEP analysis.

Because the HS analysis relies on the ordering of epenthesis operations rather than surface generalizations, it can account for the stress pattern in words like [ké̱roʔks] without any additional assumptions. In /k-rʔok-s/ 'I chop with an axe', epenthesis occurs to avoid the creation of a complex onset rather than to avoid a syllable contact violation, and as such, will precede stress assignment in order to avoid violating high-ranked *COMPLEX. Metathesis follows epenthesis, rendering stress assignment opaque:[16]

15 Michelson (1988) analyzes the metathesis process as a series of two processes, vowel copy and syncope. For ease of presentation, I will assume that that the operation is simply metathesis, a single operation in HS.

16 The relative ordering of PWDHD and *Cʔ is not possible to determine if we continue with the simplifying assumption that the metathesis operation is a single operation.

(30) a. Mohawk CCC/metathesis, Step 1: Epenthesis

/k-rʔok-s/	*Comp	PWdHd	*Cʔ	SyllCon	*Operations*
a. → ker.ʔoks		1	1		*epenthesis*
b. (krʔóks)	1 W	L	1		*stress*
c. kroʔks	1 W	1	L		*metathesis*
d. krʔoks	1 W	1	1		*no change*

b. Mohawk CCC/metathesis, Step 2: Stress

ker.ʔoks	*Comp	PWdHd	*Cʔ	SyllCon	*Operations*
a. → (kér.ʔoks)			1		*stress*
b. ker.oʔks		1 W	L		*metathesis*
c. ker.ʔoks		1 W	1		*no change*

c. Mohawk CCC/metathesis, Step 3: Metathesis

(kér.ʔoks)	*Comp	PWdHd	*Cʔ	SyllCon	*Operations*
a. → (ké.roʔks)					*metathesis*
b. (kér.ʔoks)			1 W		*no change*

The derivation will converge in Step 4 with stress correctly falling on the epenthetic vowel, even though it occurs in an open syllable. The HS analysis therefore derives the correct pattern for free, while the Head-Dep analysis cannot account for this pattern without additional assumptions regarding the interaction of the metathesis process with epenthesis and stress.

9.4.1.3 Final Cʔ epenthesis

The vowel [e] is also epenthesized into sequences of word final consonant-glottal stop (Cʔ) sequences. Like CR epenthesis, Cʔ epenthesis is opaque to stress assignment, but the way it affects stress is different: instead of pushing back stress by one syllable, as in CR clusters, stress is instead pushed back two syllables:

(31) Cʔ# [e]-epenthesis ('Type III', Michelson 1989: 43)

a.	ʌ-k-arat-ʔ	[ʌká:rate̲ʔ]	'I lay myself down'
b.	ro-kut-ot-ʔ	[rokú:tote̲ʔ]	'he has a bump on his nose'
c.	o-hnyʌʔtskwir-ʔ	[ohnyʌ́:tskwire̲ʔ]	'small branch'

As before, the HS analysis can account for this pattern by ensuring that stress assignment precedes the constraint responsible for Cʔ epenthesis. Unlike the rule-based analysis proposed by Michelson (1989), however,

CR and Cʔ# epenthesis cannot be motivated by the same epenthesis rule: I have argued that SYLLABLECONTACT is responsible for CR epenthesis, but as the final /ʔ/ is not in a different syllable, this constraint cannot motivate epenthesis here.

Instead, I propose to make use of a distinction between word-final and word-medial consonants. This distinction is attested cross-linguistically in languages such as English, where sequences of word-final consonants (restricted to coronals in English) are permitted, but where medial clusters are much more restricted in nature (for English, see Hayes 1982, on other languages that exhibit final-consonant extrametricality, see Hayes 1995). I attribute this distinction to the existence of two versions of the constraint PARSESEGMENT (PARSESEG) that exist in a stringency relationship, as defined below, where PARSESEGMENT-NONFINAL is a specific version of the general PARSESEGMENT constraint:

(32) PARSESEGMENT: assign one violation mark for every segment that is not associated with a syllable.

(33) PARSESEGMENT-NONFINAL (PARSESEG-NONFIN): assign one violation mark for every non-PWd-final segment that is not associated with a syllable.

In Mohawk, PARSESEG-NONFIN outranks PWDHD, and eliminates candidates that would leave medial consonants unparsed, as in CCC clusters like in *(wák<n>yaks)*. The general constraint, PARSESEG, however, is ranked below PWDHD, meaning that final consonants such as /ʔ/ may be left unparsed while stress is assigned (unparsed segments are set aside with <>):

(34) a. Mohawk Cʔ# cluster, Step 1: Stress

	/ʌ-k-arat-ʔ /	*COMP	PWDHD	PRSSEG	ALIGNHDR	***Operations***
a. →	ʌ(ká.rat)<ʔ>			1		*Stress*
b.	ʌ(ká.ratʔ)	1 W		L		*Stress*
c.	ʌ.ka.ra.teʔ		1 W	L		*epenthesis*
d.	ʌ.ka.ratʔ		1 W	L		*no change*

Note how, once again, high-ranking *COMPLEX intervenes to help decide between two candidates that equally satisfy PWDHD.

In Step 2, epenthesis occurs to eliminate the PARSESEG violation, rendering stress assignment opaque:

(35) Mohawk Cʔ# cluster, Step 2: Epenthesis

	ʌ(ká.rat)<ʔ>	*Comp	PWdHd	PrsSeg	AlignHdR	*Operations*
a. →	ʌ(ká.ra)teʔ				1	*epenthesis*
b.	ʌ(ká.rat)<ʔ>			1 W	L	*no change*

Finally, the derivation converges in Step 3 on the form with antepenultimate stress:

(36) Mohawk Cʔ# cluster, Step 3: Convergence

	ʌ(ká.ra)teʔ	*Comp	PWdHd	PrsSeg	AlignHdR	*Operations*
a. →	ʌ(ká.ra)teʔ				1	*no change*
b.	ʌ.ka.ra.teʔ		1 W		L	*stress deletion*

The different behavior of final Cʔ epenthesis and medial CR epenthesis with respect to stress assignment can be accounted for derivationally in a straightforward way.

Once again, the HS analysis can be compared to the Classic OT analysis using Head-Dep. Alderete (1995) analyzes the forms in (31) using a version of the Head-Dep constraint, Head(F)-Dep, that is violated by epenthetic vowels that are contained within the head foot of the word, whether or not they are the stressed vowel. For instance, in [ʌ(ká:.ra)teʔ], the foot is built such that it avoids containing the epenthetic vowel in the final syllable:

(37) Head(F)-Dep » AlignHdR

	/ʌ-k-arat-ʔ/	Head(F)-Dep	AlignHdR
a. →	ʌ(ká:.ra)teʔ		1
b.	ʌka(rá.teʔ)	1 W	L

However, this analysis of final epenthesis makes predictions regarding the behavior of other epenthetic vowels in the language that do not appear to be generally true. First, in order to affect stress assignment in words with final epenthesis, Head(F)-Dep must outrank the relevant stress constraints in Mohawk (e.g. AlignHdR), meaning that it is expected that other epenthetic vowels should also avoid being parsed as part of the head foot. Alderete (1995) proposes that such forms are excluded on the basis of a three-syllable stress window which blocks forms where stress is more than three syllables from the right edge of the word; however, as the following examples show, stress falls outside of this window when the antepenultimate syllable contains an epenthetic vowel that is opaque to stress assignment (as from CR epenthesis):

(38) Cʔ# [e]-epethesis: Pre-antepenultimate stress (Michelson 1989: 40–43)

a.	ʌ-k-rahkw-ʔ	[ʌ́kerahkweʔ]	'I will use it as a container'
b.	waʔ-k-riht-ʔ	[wà:kerihteʔ]	'I will cook it'
c.	t-ʌ-k-rik-ʔ	[tʌ́kerikeʔ]	'I will put together side by side'

Further, as can be seen in the following example, while epenthetic vowels from CR epenthesis avoid being stressed, they do not avoid being part of the head foot. Because there is no three-syllable stress-window in Mohawk, there does not appear to be any constraint that would disfavor stressing the fifth-to-last syllable in the following word, as would be predicted by high-ranking HEAD(F)-DEP:

(39) HEAD(F)-DEP is violated gratuitously

a.	waʔ-t-k-atat-nak-ʔ	[waʔtka(táte)nakeʔ]	'I scratched myself'
b.	*[waʔt(káta)tenakeʔ]	(satisfies HEAD(F)-DEP)	

The HS account does not require the assumption of a three-syllable stress-window, nor does it predict that stress will shift to avoid footing an epenthetic vowel, as in (39). In this form, the HS analysis correctly predicts that stress will fall on the pre-antepenultimate syllable, as can be illustrated by considering the derivation for this word. In Step 1, stress is assigned before epenthesis because there are no violations of *COMPLEX:[17]

(40) Mohawk CR/Cʔ epenthesis, Step 1: Epenthesis

/waʔ-t-k-atat-nak-ʔ/	*COMP	PWDHD	SYLLCON	PRSSEG	***Operations***
a. → waʔt.ka(tát.nak)<ʔ>			1	1	*stress*
b. waʔt.ka.ta.te.nak<ʔ>		1 W	L	1	*epenthesis*
c. waʔt.ka.tat.na.keʔ		1 W	1	L	*epenthesis*
d. waʔt.ka.tat.nak<ʔ>		1 W	1	1	*no change*

When CR and Cʔ epenthesis occur, stress will remain stranded on the syllable *(tá)*, which will eventually end up in pre-antepenultimate position. Based on the data considered so far, there is no ranking argument for SYLLCON and PARSESEG, such that the two epenthesis operations can occur in either order; however, both orders will result in an output consistent with the attested form.

17 This form contains a CCC sequence that is not broken up by epenthesis (/ʔtk/), indicating that *COMPLEX is too broad to capture all of the data. However, for the purposes of this analysis, I will simply assume that these forms do not violate *COMPLEX rather than explore other formulations of this constraint.

As shown in this section, the HS analysis of various stress-epenthesis interactions found in Mohawk captures the contrast between transparency and opacity in epenthesis operations by assuming that epenthesis occurs in response to three markedness constraints, *COMPLEX, SYLLCON, and PARSESEG, which are ranked independently with respect to PWDHD. In the language's constraint hierarchy, *COMPLEX outranks PWDHD, with the result that epenthesis will precede stress assignment when it is necessary to avoid the creation of a complex onset or coda. This was seen to be true both for medial CCC clusters and for the initial underlying CCC cluster in /k-rʔok-s/. In these environments, the epenthetic vowel is present when stress is assigned, and stress falls transparently on the epenthetic vowel when is in the penultimate syllable. However, SYLLCON and PARSESEG are ranked below PWDHD, with the result that epenthesis to avoid CR clusters and final Cʔ clusters follows stress assignment. Epenthetic vowels in these environments are thus not visible when stress is assigned, resulting in opaque stress. This was seen to be true both for medial and initial CR clusters, as well as final Cʔ clusters and words with both CR and Cʔ epenthesis. This analysis provides a clear advantage over the HEAD-DEP analysis because it is able to account for a wider range of data, including those cases which are problematic for the HEAD-DEP account.

9.4.2 *Levantine Arabic*

Another language with a complex array of transparent and opaque stress-epenthesis interactions is Levantine Arabic, a subset of which were discussed in Section 9.2 as they apply to closely-related Iraqi Arabic.[18] The basic pattern of stress assignment is Latin-like, where stress falls on the penultimate syllable if it is heavy or if the word contains only two syllables, but on the antepenultimate syllable if the penult is light (Abu-Salim 1982):

18 Levantine and Iraqi Arabic show the same basic stress pattern and share the opaque stress-epenthesis interaction in final CC clusters (see Section 9.2 and discussion below). They differ in their treatment of open syllables with low vowels, where such vowels are raised to [i] in Iraqi Arabic (e.g. [kítab] 'he writes') but not in Levantine Arabic ([kátab] 'he writes'). They also differ with respect to their treatment of the stress-epenthesis interaction in CCC versus CCCC clusters; this will be discussed later in this section. See Farwaneh (1995) for further details.

(41) Penultimate stress: heavy penult

a. darásna 'we studied'
b. samá:na 'our sky'
c. katáblak 'he wrote to/for you (m.sg)'
d. maká:tib 'offices'
e. maktábna 'our office'
f. katabú:ha 'they wrote it (f.)'

(42) Initial stress: disyllabic word

a. ʔána 'I'
b. kátab 'he wrote'

(43) Antepenultimate stress: light penult

a. kátabu 'they wrote'
b. mádrasa 'school'
c. ʕálamat 'she taught'
d. ʕallámato 'she taught him'

Epenthetic vowels are inserted to avoid violations of *Complex in both CCC and CCCC clusters; however, the epenthetic vowels differ with respect to whether or not they are stressed when in penultimate position. More specifically, a penultimate epenthetic vowel is only stressed if it is inserted to break up a CCCC cluster. If it instead breaks up a CCC cluster, stress falls on the antepenultimate syllable even though the penult is heavy:

(44) Opaque Pattern: Medial CCC clusters

/katab-l-ha/	ka.tá.bi̲l.ha	'he wrote to her'	*ka.ta.bí̲l.ha
cf. /katab-na/	ka.táb.na	'we wrote'	

(45) Transparent Pattern: Medial CCCC clusters

/katab-t-l-ha/ ka.tab.tí̲l.ha 'I wrote to her'

A third pattern, mentioned previously in Section 9.2, shows opaque stress assignment when a vowel is epenthesized to break up a final CC cluster. In this case, stress falls on the penult even though it is light:

(46) Levantine Arabic, Opaque Pattern: Final CC clusters

/katab-t/	katábi̲t	'I wrote'	*kátabi̲t
cf. /katab-u/	kátabu	'they wrote'	

This analysis will first consider the contrast between CCC and CCCC clusters, and then move on to account for final CC clusters.

9.4.2.1 CCC and CCCC clusters

While epenthesis in both CCC and CCCC clusters occurs to avoid violating *Complex, it is possible to distinguish the two types of epenthesis by assuming that the medial consonants are left unparsed in an intermediate step, rather than parsed into complex onsets or codas (for an analysis using Stratal OT, see Kiparsky 2003). For example, in words with CCC clusters, if stress were to precede epenthesis in the first step, this would result in a single unparsed segment because there would be no way to syllabify the middle consonant in the sequence without violating high-ranked *Complex. If ParseSeg is ranked below PWdHd, stress will be assigned before epenthesis, and will fall on what will eventually become the antepenultimate syllable.

(47) Levantine Arabic CCC cluster, Step 1: Stress

	/katab-l-ha/	*Comp	PWdHd	PrsSeg	*Operations*
a. →	ka(táb)<l>ha			1	*stress*
b.	ka.ta.bil.ha		1 W	L	*epenthesis*
c.	ka(tábl)ha	1 W		L	*stress*
d.	ka(táb)lha	1 W		L	*stress*
e.	ka.tab<l>ha		1 W	1	*no change*

When epenthesis occurs to eliminate the violation of ParseSeg, stress becomes opaque because it is now antepenultimate, and violates constraints responsible for stress assignment in the language (for instance, FootBinμ, which is violated by feet which do not dominate exactly two moras):

(48) Levantine Arabic CCC cluster, Step 2: Epenthesis

	ka(táb)<l>ha	*Comp	PWdHd	PrsSeg	FtBinμ	*Operations*
a. →	ka(tá)bil.ha				1	*epenthesis*
b.	ka(táb)<l>ha			1 W	L	*no change*

Finally, the derivation converges on the opaque form because, as in Dakota and Mohawk, foot deletion is not harmonically improving:

(49) Levantine Arabic CCC cluster, Step 3: Convergence

	ka(tá)bil.ha	*Comp	PWdHd	PrsSeg	FtBinμ	*Operations*
a. →	ka(tá)bil.ha				1	*no change*
b.	ka.ta.bil.ha		1 W		L	*stress deletion*

In contrast, the cost of not epenthesizing a vowel in a word with a CCCC cluster is a sequence of two unparsed consonants, rather than one (*ka.tab<l>ha* vs. *ka.tab<tl>ha*). We can conjecture that while epenthesis to avoid a single unparsed segment follows stress assignment, epenthesis to avoid a sequence of two unparsed segments, precedes stress assignment. As proposed in Kiparsky (2003), there is reason to believe that sequences of two unparsed segments are more marked than single unparsed segments, and that this is reflected in the preference for early epenthesis in words with CCCC clusters. This dispreference for sequences of two unparsed consonants may be captured by assuming the existence of a specific version of ParseSeg which is violated by sequences of two or more unparsed segments:

(50) ParseCC: Assign one violation mark for every sequence of two or more unparsed consonants.

This constraint can be thought of as a segment-level counterpart of the Parse-2 constraint proposed in Kager (1999), which punishes sequences of two unparsed syllables.

If ParseCC is ranked above PWdHd, we will correctly derive the transparent interaction found in words with CCCC sequences: epenthesis will precede stress in order to satisfy the higher-ranked markedness constraint that is not violated in CCC clusters:

(51) Levantine Arabic CCCC cluster, Step 1: Epenthesis

	/katab-t-l-ha/	*Comp	ParseCC	PWdHd	PrsSeg	***Operations***
a. →	ka.tab.til.ha			1		*epenthesis*
b.	ka(táb)<tl>ha		1 W	L	2 W	*stress*
c.	ka.tab<tl>ha		1 W	1	2 W	*no change*

When stress is assigned in Step 2, it is assigned transparently to the heavy penultimate syllable containing the epenthetic vowel:

(52) Levantine Arabic CCCC cluster, Step 2: Stress

	ka.tab.til.ha	*Comp	ParseCC	PWdHd	PrsSeg	***Operations***
a. →	ka.tab(tíl)ha					*stress*
b.	ka.tab.til.ha			1 W		*no change*

The derivation will converge in Step 3. Note that ParseCC and ParseSeg are in a stringency relationship with one another: violations of ParseCC are also violations of ParseSeg, while the reverse is not true. The HS analysis makes a typological prediction here, namely that when two constraints are in a stringency relation, regardless of their ranking, there is no language in which the less marked structure (here, forms with a single unparsed segment) will interact transparently with stress assignment while the more marked structure (forms which violate ParseCC) will interact opaquely. The range of opaque interactions that are predicted to occur cross-linguistically are in this way grounded in the relationships between markedness constraints, a characteristic that distinguishes HS from rule-based serial frameworks. This topic is taken up again in the discussion of typology in section 9.4.3.

While descriptively adequate, the ParseCC constraint has the disadvantage of being stipulative in nature because it encodes in its definition the observation that two unparsed segments are more marked than one, something that should ideally arise from the system rather than be encoded in the constraint itself. A different solution is to use weighted rather than ranked constraints, as in Harmonic Grammar (HG, Jesney 2011a, Legendre, Miyata and Smolensky 1990, Pater 2009a, b, Potts *et al.* 2010, Smolensky and Legendre 2006), which can be implemented in a serial framework like HS (Serial HG, Pater 2012).[19] For an introduction to HG, see Pater (this volume).

For example, the first step in a Serial HG analysis of the stress-epenthesis interaction in a CCC cluster can be represented in the following tableau, where I have assumed weight settings for PWdHd and PrsSeg consistent with the present analysis. In this tableau, the stress candidate has a higher overall H score than its competitors because it violates only PrsSeg, the constraint with a relatively lower weight. The epenthesis and *no change* candidates both incur a violation of PWdHd, and have correspondingly lower H scores. The stress candidate is therefore the winner in this step:

(53) Serial HG, Levantine Arabic CCC cluster, Step 1: Stress

	Weights	1.5	1	H	
	/katab-l-ha/	PWdHd	PrsSeg		***Operations***
a. →	ka(táb)<l>ha		−1	−1	*Stress*
b.	ka.ta.bil.ha	−1		−1.5	*epenthesis*
c.	ka.tab<l>ha	−1	−1	−2.5	*no change*

19 It should be noted that any of the above arguments provided for HS may also be analyzed in Serial HG, as basic ranking arguments in OT with strict ranking may also be made using weighted constraints.

One consequence of assuming that constraints are weighted rather than ranked is that it predicts the existence of 'gang effects' where there are asymmetric trade-offs in constraint violations, which occur when violations of two constraints can be avoided by a single violation of a conflicting constraint (Pater 2009b). A gang effect occurs when multiple violations of one or more lower-weighted constraints result in a lower overall H score for a candidate as compared to a single violation of a higher-weighted constraint for a different candidate. In Levantine Arabic, we can use a gang effect to model the distinction between a single ParseSeg violation and two ParseSeg violations. Here, a word with two unparsed consonants will violate ParseSeg twice. Given the right weightings, the sum of the two violations will result in a lower H score than the epenthesis candidate, which incurs a single violation of PWdHd:

(54) Serial HG, Levantine Arabic CCCC cluster, Step 1: Epenthesis

	1.5	1	H	
/katab-t-l-ha/	PWdHd	PrsSeg		***Operations***
a. → ka.tab.til.ha	−1		−1.5	*Epenthesis*
b. ka(táb)<tl>ha		−2	−2	*Stress*
c. ka.tab<tl>ha	−1	−2	−3.5	*no change*

In this way, HG coupled with HS can provide an alternative way of accounting for the increased dispreference for two unparsed segments as opposed to one. The advantage of this analysis is that it derives the difference in markedness between one and two unparsed segments through the tools of the framework, without need for a constraint where that markedness relation is built-in.

Another alternative is to analyse the dispreference for sequences of two unparsed consonants in CCCC clusters using local constraint conjunction (Smolensky 2006). An analysis using constraint conjunction would rank the self-conjoined version of ParseSeg (ParseSeg&ParseSeg) above PWdHd, and the simple version below PWdHd, with the result that two (or more) violations of ParseSeg would be satisfied before stress assignment, and a single violation would be resolved after stress assignment.

Local constraint conjunction, however, requires the assumption that the conjoined constraint applies within a specific domain. In the above example, the two unparsed segments are adjacent, but because they are not yet syllabified, do not both occupy any phonologically-motivated sub-domain smaller than the prosodic word. McCarthy (1999) argues that when constraint conjunction is used to account for opaque processes, locally conjoined constraints specifying a domain which is too large run the risk of eliminating non-local transparent patterns along with the opaque process. The following sketch shows that this is true of a hypothetical analysis of the Levantine Arabic epenthesis patterns using local constraint conjunction.

For instance, if we assume the domain of PARSESEG&PARSESEG to be the prosodic word, we correctly predict the transparent behavior of CCCC clusters. In addition, however, we incorrectly predict the existence of a language where words with two non-adjacent unparsed segments (hypothetical *katab<l>hab<l>ha*) would eliminate the violation of the conjoined constraint by eliminating one of the PARSESEG violations before stress is assigned. As shown in the following HS tableau for Step 1 of the derivation of this word, epenthesis is predicted to precede stress assignment, but the constraint ranking equally prefers either site for epenthesis.

(55) HS with constraint conjunction, hypothetical word with two non-adjacent violations of PARSESEG, Step 1: Stress

	/katablhablha/	(PRSSEG&PRSSEG)$_{PWd}$	PWDHD	PRSSEG	***Operations***
a. →	ka.ta.bil.hab<l>ha		1	1	*epenthesis*
b. →	ka.tab<l>ha.bil.ha		1	1	*epenthesis*
c.	ka.tab<l>(háb)<l>ha	1 W	L	2 W	*Stress*
d.	ka.tab<l>hab.<l>ha	1 W	1	2 W	*no change*

If we assume that ties in HS are resolved by randomly selecting one of the tied candidates,[20] we predict that this language will show transparent stress in one of the vowels but opaque stress in the other. More specifically, if candidate (a) is chosen, stress (motivated by the next-highest-ranked constraint PWDHD) will be assigned in Step 2 to the heavy penultimate syllable as in *ka.ta.bil(háb)<l>ha*, while Step 3 will see epenthesis of the second vowel, *ka.ta.bil(há)bil.ha*, which renders the placement of stress on the antepenultimate syllable opaque. On the other hand, if candidate (b) is chosen, Step 2 will assign stress transparently to the penultimate

20 An alternative hypothesis is that ties are always resolved by constraint ranking, such that low-ranked constraints may emerge to play a role in determining the winner. For example, in the case of epenthesis, the make-up of the consonant cluster triggering epenthesis might be expected to play a role in such cases. However, this attention to segmental detail is atypical of epenthesis processes (e.g. Blumenfeld 2006, Broselow 1982).

syllable, as in *ka.tab<l>ha(bíl)ha*. Epenthesis in Step 3 will not affect the positioning of stress assignment in this case.

Interestingly, neither the HS analysis assuming ParseCC nor the serial HG analysis share this typological prediction. In the former case, ParseCC will not be violated by non-adjacent ParseSeg violations, and we predict that stress assignment will precede epenthesis of both vowels:

(56) HS with ParseCC, hypothetical word with two non-adjacent violations of ParseSeg, Step 1: Stress

	/katablhablha/	PrsCC	PWdHd	PrsSeg	*Operations*
a. →	ka.tab<l>(háb)<l>ha			2	*Stress*
b.	ka.ta.bil.hab<l>ha		1 W	1 L	*Epenthesis*
c.	ka.tab<l>ha.bil.ha		1 W	1 L	*Epenthesis*
d.	ka.tab<l>hab.<l>ha		1 W	2	*no change*

After epenthesis occurs in Steps 2 and 3, stress will be stranded on the antepenultimate syllable. This result holds regardless of which epenthetic vowel is inserted first.

Similarly, the non-local interaction described above is impossible to derive in Serial HG under any possible weighting of the constraints PWdHd and ParseSeg.[21] In the hypothetical word with two non-adjacent unparsed consonants, the choice in the first step will be between epenthesis to eliminate one of the violations of ParseSeg, and stress, which will leave both violations intact. Because harmony is calculated cumulatively, any candidate that shows epenthesis will violate both ParseSeg and PWdHd. Under the weighting assumed here, the cumulative weight of one violation of each of these constraints would result in a lower H-score than two violations of ParseSeg, resulting in stress assignment in the first step:

(57) Serial HG, hypothetical word with two non-adjacent violations of ParseSeg, Step 1: Stress

		1.5	1	H	
	/katablhablha/	PWdHd	PrsSeg		*Operations*
a. →	ka.tab<l>(háb)<l>ha		−2	−1.5	*stress*
b.	ka.ta.bil.hab<l>ha	−1	−1	−2.5	*epenthesis*
c.	ka.tab<l>ha.bil.ha	−1	−1	−2.5	*epenthesis*
d.	ka.tab<l>hab.<l>ha	−1	−2	−3.5	*no change*

21 The HG typology was checked using OT-Help2 (Staubs *et al.* 2010).

This tableau shows that each of the violations of ParseSeg, when non-adjacent, will be satisfied after stress assignment, meaning that either epenthesis order will have the same effect on stress assignment: stress is predicted to be consistently opaque.

While I do not have access to data that would bear on this question, it is interesting to observe that HS with constraint conjunction differs in its predictions from both the HS analysis assuming ParseCC and the Serial HG analysis, and that these are the least restrictive accounts in this particular case. Further empirical work is needed to determine which analysis is preferable. See Pater (2009b) and Potts *et al.* (2010) for further discussion of HG as compared to OT frameworks with respect to traditional constraint ranking and local constraint conjunction.

9.4.2.2 Final CC Clusters

The analysis developed above may be extended to account for the opaque stress assignment in final CC clusters, as is found in both the Levantine and Iraqi dialects of Arabic. In these forms, stress falls on the penultimate syllable even though it is light:

(58) Opaque Pattern: Final CC clusters (Levantine Arabic)

/katab-t/	ka.tá.bi̱t	'I wrote'	*ká.ta.bi̱t
cf. /katab-at/	ká.ta.bat	'she wrote'	

While these patterns posed a direct problem for Head-Dep and Classic OT (see discussion in Section 9.2), these patterns can be accounted for straightforwardly under an HS framework using the analysis developed above.

As in medial consonant clusters, epenthesis occurs to avoid a violation of *Complex; however, syllabification without epenthesis leaves only a single segment unparsed, which violates the constraint ParseSeg. Since this constraint is also ranked below PWdHd, the stress candidate will win in Step 1, as shown below. Stress is assigned to the final syllable, which is heavy and non-final, owing to the extrametricality of the final consonant (McCarthy 2007):

(59) Levantine Arabic CC cluster, Step 1: Stress

	/katab-t/	*Comp	PWdHd	PrsSeg	AlignHdR	***Operations***
a. →	ka(táb)<t>			1		*stress*
b.	ka.ta.bit		1 W	L		*epenthesis*
c.	ka(tábt)	1 W		L		*stress*
d.	ka.tabt	1 W	1 W	L		*no change*
e.	ka.tab<t>		1 W	1		*no change*

As for CCC clusters, epenthesis occurs in Step 2, making stress assignment opaque, and the derivation converges in Step 3:

(60) a. Levantine Arabic CC cluster, Step 2: Epenthesis

	ka(táb)<t>	*Comp	PWdHd	PrsSeg	AlignHdR	***Operations***
a. →	ka(tá)bit				1	*epenthesis*
b.	ka(táb)<t>			1 W	L	*no change*

b. Levantine Arabic CC cluster, Step 3: Convergence

	ka(tá)bit	*Comp	PWdHd	PrsSeg	Stress	Dep	***Operations***
a. →	ka(tá)bit				1		no change
b.	ka.ta.bit		1 W		L		stress deletion

The HS analysis derives the final CC patterns using the same constraint ranking necessary to account for medial clusters.

The analysis can also be extended to account for a difference between Iraqi Arabic and Levantine Arabic. Iraqi Arabic behaves like Levantine Arabic with respect to medial CCCC and final CC clusters, but epenthesis in medial CCC clusters is transparent rather than opaque to stress assignment (Farwaneh 1995):

(61) Opaque stress: final CC clusters:
/kitab-t/ kitábi̱t 'I wrote'
(as in Levantine Arabic)

(62) Transparent stress: medial CCC clusters
/kitab-l-ha/ kitabí̱lha 'he wrote to her'
(*cf.* Levantine Arabic katábilha)

(63) Transparent stress: medial CCCC clusters
/kitab t l ha/ kitabtí̱lha 'I wrote to her'
(as in Levantine Arabic)

This contrast can be accounted for by assuming the distinction between the specific constraint PARSESEG-NONFIN and the general constraint PARSESEG that was assumed above for Mohawk. In Levantine Arabic, both versions of the constraint are ranked below PWDHD, resulting in opaque interactions except when PARSECC is violated. In Iraqi Arabic, however, PARSESEG-NONFIN is ranked above PWDHD, which derives transparent interactions in all medial clusters. PARSESEG, on the other hand, is ranked below PWDHD, as in Levantine Arabic, resulting in an opaque interaction only in word-final clusters.

9.4.3 Typological Predictions

Like Classic OT, HS makes clear typological predictions that can be derived from constraint permutation. In this paper, I have shown that the contrast between epenthetic vowels that are transparent to stress assignment with those that are opaque results from the ranking of the constraints driving epenthesis with respect to the constraint driving stress assignment: when a syllable structure markedness constraint outranks PWDHD, epenthesis precedes stress and the vowel is transparent for the purpose of stress assignment, while when the opposite ranking is true, stress precedes epenthesis and the stress pattern is opaque.

The HS analysis predicts that any constraint driving epenthesis may interact opaquely with stress in some language, and that different syllable structure markedness constraints can be independently ranked above or below PWDHD within a single language, leading to 'mixed' languages like Mohawk and Levantine Arabic. Further, as pointed out in the discussion of Levantine Arabic, markedness constraints that are in a stringency relationship (such as PARSECC and PARSESEG) are predicted to behave differently with respect to factorial typology as compared to markedness constraints that are not in a stringency relationship. For example, the markedness constraints responsible for CCC and CR epenthesis in Mohawk (*COMPLEX and SYLLCON) are not in a stringency relationship with one another. Thus, while Mohawk has the ranking *COMPLEX » PWDHD » SYLLCON, factorial typology predicts the existence of four distinct languages with epenthesis to avoid violations of *COMPLEX and SYLLCON, as follows:

(64) Factorial typology for *COMPLEX, PWDHD, and SYLLCON
(a) Language 1: *COMPLEX, SYLLCON » PWDHD (completely transparent pattern)
(b) Language 2: *COMPLEX » PWDHD » SYLLCON (Mohawk pattern)

(c) Language 3: SyllCon » PWdHd » *Complex (CR transparent, CCC opaque)
(d) Language 4: PWdHd » *Complex, SyllCon (completely opaque pattern)

On the other hand, the permutation of markedness constraints that are in a stringency relationship, such as ParseCC and ParseSeg, predicts only three distinct patterns:

(65) Factorial typology for ParseSeg, ParseCC, and PWdHd
(a) Language 1: ParseCC, ParseSeg » PWdHd (completely transparent pattern)
(b) Language 2: ParseCC » PWdHd » ParseSeg (Levantine Arabic pattern)
(c) Language 3: ParseSeg » PWdHd » ParseCC (completely opaque pattern)
(d) Language 4: PWdHd » ParseSeg, ParseCC (completely opaque pattern)

The typology above that results from the permutation of ParseSeg and ParseCC appears to be at least partially justified based on attested Arabic dialects: Palestinian Arabic optionally stresses epenthetic vowels, exhibiting the transparent pattern (Gouskova and Hall 2009). No dialect shows the completely opaque pattern, but this is justified typologically by the existence of languages like Dakota where stress-epenthesis interactions are consistently opaque. Notably, however, no dialect appears to show a pattern where CCC clusters are transparent but CCCC clusters are opaque, as predicted from the typology in (65).

In this way, HS differs from rule-based serialism because it makes specific predictions regarding factorial typology based on how markedness constraints are related to each other. Rules that make reference only to the environment triggering epenthesis miss this generalization.

Finally, the HS analysis avoids the pathological prediction of Head-Dep discussed in Section 9.2, where it was shown that Head-Dep predicts the existence of languages that avoid epenthesizing vowels in the heads of prosodic constituents, but not in non-heads. The HS analysis obviates the need for a positional faithfulness constraint of this type by deriving the language contrasts derivationally, using only basic markedness and faithfulness constraints. Based on the above discussion, the typological predictions of the HS analysis are reasonable, and appear to be an improvement on the predictions of both rule-based serialism and Head-Dep.

9.4.4 Comparison with other serial OT frameworks

In this paper, I have argued that the array of possible interactions between stress and epenthesis can be captured using Harmonic Serialism, a derivational variant of OT that retains many of the advantages of OT. Like Classic OT, the HS account provides clear typological predictions while capturing a range of data that were problematic for the HEAD-DEP account.

HS can also be compared with other serial variants of OT, including Stratal OT (Bermúdez-Otero to appear, Kiparsky 2003, to appear) and OT-with-Candidate-Chains (McCarthy 2007, Wolf 2008, this volume). Like HS, Stratal OT and OT-CC provide a constraint-based alternative to rule ordering and a serial alternative to Classic OT. Analyses of opaque and mixed systems of stress-epenthesis interactions have been developed in Stratal OT (Kiparsky 2003, to appear) and OT-CC (McCarthy 2007).

Stratal OT is a constraint-based framework that adopts many of the assumptions of Lexical Phonology (among others, Kiparsky 1982, Mohanan 1986). There are assumed to be three distinct levels of representation, stem-level, word-level, and phrase-level (also referred to as stem/lexical/postlexical). A parallel OT evaluation occurs at each of these levels; however, unlike HS, each level can have a different constraint ranking. Thus, a language may rank DEP over NOCODA at the word level, but have the opposite ranking at the phrase level. As proposed by Kiparsky (2003), when stress is a word-level process, epenthesis is predicted to interact transparently with stress if it occurs at the word or stem level, but opaquely if it occurs at the phrase level. In this way, the same principle of ordering stress and epenthesis operations can be used in Stratal OT to achieve many of the same results.

As far as the data discussed in this paper, Stratal OT can also provide an account of the patterns. However, the two frameworks are quite different in their architecture and make different predictions with respect to the types of interactions that are expected to occur. For instance, Stratal OT makes the strong prediction that there are exactly three levels of representation, and that opaque interactions are expected to occur between processes that apply at different levels, and never among processes that apply at the same level. This property of Stratal OT has been criticized as overly restrictive, as there are opaque processes that appear to occur within one of the levels assumed by the theory (McCarthy 2007). More specifically with respect to the data discussed in this paper, McCarthy (2007) shows that Arabic dialects show different patterns with respect to whether or not epenthetic vowels in CCC clusters are opaque: provided that these languages show the same process of CCC epenthesis, there is no reason to believe that the process should be active in different strata in different dialects. HS, on the other hand, does not set any specific limit with respect to the number of possible steps in a derivation, nor that the steps are tied to specific strata. A derivation may have anywhere from one step (a faithful input to output mapping) upwards to the number of possible operations that may be applied to a given input in a harmonically-improving manner.

Another difference between the two theories relates to constraint ranking. In Stratal OT, each level has a distinct ranking, and therefore, its own different grammar. HS, on the other hand, retains from Classic OT the assumption that each language has a single ranking. In HS, this constraint ranking remains constant throughout the derivation. This aspect of Stratal OT has also been criticized because it makes overly-powerful typological predictions. In this respect, with only a single ranking per language, the types of opaque interactions that are predicted by HS are greatly reduced. See McCarthy (2007) for discussion.

HS also differs from closely-related OT-CC (McCarthy 2007, Wolf 2008, this volume). Like HS, OT-CC assumes that derivations are built by a series of single operations that are applied to an input. Unlike HS, OT-CC compares possible derivations for optimality, rather than intermediate steps. One of the main innovations of OT-CC is a family of Prec constraints that impose precedence relationships on unfaithful mappings. Each step in the derivation is limited to a single unfaithful mapping, which is analogous to the use of operations as a check on gradualness as assumed in HS. Prec(A,B) constraints take as their argument any two faithfulness constraints (A and B). The typological predictions of these constraints are kept in check by a requirement on harmonic improvement and a meta-constraint that requires the faithfulness constraint B to outrank a corresponding constraint Prec(A,B). As shown in the above analysis, HS can account for stress-epenthesis interactions using only constraint ranking, without Prec constraints. As well, harmonic improvement is an integral part of HS, as each step in the derivation represents an output form that is locally optimal.

It must be taken into consideration, however, that HS is limited with respect to the types of opaque processes that it can account for. McCarthy (2000) argues that HS fares no better than Classic OT in its ability to account for counter-feeding opacity and most types of counter-bleeding opacity: in HS, as in Classic OT, constraint ranking is durable (i.e. does not change throughout the derivation, as in Stratal OT). This means that non-optimal candidates cannot result in convergence of the derivation because the derivation will continue to proceed until no further harmonic improvement is possible. Because counter-feeding opacity by definition allows non-optimal forms to surface, such derivations cannot be captured using HS. Further, as argued in this paper, stress-epenthesis interactions, a type of counter-bleeding opacity, can indeed be accounted for using HS provided that stress assignment and epenthesis are assumed to be independent operations: forms with opaque stress assignment were able to surface only because stress-shift was assumed to be a two-step operation involving stress deletion and reassignment. This same assumption must be true in order to derive other opaque processes using HS. However, as McCarthy (2000) argues, this is not necessarily the case with other types of counter-bleeding opacity, particularly cases where the processes in question may be reversed in a single step. While this paper shows that HS is indeed capable of accounting for certain types of opaque processes, it does not provide a solution for the larger array of opaque processes. For further discussion of specific and hypothetical examples, see McCarthy (2000).

In all, while more research remains to be done comparing the predictions of HS with Stratal OT and OT-CC, this paper has shown that HS is capable of accounting for the array of stress-epenthesis interactions that are found cross-linguistically, including languages with mixed systems. The analyses presented here require only the assumption of serialism, and does so using a single constraint ranking per language. As noted by Ito and Mester (2003), there is no reason to believe that there needs to be a single solution to opacity within OT, and other types of opaque processes will require solutions beyond the theory proposed in this paper.

9.5 Conclusion

This paper provides an analysis of stress-epenthesis interactions using the framework of Harmonic Serialism, a serial variant of OT that combines constraint interaction and serial derivation. I showed that HS can be used to account for a range of attested interactions between stress and epenthesis, including transparent, opaque, and mixed languages and patterns that cannot be accounted for using the strictly parallel HEAD-DEP account. Because the order of operations is determined by constraint ranking, it becomes possible in these frameworks to account for typological variation using constraint permutation, just as in Classic OT. Languages differ in the ranking of constraints, and this ranking is responsible for determining whether stress and epenthesis interact transparently or opaquely. In addition, this paper shows that the framework differs fundamentally from rule-based serialism by deriving patterns using the interaction of violable constraints.

This paper illustrates that HS can be used to account for attested stress-epenthesis interactions which show a type of counter-bleeding opacity, and does so with only the assumptions of serialism under the framework of HS. HS presents a viable alternative to other accounts of opaque stress-epenthesis interactions within OT, including the HEAD-DEP account and serial alternatives like Stratal OT and OT-CC. HS differs from other approaches to opacity within OT because serialism is an integral part of the theory, rather than a mechanism introduced to account solely for opacity, as HS has been used to account for a range of phonological patterns aside from opacity (for examples and citations, see the other papers in this volume). As shown in this paper, the account of stress-epenthesis interactions proposed here derives naturally from the assumptions of the framework, and its potential to account for other types of patterns that are difficult to analyze in Classic OT deserves further attention.

References

Abu-Salim, Issam M. (1982) *A Reanalysis of Some Aspects of Arabic Phonology: A Metrical Approach*. Doctoral dissertation, University of Illinois, Urbana, IL.

Alderete, John. (1995) Faithfulness to prosodic heads. Unpublished paper, University of Massachusetts, Amherst.

Alderete, John (1999) Head Dependence in Stress-Epenthesis Interaction. In Ben Hermans and Marc van Oostendorp (eds), *The Derivational Residue in Optimality Theory*, 29–50. Amsterdam: Benjamins.

Ashton, E. O. (1944) *Swahili Grammar (including Intonation)*. Burnt Hill, Harlow: Longman Group Ltd.

Bermúdez-Otero, Ricardo (to appear) *Stratal Optimality Theory*. Oxford: Oxford University Press.

Blevins, Juliette (1995) The syllable in phonological theory. In John A. Goldsmith (ed.), *The Handbook of Phonological Theory*, 206–244. Cambridge, MA, and Oxford: Blackwell.

Blumenfeld, Lev (2006) *Constraints on Phonological Interactions*. Doctoral dissertation, Stanford University, Stanford, CA. Available on Rutgers Optimality Archive, ROA-877.

Broselow, Ellen (1982) On predicting the interaction of stress and epenthesis. *Glossa* 16: 115–132.
Broselow, Ellen (2008) Stress-epenthesis interactions. In Bert Vaux and Andrew Nevins (eds), *Rules, Constraints, and Phonological Phenomena*, 121–148. Oxford: Oxford University Press.
Clements, G. N. (1986) Syllabification and epenthesis in the Barra dialect of Gaelic. In Koen Bogers, Harry van der Hulst and Maarten Mous (eds), *The Phonological Representation of Suprasegmentals*, 317–336. Dordrecht: Foris.
Davis, Stuart and Seung-Hoon Shin (1999) The syllable contact constraint in Korean: An Optimality-Theoretic analysis. *Journal of East Asian Linguistics* 8: 285–312.
Elfner, Emily. (2009) Syllabification and stress-epenthesis interactions in Harmonic Serialism. Unpublished paper, University of Massachusetts Amherst. Available on Rutgers Optimality Archive, ROA-1047.
Farwaneh, Samira (1995) *Directionality Effects in Arabic Dialect Syllable Structure.* Doctoral dissertation, University of Utah, Salt Lake City, UT.
Gouskova, Maria (2004) Relational hierarchies in Optimality Theory: The case of syllable contact. *Phonology* 21: 201–250.
Gouskova, Maria and Nancy Hall (2009) Acoustics of unstressable vowels in Lebanese Arabic. In *Phonological Argumentation: Essays on Evidence and Motivation.* London: Equinox Publishing.
Hayes, Bruce (1982) Extrametricality and English stress. *Linguistic Inquiry* 13: 227–276.
Hayes, Bruce (1989) Compensatory lengthening in moraic phonology. *Linguistic Inquiry* 20: 253–306.
Hayes, Bruce (1995) *Metrical Stress Theory: Principles and Case Studies.* Chicago, IL: The University of Chicago Press.
Ito, Junko and Armin Mester (2003) On the sources of opacity in OT: Coda processes in German. In *The Syllable in Optimality Theory*, 271–304. Cambridge: Cambridge University Press.
Jesney, Karen (2011a) *Cumulative Constraint Interaction in Phonological Acquisition and Typology.* Doctoral dissertation, University of Massachusetts Amherst, Amherst, MA.
Jesney, Karen (2011b) Positional faithfulness, non-locality, and the Harmonic Serialism solution. In Suzi Lima, Kevin Mullin and Brian Smith (eds), *Proceedings of the 39th Annual Meeting of the North East Linguistic Society*, 239–440. Amherst, MA: GLSA Publications. Available on Rutgers Optimality Archive, ROA-1018.
Kager, René (1999) Surface opacity of metrical structure in Optimality Theory. In Ben Hermans and Marc van Oostendorp (eds), *Derivational Residue in Phonological Optimality Theory*, 207–245. Amsterdam: John Benjamins.
Kiparsky, Paul (1982) Lexical phonology and morphology. In The Linguistic Society of Korea, *Linguistics in the Morning Calm*, 3–91. Seoul: Hanshin.
Kiparsky, Paul (2003) Syllables and moras in Arabic. In Caroline Féry and Ruben van de Vijver (eds), *The Syllable in Optimality Theory*, 147–182. Cambridge: Cambridge University Press.
Kiparsky, Paul (to appear) *Paradigmatic Effects.* Stanford, CA: CSLI Publications.
Legendre, Géraldine, Yoshiro Miyata and Paul Smolensky (1990) Can connectionism contribute to syntax? Harmonic Grammar, with an application. In M. Ziolkowski,

M. Noske and K. Deaton (eds), *Proceedings of the 26th Regional Meeting of the Chicago Linguistic Society*, 237–252. Chicago, IL: Chicago Linguistic Society.

McCarthy, John J. (1999) Sympathy and phonological opacity. *Phonology* 16: 331–399.

McCarthy, John J. (2000) Harmonic serialism and parallelism. In *Proceedings of the North East Linguistics Society 30*, pp. 501–524. Amherst, MA: GLSA Publications.

McCarthy, John J. (2002) *A Thematic Guide to Optimality Theory*. Cambridge: Cambridge University Press.

McCarthy, John J. (2003) Sympathy, cumulativity, and the Duke-of-York gambit. In Caroline Féry and Ruben van de Vijver (eds.), *The Syllable in Optimality Theory*, 23–76. Cambridge: Cambridge University Press.

McCarthy, John J. (2007) *Hidden Generalizations: Phonological Opacity in Optimality Theory*. London: Equinox Publishing.

McCarthy, John J. (2008a) The gradual path to cluster simplification. *Phonology* 25: 271–319.

McCarthy, John J. (2008b) The serial interaction of stress and syncope. *Natural Language & Linguistic Theory* 26: 499–546.

McCarthy, John J. (2010) Studying Gen. *Journal of the Phonetic Society of Japan* 13: 3–12.

McCarthy, John J. (this volume) The theory and practice of Harmonic Serialism. In John McCarthy and Joe Pater (eds), *Harmonic Grammar and Harmonic Serialism*. London: Equinox Press.

McCarthy, John J. and Alan Prince (1995) Faithfulness and Reduplicative Identity. In Jill Beckman, Laura Walsh Dickey and Suzanne Urbanczyk (eds), *Papers in Optimality Theory*, 249–384. Amherst, MA: GLSA Publications.

McCarthy, John J. and Alan Prince (1999) Faithfulness and identity in Prosodic Morphology. In René Kager, Harry van der Hulst and Wim Zonneveld (eds), *The Prosody-Morphology Interface*, 218–309. Cambridge: Cambridge University Press.

Michelson, Karin (1983) *A comparative study of accent in the Five Nations Iroquoian languages*. Doctoral dissertation, Harvard University.

Michelson, Karin (1988) *A comparative study of Lake Iroquoian accent*. Dordrecht: Kluwer.

Michelson, Karin (1989) Invisibility: Vowels without a timing slot in Mohawk. In Donna Gerdts and Karin Michelson (eds), *Theoretical Perspectives on Native American Languages*, 38–69. Albany: SUNY Press.

Mohanan, K. P. (1986) *The Theory of Lexical Phonology*. Dordrecht: Reidel.

Moore-Cantwell, Claire (this volume) Contexts for epenthesis in Harmonic Serialism. In John McCarthy and Joe Pater (eds), *Harmonic Grammar and Harmonic Serialism*. London: Equinox Press.

Murray, Robert W. and Theo Vennemann (1983) Sound change and syllable structure in Germanic phonology. *Language* 59: 514–528.

Pater, Joe (2009a) Review of Smolensky and Legendre 2006. *Phonology* 26: 217–226.

Pater, Joe (2009b) Weighted Constraints in Generative Linguistics. *Cognitive Science* 33: 999–1035.

Pater, Joe (2012) Serial Harmonic Grammar and Berber syllabification. In Toni Borowsky, Shigeto Kawahara, Takahito Shinya and Mariko Sugahara (eds),

Prosody Matters: Essays in Honor of Lisa Selkirk, 43–72. London: Equinox Publishing.

Pater, Joe (this volume) Universal Grammar with weighted constraints. In John McCarthy and Joe Pater (eds), *Harmonic Grammar and Harmonic Serialism.* London: Equinox Press.

Polomé, Edgar (1967) *Swahili Language Handbook.* Washington, DC: Center for Applied Linguistics.

Potts, Christopher, Joe Pater, Karen Jesney, Rajesh Bhatt and Michael Becker (2010) Harmonic Grammar with linear programming: From linear systems to linguistic typology. *Phonology* 27: 77–117.

Prince, Alan (2002) Arguing Optimality. *Papers in Optimality Theory II*, 269–304. Amherst, MA: GLSA.

Prince, Alan and Paul Smolensky (1993/2004) *Optimality Theory: Constraint Interaction in Generative Grammar.* Piscateway, NJ: Rutgers University Center for Cognitive Science: Rutgers University.

Pruitt, Kathryn (2010) Serialism and locality in constraint-based metrical parsing. *Phonology* 27: 481–526.

Rose, Sharon (2000) Epenthesis positioning and syllable contact in Chaha. *Phonology* 17: 397–425.

Selkirk, Elisabeth (1981) Epenthesis and degenerate syllables in Cairene Arabic. In *Theoretical Issues in the Grammar of the Semitic Languages (MIT Working Papers in Linguistics 3)*, 111–140. Cambridge, MA.: Department of Linguistics and Philosophy, MIT.

Shaw, Patricia (1976) *Theoretical issues in Dakota Phonology and Morphology.* University of Toronto, Toronto.

Shaw, Patricia (1985) Modularisation and substantive constraints in Dakota Lexical Phonology. *Phonology Yearbook* 2: 171–202.

Smolensky, Paul (2006) Optimality in Phonology II: Harmonic Completeness, Local Constraint Conjunction, and Feature Domain Markedness. In Paul Smolensky and Géraldine Legendre (eds) *The Harmonic Mind*, 27–160. Cambridge, MA: MIT Press.

Smolensky, Paul and Géraldine Legendre (2006) *The Harmonic Mind: From Neural Computation to Optimality-Theoretic Grammar.* Cambridge, MA: MIT Press.

Staubs, Robert, Michael Becker, Christopher Potts, Patrick Pratt, John McCarthy and Joe Pater (2010) *OT-Help 2.0.* Software package, University of Massachusetts Amherst.

Torres-Tamarit, Francesc (this volume) Compensatory and opaque vowel lengthening in Harmonic Serialism. In John McCarthy and Joe Pater (eds), *Harmonic Grammar and Harmonic Serialism*. London: Equinox Press.

Wolf, Matthew (2008) *Optimal Interleaving: Serial Phonology-Morphology Interaction in a Constraint-Based Model.* Doctoral dissertation, University of Massachusetts Amherst, Amherst, MA. Available on Rutgers Optimality Archive, ROA-996.

Wolf, Matthew (this volume) Cyclicity and non-cyclicity in Maltese: Local ordering of phonology and morphology in OT-CC In John McCarthy and Joe Pater (eds), *Harmonic Grammar and Harmonic Serialism*. London: Equinox Press.

Chapter 10

Compensatory and opaque vowel lengthening in Harmonic Serialism[1]

Francesc Torres-Tamarit

10.1 Introduction and overview

Classic compensatory lengthening (CL) occurs when a mora-bearing coda consonant is deleted and its mora reassociates with a preceding vowel (Ingria 1980, Steriade 1982, Hock 1986, Wetzels 1986, Hayes 1989, Kavitskaya 2002). In other words, the deleted consonant is replaced by a lengthened vowel. This phenomenon is usually referred to as CL via mora preservation. The derivation in (1) illustrates one case of classic CL in Latin.

(1) /kasnus/ → [kaːnus] 'gray' (Hayes 1989)

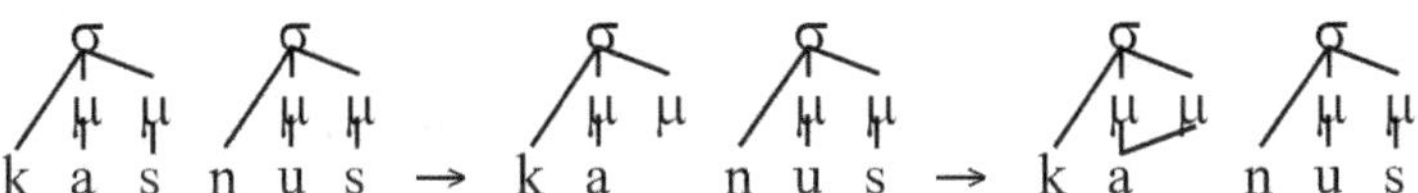

The derivation in (1) proceeds as follows. First coda consonants project their own mora because CVC syllables count as heavy in Latin. Then /s/ deletion applies when followed by a [+sonorant, +anterior] segment. Finally the first vowel of the word links to the mora left behind and surfaces as a lengthened vowel. A strong prediction follows from this

1 This research has been supported by a Rubicon postdoctoral fellowship (446-11-022) from the Netherlands Organization for Scientific Research. I am grateful to Joan Mascaró, John J. McCarthy, Joe Pater and one anonymous reviewer for helpful comments.

standard analysis: classic CL is restricted to weight-sensitive languages only when coda consonants are deleted, but not onsets, which are universally weightless (Hayes 1989).[2]

If classic CL is claimed to be a process involving mora preservation, parallel Optimality Theory (POT) fails in selecting the right output forms because the conditions that make weight by position applicable are non-surface-apparent, since the coda consonant is deleted on the surface. That is, deletion of a mora-bearing coda consonant counterbleeds the application of weight by position.[3]

The main goal of this chapter is to argue on the basis of classic CL and opaque vowel lengthening (VL) in favor of Harmonic Serialism (HS), a version of OT where input-output mappings usually require a serial derivation in which harmony is incrementally achieved. I argue that HS derives classic CL and double flop (see discussion below) only if two independently motivated proposals about the gradual nature of GEN are considered together. First, that syllabification is built serially in harmonically improving steps (Elfner 2009, Jesney 2011, Pater 2012, this volume, Elsman this volume), and second, that deletion of a coda consonant is a two-step process that begins with debuccalization (McCarthy 2008b). These theoretical assumptions are also extended to account for opaque interactions between VL and word-final obstruent devoicing in Friulian, as well as a more complex opaque interaction between VL and syllable-final obstruent devoicing, vowel epenthesis, and subsequent resyllabification in Alsatian French.

The remainder of this chapter is organized as follows. It begins with showing that POT is unable to derive classic CL and double flop, which constitute particular cases of opacity by overapplication (Section 10.2). Then it continues by presenting some theoretical background of what recent literature on HS has said about syllabification and deletion processes (Section 10.3). An explanation of the representational system assumed, as well as its implications for the theory of GEN in HS follows (Section 10.4). The HS analysis of classic CL and double flop is presented

2 See Topintzi (2006) for cases of CL triggered by onset deletion.

3 In Sprouse (1997) the opacity problem is avoided because inputs are enriched with prosodic information. Other alternatives, which do not go beyond the basic assumptions of POT, are found in Hermans (2001) and Topintzi (2006), in which classic CL is argued to be a non-mora preservation process. Campos-Astorkiza (2011) also argues against a mora-preservation approach to CL based on perceptual similarity.

in Section 10.5. This proposal is extended to cases of opaque VL in Friulian and Alsatian French in Section 10.6. Section 10.7 includes a brief excursus in which HS is compared with OT with candidate-chains and Stratal OT. Section 10.8 concludes.

10.2 Classic compensatory lengthening and double flop in POT

Classic CL is an opaque process in which weight by position overapplies because the condition that makes weight by position applicable is wiped out by the process of coda deletion. In other words, there is no coda consonant in the surface representation that justifies the presence of the second mora associated with the vowel. Using a rule-based rhetoric, deletion of a mora-contributing coda consonant counterbleeds weight by position: first coda consonants project a mora, and deletion, which is the bleeding process, follows. A reverse ordering between these two processes would give the wrong result in any rule-based derivational framework.

A CL output form is harmonically bounded in POT. This is so because the violation marks incurred by the transparent candidate, with deletion but no lengthening, are a proper subset of those incurred by the CL candidate, which cannot win under any permutation of the constraint set. The tableau (2) illustrates this fact with the example from (1). Throughout this paper I use the following notational conventions for candidates: parentheses mark syllable boundaries, a subscript <$_\mu$> stands for a mora linked to the preceding segment, a regular <μ> stands for a shared mora, and a superscript <$^\mu$> stands for a floating mora.

(2) Classic CL in POT

	/kasnus/	WBP	*s[+son,+ant]	DEP-μ	MAX-C
a. →	$(\text{ka}_\mu)(\text{nu}_\mu\text{s}_\mu)$			L	1
b. ☹	$(\text{ka}_{\mu\mu})(\text{nu}_\mu\text{s}_\mu)$			1	1
c.	$(\text{ka}_\mu\text{s}_\mu)(\text{nu}_\mu\text{s}_\mu)$		1 W	L	L
d.	$(\text{ka}\mu\text{s})(\text{nu}_\mu\text{s}_\mu)$	1 W	1 W	L	L

Candidates (a) and (b) are equally unmarked. They vacuously satisfy the top-ranked markedness constraint WEIGHT-BY-POSITION, which assigns one violation mark for every coda consonant that does not project a mora, because there is no coda consonant in the surface that can be evaluated. They also satisfy the phonotactic constraint *s[+sonorant, +anterior], which triggers /s/ deletion. However, although both candidates equally violate the anti-deletion faithfulness constraint MAX-C, candidate (b) incurs one violation of the faithfulness constraint against inserting moras that are not positionally licensed (see Bermúdez-Otero 2001), DEP-μ, as opposed to candidate (a).[4] The problem is that there is no top-ranked markedness constraint that compels the violation of DEP-μ in candidate (b), which is less faithful than candidate (a) without apparent motivation. The last two candidates lose because they violate one or both of the top-ranked markedness constraints.

Another classic example of CL extensively discussed in the literature is double flop in Ancient Greek, where CL is triggered by deletion of a segment that is not adjacent to the lengthened vowel (Steriade 1982, Wetzels 1986, Hayes 1989). East Ionic Greek illustrates double flop. As can be seen in (3), a rule of /w/ deletion applied in this dialect. What is interesting in these data is that /w/ deletion only triggers CL if a consonant intervenes between /w/ and the lengthened vowel.

(3) /w/ deletion in East Ionic Greek (Hayes 1989)

*woikos	→	oikos	'house'
*newos	→	neos	'new'
*odwos	→	o:dos	'threshold'

Steriade (1982), Wetzels (1986) and Hayes (1989) showed that this situation could be accounted for by resorting to double flop. A standard derivation of a form like /odwos/ could be as follows. First the word is syllabified as (*od*)(*wos*), where /d/ is assigned a mora because closed syllables count as heavy in Ancient Greek. Then /w/ deletion takes place giving rise to an intermediate representation (*od*)(*os*). This universally ill-formed intermediate syllabification in which a closed syllable precedes an onsetless syllable is fixed by means of resyllabification of /d/, resulting in (*o*)(*dos*) (first flop). Once /d/ has shifted from coda position to onset position, the first vowel links to the mora left behind, giving

4 Bermúdez-Otero's (2001) revised formulation of DEP-μ avoids penalizing weight by position and coda adjunction, which, as opposed to vowel lengthening or gemination, are basic syllabification operations. According to Bermúdez-Otero (2001), 'a non-syllabic segment α is positionally μ-licensed by a mora μ if, and only if, (a) α does not have an input correspondent linked to a mora, and (b) α is immediately dominated only by μ' Bermúdez-Otero (2001): 7. See also Campos-Astorkiza (2004) for a discussion of the proper formulation of DEP-μ.

rise to CL, [(oː)(dos)] (second flop). These facts are represented in the derivation in (4).

(4) 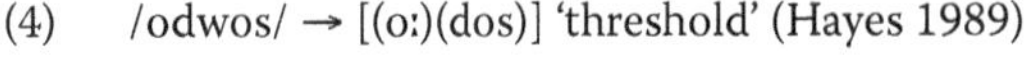

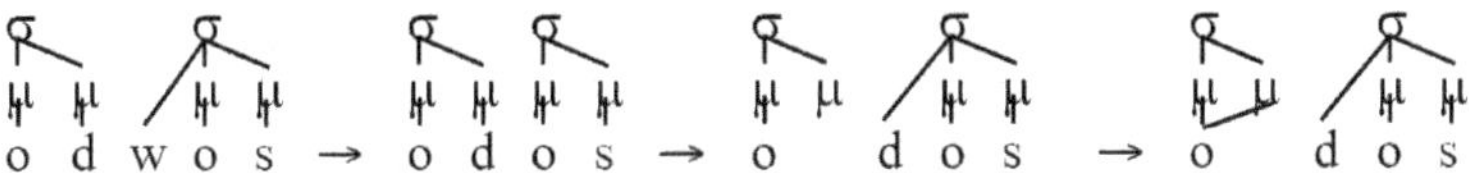

The following tableau demonstrates again that POT cannot select the opaque double flopped candidate. Candidate (a), the transparent one, is selected because it satisfies Dep-μ, and harmonically bounds candidate (b), the actual output form. Candidate (c), on the one hand, loses because it violates Onset twice. Candidate (d), on the other hand, is ruled out because it fatally violates the top-ranked markedness constraint against occurrences of /w/, *w. The constraint Weight-by-Position is omitted in tableau (5).

(5) Double flop in POT

	/odwos/	*w	Onset	Dep-μ	Max-C
a. →	$(o_\mu)(do_\mu s_\mu)$		1	L	1
b. ☹	$(o_{\mu\mu})(do_\mu s_\mu)$		1	1	1
c.	$(o_\mu d_\mu)(o_\mu s_\mu)$		2 W	L	1
d.	$(o_\mu d_\mu)(wo_\mu s_\mu)$	1 W	1	L	L

In this section it has been shown that a POT analysis of classic CL and double flop as mora preservation processes fails in selecting the actual output forms, in which deletion, or resyllabification triggered by deletion, counterbleeds weight by position. Classic CL and double flop are typical cases of non-surface-apparent linguistic generalizations, which cannot be accommodated within a parallel constraint-based model.

The next section presents some theoretical background in HS with respect to syllabification and deletion processes. It will serve as the basis for further discussion.

10.3 Theoretical background

POT is mostly concerned with the study of Con. One of its aims is to establish the set of universal, but violable constraints in order to predict phonological patterns, which are the result of factorial typology.

Given that in POT input-output mappings are accomplished at once, with intermediate representations excluded, the exact nature of the GEN component of Universal Grammar is irrelevant. Actually GEN is a blind brute force that generates an infinite set of candidates that may differ in many ways with respect to the input. Thus the only source of language variation and language typology is constraint ranking.

In HS, on the contrary, GEN plays a paramount role. It is restrained by a gradualness condition on candidate generation by which candidates only introduce one single modification with respect to the input. Defining gradualness, that is, exploring what it means to introduce a single change at a time is one of the main research interests in HS, a theory which has already been proved to be adequate both in the study of language typology and phonological opacity.[5]

This paper aims to contribute to the study of opacity in HS in the light of classic CL and double flop. When facing the problems posed by classic CL and double flop, two main questions arise within the context of HS: how is syllable structure built by GEN; what is the exact nature of deletion operations? The next subsections briefly explore these questions.

10.3.1 Syllabification in HS

Two different conceptions are plausible in HS with respect to syllabification. In one of them, rooted in a faithfulness-based definition of gradualness, syllable-building operations are not subject to the gradualness requirement on GEN because a single modification is defined in terms of a single violation of a basic faithfulness constraint (McCarthy 2007a). Given that syllabification is never contrastive in tautomorphemic

5 Given the same assumptions about CON, the typological predictions of HS have been shown to be more restrictive than those derived from POT (McCarthy 2007b, 2008a, b, 2010, 2011; Pruitt 2010; Jesney 2011; Kimper 2011; Elfner this volume). This restrictiveness is a natural consequence of gradualness, which confers to HS the property of local optimality or the impossibility to 'look ahead' towards global optimality. If a mapping /A/ → [C] requires an intermediate form *B* in HS given gradualness, but there is no plausible universal constraint in CON favoring *B* over /A/, or there is one but the necessary ranking to get *B* from /A/ contradicts known constraint rankings in a given language, then [C] is not a possible output derived from /A/, even though [C] is globally more harmonic than /A/. This is different in POT, where the relative harmony between *B* and /A/ or [C] does not matter and the mapping /A/ → [C] can be obtained just if [C] is more harmonic than /A/. Under these circumstances, when the mapping /A/ → [C] is unattested, the explanatory power of HS is superior to that of POT (McCarthy 2010).

sequences in a given language, faithfulness constraints protecting syllabification must be excluded from CON (but see Elfner (2006) for arguments in favor of contrastive syllabification). It follows from this that syllabification can be accomplished simultaneously with another single operation without violating gradualness (McCarthy 2008a, 2010, Elfner 2009, Pruitt 2010). As opposed to syllabification, stress assignment does count as a single modification because it can be contrastive. Metrical foot building is thus an unfaithful operation that violates a constraint like IDENT(stress) (McCarthy 2008a) or DEP-PROMINENCE (Elfner 2009).

The other approach to syllabification departs from an operation-based definition of gradualness, in which all prosody-building operations, including syllabification, count as autonomous operations (Jesney 2011, Pater 2012, this volume, Elfner this volume, Elsman this volume). I will call this the operation-based definition of gradualness. Under this view, syllabification cannot coexist with other phonological operations within the same candidate set. I will not discuss whether syllables are built one at a time or whole syllabification is reached in a separate single step because both alternatives make no distinction with respect to classic CL or double flop, although I will make use of Elfner's (this volume) serial syllabification operations to be consistent with previous work. McCarthy (2008a) also discusses this issue and concludes in favor of a model of serial syllabification in which syllables are built one at a time. This topic connects with the 'multiple application problem', which was a controversial issue in the context of rule-based generative phonology in the early 1970s: are phonological rules applied at once at different loci if the structural description is met or do they apply iteratively at different derivational stages?

In this chapter I will argue in favor of the operation-based definition of gradualness, according to which syllabification is subject to the gradualness requirement on GEN.

10.3.2 Deletion in HS

In McCarthy (2008b) a serial theory of consonantal cluster reduction in HS is proposed. Processes of deletion and place assimilation are split into two serially ordered single-step processes. In deletion, the first step is debuccalization (i.e., deletion of oral place features), which introduces a MAX(place) violation, followed by deletion of the root node, which correlates with a MAX-C violation. In place assimilation, the first step is also debuccalization, and the second one, insertion of a new association line between the placeless root node and the place feature autosegment associated with the onset consonant, which constitutes a violation of NO-LINK(place). The markedness constraint that triggers debuccalization is CODA-CONDITION, which prohibits place features from being associated with segments that are parsed in coda position. In a NO-LINK(place)-violating representation, CODA-CONDITION is satisfied because the place feature associated with the coda consonant, which is also associated with the onset, becomes licensed. The main argument in favor of splitting deletion and place assimilation into two separate autosegmental processes comes from the coda/onset asymmetry. The coda/onset asymmetry refers to the fact that codas, but not onsets, are generally affected by deletion and place assimilation processes. A HS derivation of the mapping /patka/ → [pa.ka] is thus as follows: /patka/ → *paH.ka* → [pa.ka], where capital *H* represents a debuccalized, placeless stop consonant. This derivation is gradual and harmonically improving when evaluated by a constraint hierarchy in which CODA-CONDITION dominates HAVE-PLACE and MAX(place), and HAVE-PLACE dominates MAX-C. However, a derivation that results in onset deletion like **/patka/ → *pat.Ha* → [pa.ta] (where ** marks a non-harmonically improving derivation), is harmonically bounded at the first step of the derivation, where it gets stuck. Debuccalization of the second consonant in *pat.Ha* is not a harmonically improving step under any permutation of the constraint set because *pat.Ha* incurs a superset of the violations incurred by the fully faithful parse of the input *pat.ka*. Both *pat.Ha* and *pat.ka* violate CODA-CONDITION, because there is a coda consonant associated with its own place features, but *pat.Ha* adds violations of HAVE-PLACE and MAX(place). POT, on the contrary, cannot predict the coda/onset asymmetry because both [pa.ka] and [pa.ta] tie under the same assumptions of CON.

10.4 Representations and GEN operations

10.4.1 Representational assumptions

I follow Hayes' (1989) moraic theory of syllable structure. Vowels project one mora; onsets are exclusively dominated by the syllable node; long vowels are linked to two moras; and coda consonants are linked to their own mora in languages where closed syllables count as heavy, or share the mora associated with the vowel in languages where closed syllables count as light.[6]

(6) Moraic syllabic representations (Hayes 1989)

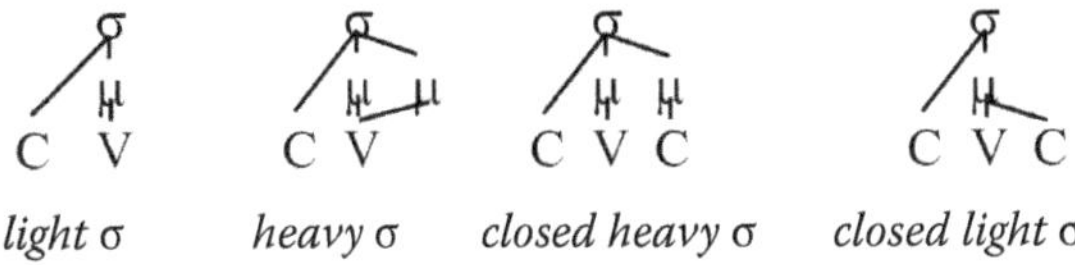

light σ *heavy σ* *closed heavy σ* *closed light σ*

With respect to weight sensitivity, in languages where closed syllables count as heavy, the markedness constraint WEIGHT-BY-POSITION dominates the faithfulness constraint against moras that exclusively dominate consonants, *μ/C (Broselow *et al.* 1997). In languages where closed syllables count as light, the reverse ranking, where *μ/C dominates WEIGHT-BY-POSITION, is observed.

10.4.2 Syllable-building operations

In Elfner (2009) a serial theory of syllabification in HS is proposed, based on a set of three basic syllable formation operations, listed in (7).

(7) Syllable formation operations (Elfner 2009)
- (a) Project syllable: from a segment *x*, create a syllable (x), where *x* can be either moraic, (x_μ), or not, (x).
- (b) Adjunction: given a syllable (x_μ) or (x), adjoin a segment *y*, where *y* can be moraic or not, to the right (coda adjunction) or to the left (onset adjunction).
- (c) Core syllabification: from segments *x* and *y*, create a binary syllable (xy_μ), where *x* is the dependent and *y* the moraic head.

Another operation that will be relevant to this analysis is resyllabification. I assume that this operation is accomplished at once as a separate step (*i.e.*, as a delinking-cum-association single-step process), and is not correlated with any violation of a faithfulness constraint (see McCarthy 2008a for arguments in favor of considering resyllabification an operation that co-occurs with syncope).

With respect to epenthesis, I also follow Elfner (2009) in that vowel epenthesis is an operation triggered by the satisfaction of a markedness constraint against moraless syllables, SYLLABLE-HEAD. First an

6 See Rosenthall and van der Hulst (1999) for a different approach to weightless coda consonants, according to which they adjoin directly to the syllable node.

unsyllabifiable consonant projects its own degenerate syllable, because PARSE-SEGMENT dominates SYLLABLE-HEAD. Then a moraic vowel is inserted in order to fix that minor syllable because SYLLABLE-HEAD dominates DEP-V.[7]

10.5 Classic compensatory lengthening and double flop in HS

Tableaux (8–12) illustrate that a harmonically improving derivation towards classic CL is possible if syllabification is gradual and deletion counts as a two-step process.

At the first step of the derivation, the candidate in which a core syllable has been projected, $(CV_{\mu})C$, is selected as the most harmonic candidate because this is the only operation that maximally satisfies PARSE-SEGMENT.

(8) Step 1: core syllabification

	/CVC/	PARSE-SEG	MAX(place)	HAVE-PLACE
a. →	$(CV_{\mu})C$	1		
b.	$C(V_{\mu})C$	2 W		
c.	CVC	3 W		
d.	CVH	3 W	1 W	1 W

At the second step, candidate (a) in tableau (9) satisfies PARSE-SEGMENT because of the adjunction of the last consonant to the already existing syllable, $(CV_{\mu}C_{\mu})$.[8] At this step, the transparent candidate with complete

7 See also Moore-Cantwell (this volume) for a theory of epenthesis in HS that solves a too-many-solutions problem in POT.

8 One anonymous reviewer points out that this analysis assumes that mora insertion is a free operation that accompanies consonant adjunction and asks whether the analysis predicts moras to be inserted at any point in the derivation, regardless of what other changes are being performed at that point. The answer is no. One specific reason for assuming that mora insertion comes for free only when core syllables are projected or consonants are adjoined to already existing syllables is the fact that these operations are basic syllabification operations that do not violate DEP-μ, according to Bermúdez-Otero's (2001) reformulation of this faithfulness constraint based on positional licensing. The same anonymous reviewer points out the asymmetry I assume between consonant deletion as a two-step process and vowel insertion as a single-step process. I agree the

deletion is not a possible GEN-generated candidate because deletion is a two-step process. Opacity is then the result. The constraint PARSE-SEGMENT can also be satisfied by inserting a mora link between the last consonant and the mora headed by the vowel, as candidate (b) illustrates, but this possibility is ruled out because of the activity of WEIGHT-BY-POSITION. The fully faithful candidate (c), and candidate (d), which shows debuccalization, are ruled out because they fatally violate the top-ranked constraint PARSE-SEGMENT.

(9) Step 2: coda adjunction

	$/(CV_\mu)C/$	PARSE-SEG	WBP	μ/C	CODA-COND	MAX (place)	HAVE-PLACE
a. →	$(CV_\mu C_\mu)$			1	1		
b.	(CVμC)		1 W	L	1		
c.	$(CV_\mu)C$	1 W		L	L		
d.	$(CV_\mu)H$	1 W		L	L	1 W	1 W

It is at the third step of the derivation when debuccalization applies, $(CV_\mu H_\mu)$. Candidate (a) is more harmonic than candidate (b) in tableau (10) because it satisfies CODA-CONDITION by deleting the place features associated with the coda consonant.

(10) Step 3: debuccalization

	$/(CV_\mu C_\mu)/$	CODA-COND	MAX(place)	HAVE-PLACE
a. →	$(CV_\mu H_\mu)$		1	1
b.	$(CV_\mu C_\mu)$	1 W	L	L

At step 4, deletion of the root node takes place because HAVE-PLACE dominates MAX-C and a markedness constraint against moras that are not linked to any segment, *FLOATINGμ. (Recall that superscript <μ> represents a floating mora.)

insertion of a segment could also be the result of two operations: first insertion of a root node and then insertion of subsegmental features. This paper, however, assumes vowel insertion to be accomplished at a single step because this issue is tangential to the cases of opacity addressed here. Moreover, although insertion of a vowel requires one violation of DEP-V, it does not constitute a violation of DEP-μ because the inserted mora is positionally-licensed, i.e., headed by the syllable nucleus.

(11) Step 4: deletion

	$/(CV_{\mu}H_{\mu})/$	Have-Place	Max-C	*Floatingμ
a. →	$(CV_{\mu}{}^{\mu})$		1	1
b.	$(CV_{\mu}H_{\mu})$	1 W	L	L

At the fifth step of the derivation, two operations are at play. Deletion of the floating mora is excluded because of the higher ranked position of Max-μ. Inserting a mora link is the most harmonic operation because *Floatingμ is satisfied, and only low-ranked Dep-Link-μ is violated. This is why candidate (a) in (12) emerges as the optimal candidate. Convergence is reached at the next step of the derivation, omitted here, where Max-μ and *Floatingμ also dominate the markedness constraint against long vowels, *V:.

(12) Step 5: lengthening

	$/(CV_{\mu}{}^{\mu})/$	Max-μ	*Floatingμ	Dep-Link-μ	*V:
a. →	$(CV_{\mu\mu})$			1	1
b.	$(CV_{\mu}{}^{\mu})$		1 W	L	L
c.	(CV_{μ})	1 W		L	L

The ranking arguments that derive classic CL in HS if syllabification is gradual and deletion counts as a two-step process appear as a Hasse diagram in (13).

(13) Hasse diagram for classic CL

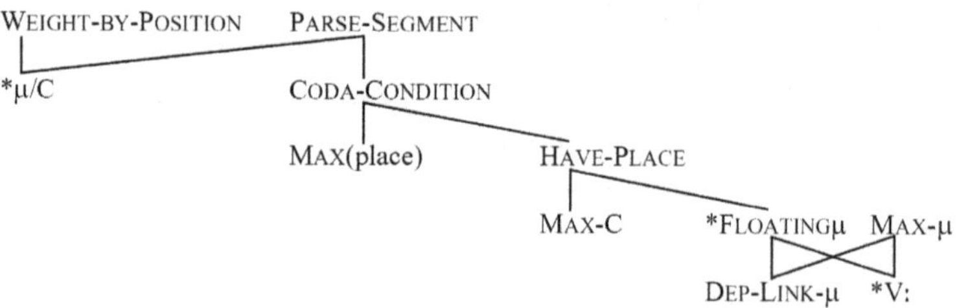

With respect to double flop, a HS derivation for the East Ionic Greek form /odwos/ would be as follows: /odwos/ → ... → $(o_{\mu}d_{\mu})(wo_{\mu}s_{\mu})$ → $(o_{\mu}d_{\mu})(o_{\mu}s_{\mu})$ → $(o_{\mu}{}^{\mu})(do_{\mu}s_{\mu})$ → $(o_{\mu\mu})(do_{\mu}s_{\mu})$. Here I treat deletion as a one-step process for the sake of clarity. Notice that in double flop serial syllabification is enough to capture the CL facts. Once Parse-Segment is satisfied, the word is syllabified as $(o_{\mu}d_{\mu})(wo_{\mu}s_{\mu})$, given that a rising intersyllabic contact is preferable to a complex onset. This syllabification pattern could be derived from a ranking in which *COMPLEX-ONSET outranks CONTACT. At that stage of the derivation the coda consonant /d/ has projected its own mora because of the top-ranked constraint WEIGHT-BY-POSITION. Later on /w/ deletion applies, yielding $(o_\mu d_\mu)(o_\mu s_\mu)$, because the markedness constraint *w outranks MAX-C. When /w/ deletion has applied, ONSET can be minimally satisfied if the coda consonant /d/ is resyllabified, $(o_\mu{}^\mu)(do_\mu s_\mu)$. This operation leaves a floating mora that will be reassociated with the first vowel, $(o_{\mu\mu})(do_\mu s_\mu)$.

If syllabification coexisted with the candidate showing deletion and deletion was considered a single step process, the wrong result would be predicted. In this hypothetical scenario, the first generated set of candidates would contain the transparent candidate $(o_\mu)(do_\mu s_\mu)$, with /w/ deletion and unmarked syllabification. This candidate vacuously satisfies the top-ranked constraint WEIGHT-BY-POSITION and minimally violates ONSET and DEP-μ. For this reason it harmonically bounds the desired candidate in which /d/ is syllabified in coda position and has projected a mora, $(o_\mu d_\mu)(o_\mu s_\mu)$, as can be seen in (14).

(14) Step 1: syllabification not subject to gradualness

	/odwos/	WBP	ONSET
a. →	$(o_\mu)(do_\mu s_\mu)$		L 1
b. ☹	$(o_\mu d_\mu)(o_\mu s_\mu)$		2

Following the rationale of McCarthy (2007b), I have demonstrated that this particular case of counterbleeding opacity can be accounted for in HS if GEN operations are able to be decomposable into more primitive operations. To deal with CL and double flop, satisfying at once both of the top-ranked markedness constraints involved in the opaque interaction, WEIGHT-BY-POSITION and CODA-CONDITION, must be blocked. The theory of gradual syllabification and deletion as a two-step process confer on HS a myopic nature. In this analysis, HS cannot 'look ahead' to the global minimum of harmonic improvement, which would correspond to the transparent candidate. By allowing the satisfaction of WEIGHT-BY-POSITION to happen first, opacity is evaded. If deletion must be preceded by debuccalization, then complete deletion is not a possible generated candidate at the derivational step where only WEIGHT-BY-POSITION can be satisfied. The strategy of simulating ordering relations with constraint ranking by looking closely at the nature of GEN is the strategy to follow in order to accommodate counterbleeding opacity in HS.[9]

In the next section, I will also argue that opaque VL in Friulian and Alsatian French can only be accommodated in HS if syllabification is gradual, meaning that it is able to precede final obstruent devoicing.

10.6 Opaque VL in Friulian and Alsatian French

An HS analysis of opaque VL in two romance varieties, Friulian[10] and Alsatian French[11], is presented in this section. As opposed to classic CL, in these two languages VL is not triggered by deletion of a mora-bearing coda consonant and subsequent association of the vowel with the mora left behind, but by a process of VL that applies before underlying voiced obstruents in coda position before undergoing final devoicing.

10.6.1 Friulian

Friulian presents a fully productive synchronic alternation between long and short vowels in some morphologically related words (Hualde 1990).[12] As can be seen in (15), when a stressed vowel precedes an underlying voiced obstruent that surfaces in word-final coda position as a voiceless obstruent due to word-final obstruent devoicing, the vowel is lengthened. However, in the presence of a vowel-initial inflectional or derivational suffix, the vowel remains short and the obstruent faithfully surfaces as voiced. When post-vocalic obstruents are underlyingly voiceless, as can be seen in (16), vowels do not undergo lengthening.

(15) Vowel length alternations in Friulian (Hualde 1990)

[ˈloːf]	'wolf.masc'	[ˈlo.ve]	'wolf.fem'
[ˈfreːt]	'cold.masc'	[ˈfre.de]	'cold.fem'
[saˈvuːt]	'known.masc'	[sa.ˈvu.de]	'known.fem'

9 Counterfeeding opacity, on the contrary, seems harder to account for in HS (McCarthy 2000).

10 Friulian belongs to the Rhaeto-Romance subfamily of Romance languages and is spoken in the Friuli region of northeastern Italy.

11 Alsatian French must not be confused with Alsatian, a Low Alemannic German dialect spoken also in Alsace, a region of northeastern France.

12 Long vowels in Friulian are not only the result of a predictable VL process. Long vowels are also phonemic. See Prieto (1992) for a diachronic study of the origin of underlying long vowels in Friulian.

[fiˈniːt]	'finished.masc'	[fi.ˈni.de]	'finished.fem'
[ˈnaːf]	'ship'	[na.vi.ˈgaː][13]	'to navigate'
[faˈmoːs]	'famous.masc'	[fa.ˈmo.ze]	'famous.fem'
[ˈlaːk]	'lake'	[la.ˈgu.ne]	'lagoon'

(16) Non-alternating forms in Friulian (Hualde 1990)

[ˈskrit]	'written.masc'	[ˈskri.te]	'written.fem'
[ˈfat]	'made.masc'	[ˈfa.te]	'made.fem'
[ˈmat]	'crazy.masc'	[ˈma.te]	'crazy.fem'
[ˈpak]	'package'	[pa.ˈkut]	'package.dim'
[ˈklap]	'stone'	[kla.pa.ˈdaː]	'to stone'

I interpret VL in Friulian as a sonority-related mora-sharing process (see Zec 2003, Montreuil 2010 for the same process in Lorraine French, Iosad 2012 for a POT analysis of Friulian based on a substance-free phonology approach). In (17) the output moraic representations of the pair [ˈloːf] ~ [ˈlo.ve] are illustrated. VL is represented as a mora-sharing configuration, where the vowel is doubly linked to its own mora and to the extra mora associated with the coda consonant. When the root-final voiced consonant is syllabified in onset position due to the presence of a vowel-initial inflectional or derivational suffix, the vowel is not lengthened because onsets do not contribute to syllabic weight.

(17) Surface representation of the pair [ˈloːf] ~ [ˈlo.ve]

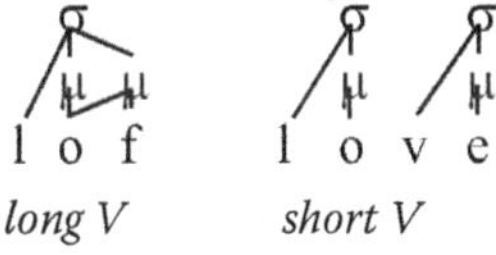

I propose that VL is triggered by the satisfaction of WEIGHT-BY-POSITION and the markedness constraint *μHEAD/OBSTRUENT$_{[+voice]}$, a more specific version of *μ/C, which militates against voiced obstruents that are mora heads. The representation in (18) violates *μHEAD/OBSTRUENT$_{[+voice]}$, although it satisfies WEIGHT-BY-POSITION.

13 The source of world-final long vowels in infinitives is the diachronic loss of the infinitive morph /r/, which triggered classic CL. This specific case of classic CL in infinitive forms is synchronically opaque (*i.e.*, it became lexicalized) because /r/ does not surface under any circumstance in Modern Friulian (Paolo Roseano p.c.).

(18) *μHEAD/OBSTRUENT$_{[+voice]}$-violating representation

l o v

In (19) I show the derivation of the mapping /lov/ → [lo:f], which is discussed below.

(19) Derivation of the mapping /lov/ → [lo:f]

In the first representation in (19), WEIGHT-BY-POSITION is satisfied, resulting in a *μHEAD/OBSTRUENT$_{[+voice]}$-violating configuration, meaning that the former constraint dominates the latter. Then *μHEAD/OBSTRUENT$_{[+voice]}$ is satisfied by means of adding a link between the vowel and the mora already associated with the voiced obstruent, meaning that *μHEAD/OBSTRUENT$_{[+voice]}$ dominates DEP-LINK-μ. Finally the voiced consonant undergoes word-final obstruent devoicing. The last representation in (19) does not violate the constraint *μHEAD/OBSTRUENT$_{[-voice]}$, the counterpart constraint of *μHEAD/OBSTRUENT$_{[+voice]}$, because the mora linked to the consonant is licensed by being also linked to the vowel. The second representation in (19) does not violate *μHEAD/OBSTRUENT$_{[+voice]}$ for the same reason.

Following Hualde (1990), I assume that voiceless obstruents are not weight contributing. I therefore interpret the lack of VL in the left column in (16) as the result of the constraint ranking *μHEAD/OBSTRUENT$_{[-voice]}$ » WEIGHT-BY-POSITION. A weightless voiceless coda is thus represented as in (20). Lengthening is then blocked in inputs like /fat/ because there is only one mora to start with.

(20) Weightless coda

The most interesting aspect of the data presented so far is that VL, as triggered by the presence of an underlying voiced coda consonant, becomes a non-surface-apparent generalization due to word-final obstruent devoicing. The data in the left column in (15) exemplify cases of overapplication of VL because at the surface there is no voiced coda consonant, which is the triggering environment for VL, due to the interaction of word-final obstruent devoicing. Word-final obstruent devoicing thus counterbleeds VL. I will argue below that these facts can only be accounted for in HS if gradual syllabification is assumed.

In order to derive VL before underlying voiced coda consonants in POT, the constraint Weight-by-Position should dominate Dep-Link-μ, as can be seen in tableau (21). (Recall that a regular <μ> represents a doubly linked mora.)

(21) VL in POT

/lov/	WbP	Dep-Link-μ
a. → (lo$_{\mu}$μf)		1
b. (loμf)	1 W	L

However, as can be seen in tableau (22), a reverse ordering between Weight-by-Position and Dep-Link-μ would be needed in order to account for the absence of VL when the coda consonant is voiceless in the underlying representation. This ranking paradox is a strong argument against a POT analysis of opaque VL in Friulian.

(22) VL in POT (ranking paradox)

/fat/	WbP	Dep-Link-μ
a. → (fa$_{\mu}$μt)	L	1 W
b. ☹ (faμt)	1	

As opposed to POT, HS with gradual syllabification can account for the vowel length alternations in (15) and the lack of VL in (16) with the same ranking.

(23) Partial ranking for Friulian
*μHead/Obstruent$_{\text{[-voice]}}$ » Weight-by-Position » *μHead/Obstruent$_{\text{[+voice]}}$ » Dep-Link-μ

To begin the analysis, let's first take the input /lov/. Core syllabification is the most harmonic candidate at step 1, as shown in all previous examples. At the second step of the derivation, the voiced consonant following the vowel is adjoined to the existing syllable as a mora-bearing coda, as candidate (a) in tableau (24) shows. The selection of candidate (a) instead of candidate (b) derives from a crucial ranking, in which Weight-by-Position outranks *μHead/Obstruent$_{\text{[+voice]}}$. Candidate (c) fatally violates the top-ranked constraint Parse-Segment.

(24) Step 2: coda adjunction

/(lo$_\mu$)v/	Parse-Seg	WbP	*μHead/Obs$_{[+voice]}$	*VcdObs/Coda
a. → (lo$_\mu$v$_\mu$)			1	1
b. (loμv)		1 W	L	1
c. (lo$_\mu$)v	1 W		L	L

In a theory of HS in which syllabification coexists with word-final obstruent devoicing, the first set of candidates of /lov/ includes a candidate (loμf), preventing /lov/ from mapping to an intermediate form (lo$_\mu$v$_\mu$) in which the underlying voiced obstruent projects its own mora, which is the source of VL. Parallel syllabification must therefore be discarded.

At the third step of the derivation, lengthening takes place. Candidate (b) is eliminated because it violates *μHead/Obstruent$_{[+voice]}$. Candidate (c), with devoicing, fatally violates the top-ranked constraint *μHead/Obstruent$_{[-voice]}$, which must dominate Weight-by-Position, and thus dominates *μHead/Obstruent$_{[+voice]}$ by transitivity. Candidate (a) is the optimal candidate, in which the vowel links to the second mora, because it satisfies *μHead/Obstruent$_{[+voice]}$ at the expense of violating the low-ranked constraint Dep-Link-μ.

(25) Step 3: lengthening

/(lo$_\mu$v$_\mu$)/	*μHead/ Obs$_{[-voice]}$	*μHead/ Obs$_{[+voice]}$	Dep-Link-μ	*VcdObs/ Coda
a. → (lo$_\mu$μv)			1	1
b. (lo$_\mu$v$_\mu$)		1 W	L	1
c. (lo$_\mu$f$_\mu$)	1 W		L	L

Devoicing applies at the next step of the derivation, omitted here. The positional markedness constraint *VoicedObstruent/Coda (Prince and Smolensky 1993/2004) dominating Ident(voice) is responsible for that. Notice that when devoicing has the chance to apply after lengthening, the top-ranked markedness constraint *μHead/Obstruent$_{[-voice]}$ is vacuously satisfied because the mora linked to the voiceless obstruent is also linked to the vowel.[14]

14 One anonymous reviewer suggests an alternative analysis in which VL is interpreted not as a mora-sharing process triggered by the satisfaction of *μHead/Obstruent$_{[+voice]}$, but instead by a process of delinking, after word-final obstruent devoicing, of the mora associated to the devoiced obstruent as a way to

Let's take now the input /fat/. After applying core syllabification at step 1, the underlying voiceless obstruent adjoins to the previously built syllable as a weightless coda at step 2 because of the ranking *μHEAD/OBSTRUENT$_{[-voice]}$ » WEIGHT-BY-POSITION, as can be seen in (26).

(26) Step 2: coda adjunction

	$/(fa_\mu)t/$	*μHEAD/OBS$_{[-voice]}$	WBP
a. →	(faμt)		1
b.	$(fa_\mu t_\mu)$	1 W	L

At the next step of the derivation lengthening has no chance to apply and the derivation gets stuck because no harmonic improvement is possible.

The ranking arguments that derive opaque VL in Friulian appear as a Hasse diagram in (27).

(27) Hasse diagram for Friulian

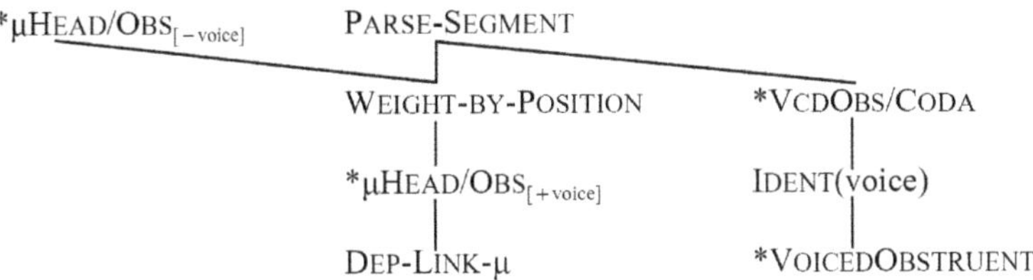

In this subsection it has been demonstrated that overapplication of VL in Friulian is straightforwardly accounted for in HS because syllabification, VL (*i.e.*, link insertion) and word-final obstruent devoicing must apply at different derivational steps given gradualness.

satisfy *μHEAD/OBSTRUENT$_{[-voice]}$, which dominates WEIGHT-BY-POSITION and MAX-LINK-μ. This delinking operation leaves the mora afloat, which reassociates with the preceding vowel in order to satisfy *FLOATINGμ. The HS derivation would be as follows: $(lo_\mu)v \rightarrow (lo_\mu v_\mu) \rightarrow (lo_\mu f_\mu) \rightarrow (lo_\mu {}^\mu f) \rightarrow (lo_{\mu\mu} f)$. This interpretation of the facts is parallel to the rule-based derivational analysis defended in Hualde (1990), in which devoicing automatically triggers delinking, which is the source of VL. This analysis seems at first sight empirically correct. The only drawback I can see in this account has to do with the consequences of delinking regarding the syllabic affiliation of the devoiced obstruent. If the devoiced coda obstruent is parsed into the syllable by means of being linked to a mora, once delinking applies there is no reason to believe that syllabic affiliation is still preserved. This situation would prevent the candidate showing delinking from winning because of the top-ranked position of PARSE-SEGMENT.

10.6.2 Alsatian French

In Alsatian French, root-final stop plus lateral sequences, which show a rising sonority profile, are repaired by means of word-medial schwa epenthesis: /sikl/ → [ˈsi.kəl] 'cycle'; /kupl/ → [ˈku.pəl] 'couple' (Montreuil 2010). The interesting data come from root-final /bl/ clusters. I assume that in Alsatian French, as in Friulian, the source of vowel length is the mora projected by a voiced coda consonant.

Syllable-final obstruent devoicing applies in Alsatian French, giving rise to the opaque mapping /tabl/ → [(ˈta:)(pəl)] 'table', where syllable-final obstruent devoicing counterbleeds VL. As Montreuil (2010) points out, epenthesis also needs to follow devoicing. If epenthesis were to precede devoicing, there would be no reason to devoice the obstruent since the stop would be syllabified in onset position, and devoicing only applies in coda position, yielding transparent *[(ˈta)(bəl)]. This double counterbleeding ordering relation between VL and devoicing, on the one hand, and devoicing and epenthesis, on the other hand, is exemplified in (28).

(28) Rule-based derivation (Montreuil 2010)

input:	/tabl/
vowel lengthening	(ta:b)l
counterbleeds	
devoicing	(ta:p)l
counterbleeds	
epenthesis	(ta:p)(əl)
feeds	
resyllabification	[(ˈta:)(pəl)]

The same ranking that has been proposed for opaque VL in Friulian also applies in Alsatian French. However, some additional constraints must be included in order to account for schwa epenthesis (see 10.4.2). PARSE-SEGMENT must dominate SYLLABLE-HEAD in order for GEN to build a degenerate, moraless syllable, and SYLLABLE-HEAD must outrank DEP-V because the minor syllable is fixed by means of schwa epenthesis.

The HS derivation of the input /tabl/ is then as follows: /tabl/ → (ta_{μ}) bl → $(ta_{\mu}b_{\mu})$l → $(ta_{\mu}b_{\mu})$(l) → $(ta_{\mu}\mu b)$(l) → $(ta_{\mu}\mu p)$(l) → $(ta_{\mu}\mu p)(ə_{\mu}l_{\mu})$ → $(ta_{\mu\mu})$ $(pə_{\mu}l_{\mu})$. The harmonic improvement tableau in (29) illustrates this derivation by showing the winning candidates at each derivational step. For the sake of clarity, only the constraints that are violated by the winning candidates are included in (29).

(29) Harmonic improvement tableau for Alsatian French

/tabl/	PARSE-SEG	SYLL-HEAD	*μHEAD/ $\text{OBS}_{[+voice]}$	DEP-V	ONSET	*VCDOBS/ CODA]_σ	DEP-LINK-μ	ID (vc)	*VCD OBS
Step 1 $(ta_\mu)bl$	2!								1
Step 2 $(ta_\mu b_\mu)l$	1!		1			1			1
Step 3 $(ta_\mu b_\mu)$ (l)		1	1(!)			1(!)			1
Step 4 $(ta_\mu \mu b)$ (l)		1				1!	1		1
Step 5 $(ta_\mu \mu p)$ (l)		1!						1	
Step 6 $(ta_\mu \mu p)$ $(ə_\mu l_\mu)$				1(!)	1(!)				
Step 7 $(ta_{\mu\mu})$ $(pə_\mu l_\mu)$									
Step 8 $(ta_{\mu\mu})$ $(pə_\mu l_\mu)$									

At the first step of the derivation, core syllabification applies. This is the most harmonic operation because the winning candidate minimally violates the top-ranked constraint PARSE-SEGMENT. Then /b/ is adjoined to the previously existing syllable as a weighted coda, and the top-ranked constraint WEIGHT-BY-POSITION is satisfied, not included in (29). At the third step, /l/ must be parsed as a degenerate, moraless syllable because PARSE-SEGMENT dominates SYLLABLE-HEAD. The markedness constraint against mora-bearing voiced coda consonants *μHEAD/ $\text{OBS}_{[+voice]}$ is then satisfied by means of inserting a mora link between the vowel and the mora headed by the voiced coda consonant. At the next step, syllable-final obstruent devoicing applies because the positional markedness constraint *VOICEDOBSTRUENT/CODA]_σ dominates IDENT(voice). At the next step SYLLABLE-HEAD is satisfied by means of word-medial schwa epenthesis. At the seventh step resyllabification applies and ONSET is satisfied. Finally the derivation converges and the actual opaque output $[(ta_{\mu\mu})(pə_\mu l)]$, with lengthening, devoicing, and schwa epenthesis, is obtained.

10.7 Derivational constrained-based alternatives

Apart from HS, other derivational versions of OT could in principle accommodate classic CL and opaque VL, namely OT with candidate-chains (OT-CC, McCarthy 2007) and Stratal OT (Kiparsky 2000, Bermúdez-Otero in preparation).

On the one hand, opaque VL in Lorraine French, which mirrors the Friulian data discussed here, and opaque VL in Alsatian French have been analyzed in OT-CC in Montreuil (2010). Classic CL has also been analyzed in OT-CC in Shaw (2009).[15] On the other hand, in Kiparsky (2010) a Stratal OT account of non-classic compensatory lengthening in Finnish and Samothraki Greek is presented, in which the trigger of VL is not the loss of a coda consonant. It is not the purpose of this paper to review all these proposals, but to show very briefly that classic CL and the type of opaque VL discussed so far find a more simple treatment in HS.

OT-CC needs PRECEDENCE constraints to deal with opacity. This type of metaconstraint forces a particular ordering among faithfulness violations between two consecutive members of a candidate chain. In Montreuil's (2010) OT-CC analysis of Lorraine and Alsatian French, PRECEDENCE constraints forcing lengthening to apply before devoicing and devoicing before schwa epenthesis are needed. One could argue against PRECEDENCE constraints based on its *ad hoc* character. Why should a PRECEDENCE constraint state that an A faithfulness violation must precede a B faithfulness violation and not the other way around? In other words, one should be skeptical about their stipulative character,

15 In Shaw (2009) it is argued that OT-CC makes a more restricted prediction regarding the languages that allow classic CL, which are presumably those that, apart from being sensitive to weight, also undergo the process of closed syllable lengthening. This statement follows from the fact that, according to Shaw (2009), GEN does not produce stray elements, meaning that before deleting a mora-bearing coda consonant, the vowel must be linked to the mora headed by the coda, the source of closed syllable lengthening. This way, the path towards vowel lengthening is said to start with lengthening, and then followed by deletion. It is not clear to me why a classic derivation that starts with deletion is not possible in OT-CC. In fact, the floating mora is not 'produced' by GEN as a stray element if deletion precedes lengthening, but is just the natural consequence of coda deletion. If a markedness constraint such as *FLOATINGμ is included in CON, there is no reason to discard a constraint-based derivation of CL in OT-CC that starts with deletion.

the same way extrinsic rule ordering was seen as an undesirable artifact in rule-based phonology.

Regarding Stratal OT, classic CL could be easily accounted for because, although EVAL applies in parallel, ranking permutation is allowed at each morphological stratum. In order to derive classic CL, MAX must outrank NO-CODA at the stem-level phonology in order to select the candidate in which weight by position has applied. At the word-level phonology, the reverse ranking is needed, in which NO-CODA dominates MAX. The main problem with this kind of analysis is that, in many cases, no morphology at all is involved in classic CL or in the case of opaque VL in Alsatian French, in which lengthening applies in monomorphemic words.

In HS neither PRECEDENCE constraints nor ranking permutation are needed. Classic CL, double flop and opaque VL are just a natural consequence of the intrinsic derivational nature of the theory imposed by the gradualness requirement on GEN.

10.8 Conclusions

Classic CL challenges POT because the conditions that make weight by position applicable are non-surface-apparent; deletion of a mora-bearing coda consonant counterbleeds the application of weight by position. In this chapter I have argued in favor of HS, a version of OT in which input-output mappings usually require a serial derivation in which harmony is incrementally achieved. It has been shown that HS derives classic CL and double flop only if two independently motivated proposals about the gradual nature of GEN are considered together. Namely, that syllabification is built serially in harmonically improving steps (Elfner 2009, Jesney 2011, Pater 2012, this volume, Elsman this volume) and that deletion of a coda consonant is a two-step process that begins with debuccalization (McCarthy 2008b). These theoretical assumptions have also been extended to account for opaque interactions between VL and final obstruent devoicing in Friulian and Alsatian French, which also require syllable formation operations to apply before final devoicing and schwa epenthesis.

The contribution of this paper has been twofold. From an empirical point of view, it has been demonstrated that HS is able to accommodate particular cases of counterbleeding opacity in which syllabification is involved. From an internal theoretical perspective, the proposed analyses have given additional support to two hypotheses about the gradual nature of GEN that have been proposed in recent literature on HS for independent reasons: serial syllabification and deletion as a two-step process.

References

Bermúdez-Otero, Ricardo (2001) Underlying non-moraic coda consonants, faithfulness, and sympathy. Unpublished manuscript. Manchester: University of Manchester.

Bermúdez-Otero, Ricardo (in preparation) *Stratal Optimality Theory*. Oxford: Oxford University Press.

Broselow, Ellen, Chen, Su-I and Huffman, Marie (1997) Syllable weight: convergence of phonology and phonetics. *Phonology* 14: 47–82.

Campos-Astorkiza, Rebeka (2004) Faith in moras: A revised approach to prosodic faithfulness. In Keir Moulton and Matthew Wolf (eds), *Proceedings of the North East Linguistics Society* 34, 163–174. Amherst, MA: GLSA Publications.

Campos-Astorkiza, Rebeka (2011) What drives compensatory lengthening? Beyond moraic conservation. In Gilda M. Socarrás and Almitra Medina (eds), *Philological Research*, 9–24. Athens, Greece: Athens Institute for Education and Research.

Elfner, Emily (2006) Contrastive syllabification in Blackfoot. In Donald Baumer, David Montero and Michael Scanlon (eds), *Proceedings of the 25th West Coast Conference on Formal Linguistics*, 141–149. Somerville, MA: Cascadilla Press.

Elfner, Emily (2009) Syllabification and stress-epenthesis interactions in Harmonic Serialism. Unpublished paper, University of Massachusetts Amherst. Available on Rutgers Optimality Archive, ROA-1047.

Elsman, Minta (this volume) Parallelism vs. serialism, or constraints vs. rules?

Hayes, Bruce (1989) Compensatory lengthening in moraic phonology. *Linguistic Inquiry* 20: 253–306.

Hermans, Ben (2001) CL in OT. In Ton van der Wouden and Hans Broeckhuis (eds), *Linguistics in the Netherlands 2001*, 139–150. Amsterdam: John Benjamins.

Hock, Hans H. (1986) Compensatory lengthening: In defense of the concept 'mora.' *Folia Linguistica* 20: 431–460.

Hualde, José I. (1990) Compensatory lengthening in Friulian. *Probus* 2: 31–46.

Hyman, Larry M. (1985) *A Theory of Phonological Weight*. Dordrecht: Foris.

Ingria, Robert (1980) Compensatory lengthening as a metrical phenomenon. *Linguistic Inquiry* 11: 465–495.

Iosad, Pavel (2012) Final devoicing and vowel lengthening in Friulian: A representational approach. *Lingua* 122: 922–951.

Jesney, Karen (2011) Positional faithfulness, non-locality, and the Harmonic Serialism solution. In Suzi Lima, Kevin Mullin and Brian Smith (eds), *Proceedings of the North East Linguistic Society* 39, 429–440. Amherst, MA: GLSA Publications. Available on Rutgers Optimality Archive, ROA-1018.

Kavitskaya, Darya (2002) *Compensatory Lengthening: Phonetics, Phonology, Diachrony*. New York, NY: Routledge.

Kimper, Wendell (2011) Locality and globality in phonological variation. *Natural Language and Linguistic Theory* 29: 423–465.
Kiparsky, Paul (2000) Opacity and cyclicity. *The Linguistic Review* 17: 351–367.
Kiparsky, Paul (2010) Compensatory lengthening. In Charles E. Cairns and Eric Raimy (eds), *Handbook of the Syllable*, 33–70. Leiden and Boston, MA: Brill Academic Publishers.
McCarthy, John J. (2000) Harmonic Serialism and parallelism. In Masako Hirotani, Andries Coetzee and Nancy Hall (eds), *Proceedings of the North East Linguistic Society* 30, 501–524. Amherst, MA: GLSA Publications.
McCarthy, John J. (2007a) *Hidden Generalizations: Phonological Opacity in Optimality Theory*. London: Equinox.
McCarthy, John J. (2007b) Restraint of analysis. In Sylvia Blaho, Patrik Bye and Martin Krämer (eds), *Freedom of Analysis*, 203–231. New York: Mouton de Gruyter.
McCarthy, John J. (2008a) The serial interaction of stress and syncope. *Natural Language and Linguistic Theory* 26: 499–546.
McCarthy, John J. (2008b) The gradual path to cluster simplification. *Phonology* 25: 271–319.
McCarthy, John J. (2010) Studying GEN. *Journal of the Phonetic Society of Japan* 13: 3–12.
McCarthy, John J. (2011) Autosegmental spreading in Optimality Theory. In John Goldsmith, Elizabeth Hume and Leo Wetzels (eds), *Tones and Features (Clements Memorial Volume)*, 195–222. Berlin: Mouton de Gruyter.
Montreuil, Jean-Pierre Y. (2010) Multiple opacity in Eastern Regional French. In Sonia Colina, Antxon Olarrea and Ana M. Carvalho (eds), *Romance Linguistics 2009: Selected Papers from the 39th Linguistic Symposium on Romance Languages*, 153–166. Amsterdam: John Benjamins.
Morén, Bruce (1999) *Distinctiveness, coercion, and sonority: A unified theory of weight*. Doctoral dissertation. Maryland, MD: University of Maryland.
Moore-Cantwell, Claire (this volume) Contexts for epenthesis in Harmonic Serialism.
Pater, Joe (this volume) Universal Grammar with weighted constraints.
Pater, Joe (2012) Serial harmonic grammar and Berber syllabification. In Toni Borowsky, Shigeto Kawahara, Takahito Shinya and Mariko Sugahara (eds), *Prosody Matters: Essays in Honor of Elisabeth O. Selkirk*, 43–72, London: Equinox Press.
Prieto, Pilar (1992) Compensatory lengthening by vowel and consonant loss in Early Friulian. *Catalan Working Papers in Linguistics* 2: 205–244.
Prince, Alan and Smolensky, Paul (1993/2004) *Optimality Theory: Constraint Interaction in Generative Grammar*. Malden, MA, and Oxford: Blackwell. [Revision of 1993 technical report, Rutgers University Center for Cognitive Science.]
Pruitt, Kathryn (2010) Serialism and locality in constraint-based metrical parsing. *Phonology* 27: 481–526.
Rosenthall, Sam and van der Hulst, Harry (1999) Weight-by-position by position. *Natural Language and Linguistic Theory* 17: 499–540.
Shaw, Jason A. (2009) Compensatory lengthening via mora preservation in OT-CC: theory and predictions. In Muhammad Abdurrahman, Anisa Schardi and Martin

Walkow (eds), *Proceedings of the North East Linguistic Society* 38, 297–310. Amherst, MA: GLSA Publications.

Sprouse, Ronald (1997) A case for enriched input. Handout from TREND 3.

Steriade, Donca (1982) *Greek prosodies and the nature of syllabification*. Doctoral dissertation. Cambridge, MA: Massachusetts Institute of Technology.

Topintzi, Nina (2006). A (not so) paradoxical instance of compensatory lengthening. *Journal of Greek Linguistics* 7: 71–119.

Wetzels, Leo (1986) Phonological timing in Ancient Greek. In Leo Wetzels and Engin Sezer (eds), *Studies in Compensatory Lengthening*, 296–344. Dordrecht: Foris.

Zec, Draga (2003) Prosodic weight. In Caroline Féry and Ruben van de Vijver (eds), *The Syllable in Optimality Theory*, 123–143. Cambridge: Cambridge University Press.

Chapter 11

Cyclicity and non-cyclicity in Maltese: Local ordering of phonology and morphology in OT-CC[1]

Matthew Wolf

11.1 Introduction

Stress in Maltese furnishes one of the case studies in Brame's (1974) classic article arguing in support of the transformational cycle in phonology (Chomsky, Halle and Lukoff 1956). One argument for cyclic application of stress in Maltese is a phenomenon which the language shares with a number of Levantine Arabic dialects: syncope of vowels in unstressed open syllables underapplies in verb stems with pronominal (object-marking) suffixes. For instance, subject-marked [ˈħtaf-na] 'we snatched' contrasts with object-marked [ħa.ˈtaf-na] 'he snatched us' (Odden 1993: 139–140); in the latter form the first vowel fails to syncopate, which can be explained by assuming that it gets stress on an earlier cycle containing just the bare stem /ħataf/.

1 For comments, ideas, questions asked and answered, and other assistance, I am grateful to Stephen R. Anderson, Michael Becker, Mark Borg, Kyle Gorman, Maria Gouskova, Karen Jesney, John McCarthy, Paul Micallef, Neil Myler, Kathryn Pruitt, Gilbert Puech, Tamara Schembri, Michael Spagnol, an anonymous reviewer, and members of audiences at the Third Conference on Maltese Linguistics, especially Elizabeth Hume, Gilbert Puech, Alina Twist, Adam Ussishkin, and Alexandra Vella; at UMass Amherst, especially Abdelaziz Boudlal, Wendell Kimper, and Alexei Nazarov; at the University of Southern California, especially Louis Goldstein, Assaf Israel, and Michael Proctor; at the University of Pennsylvania, especially Gene Buckley, Tony Kroch, and Florian Schwarz; and at Reed College, especially Matt Pearson. The views presented here are my own, as are all errors.

This paper addresses a complication in the Maltese facts first raised in the theoretical literature by Odden (1990, 1993): this evidence for the cycle obtains only in consonant-final stems. Vowel-final (third weak radical)[2] stems show no underapplication of syncope in object-suffixed forms. The analysis I propose here builds on Odden's idea that the difference is connected with the fact that stem-final vowels lengthen before suffixes. I will argue that the facts can be understood as resulting from conflicting preferences about pairwise orderings between phonological and morphological processes: on the one hand there is a pressure for object marking to be preceded by stress (i.e., for stress to be applied on the inner cyclic domain), but on the other hand, there is a pressure for stress to be applied after vowel lengthening (which encourages deferring stress until after object suffixation and the lengthening it causes). This arrangement, which echoes in many ways Anderson's (1969, 1974) theory of Local Ordering, can be straightforwardly implemented in Optimality Theory with Candidate Chains (OT-CC: McCarthy 2007), and more specifically in a variant of OT-CC called Optimal Interleaving (OI: Wolf 2008) which includes morph-insertion among the operations which take place in the OT-CC phonology. Under this view, 'cyclicity' (the application of certain phonology before certain morphology) is not something that is hard-wired into the overall modular architecture of a language's grammar, but is instead simply a particular type of ordering relation between phonological and morphological processes which a language may select for some forms but not for others.

11.2 Stress, syncope, and the cycle in Maltese

Maltese is the national language (and, with English, the co-official language) of the Republic of Malta, a small archipelago of 445,000 inhabitants located in the center of the Mediterranean. The Maltese language is a descendant of Arabic, which was introduced to the islands during the period of Arab rule from 870 to 1091. This era concluded with the

2 Third weak radical stems are those that historically had a glide as the third root consonant. There is a second class of verbs in Maltese whose stem is vowel-final in the absence of a following suffix, but which do show cyclic stress and underapplication of syncope. This is the class of verbs with orthographic <'> as the third radical and [a] as the preceding rightmost vowel. The character <'> (in other environments written <għ>) represents historical /ɣ/ or /ʕ/, and so the fact that these stems pattern with C-final stems in showing cyclic underapplication of syncope (Aquilina 1965: 165) is consistent with Brame's (1972) arguments that the /ʕ/ remains synchronically underlying.

conquest of Malta by Norman Sicily, which politically and religiously separated the islands from the rest of the Arabic-speaking world. Arabic remained the vernacular language, but now came into close contact with Italian. The present-day Maltese language resulting from this history is often described as a 'mixed language', due to the co-existence of a large amount of vocabulary (and certain grammatical influences) borrowed from Italian alongside that inherited from Arabic. Malta was under British rule from 1800 to 1964, which has resulted more recently in heavy borrowing from English; the Maltese population today is extensively bilingual in Maltese and English.

My concern in this paper is primarily to account for vowel ~ zero alternations in verbs in the Semitic part of the vocabulary, more specifically verbs of the first measure whose stems have the shape [CVCVC] in the third-person masculine singular perfective, where there are no phonologically overt affixes. Taking /CVCVC/ to in fact be the underlying form of the stem, Brame (1972, 1974) argues that vowel ~ zero alternations in verb paradigms can be accounted for if we assume that stress is assigned as in (1), and that stress assignment is followed by a rule which syncopates vowels in unstressed, non-final open syllables:[3]

(1) Place stress:
  (a) On the ultima, if it is superheavy (or the word is monosyllabic), else
  (b) On the penult, if it is heavy (or the word is bisyllabic), else
  (c) On the antepenult.

This can be illustrated by looking at a representative inflectional paradigm of first-measure verb in the perfective:

(2) *qatel* 'kill' (Sutcliffe 1936: 68)[4]

a. /ʔitil-t/	[ˈʔtilt]	'I killed'	*e.* /ʔitil-na/	[ˈʔtil.na]	'we killed'
b. /ʔitil-t/	[ˈʔtilt]	'you.SG killed'	*f.* /ʔitil-tu/	[ˈʔtil.tu]	'you.PL killed'
c. /ʔitil-it/	[ˈʔat.let]	'she killed'	*g.* /ʔitil-u/	[ˈʔat.lu]	'they killed'
d. /ʔitil-Ø/	[ˈʔa.tel]	'he killed'			

3 The Maltese stress system closely resembles those of the Levantine Arabic dialects (Brame 1974: 45; Vella 2003: 270–271). This includes not only the basic pattern in (1) and cyclic application of stress before object suffixes, but also the lack of cyclic stress in third weak radical verbs, which is shared with some though not all Levantine dialects, as discussed by Johnson (1990: §5). I am grateful to Gilbert Puech for pointing out the existence of Johnson's paper to me.

4 The vowel-quality alternations in this paradigm are due to (a) lowering of /i/ to [a] adjacent to /ʔ, h/, and (b) reduction of /i/ to [e] before a single word-final consonant. See Hume (1994: 167–181) for details.

Final stress (clause (1a)) is seen in (2a,b); penult stress (clause (1b)) is seen in (2d,e,f); and antepenult stress (clause (1c)) is seen in (2c,g). It should be acknowledged that these assumptions about the stress-syncope relationship are not universally accepted. In particular, Puech (2011) argues that 'syncope' in Maltese is actually a matter of vowels of the vocalic melody not being incorporated into syllable structure in the first place. For forms like [ˈʔat.lu] for which Brame's (1972, 1974) analysis posits a pre-syncope antepenult stress (/ˈʔa.te.lu/), Puech's analysis has the syllabification algorithm produce /ʔat.lu/, which then is assigned penult stress. One argument for Puech's (2011) analysis is that by having 'syncope' apply before stress assignment rather than after, it avoids the need to posit the antepenult stress clause (1c). That is claimed to be advantageous on the grounds that loanwords with antepeult stress are often adapted with penult stress (e.g. English *ˈArabic* → Maltese *e.ˈre.bic*).[5, 6] Limits of space unfortunately make it impossible to do justice to Puech's alternative approach in this chapter, so from here on I will adhere to the probably more familiar approach from Brame (1972, 1974) that stress precedes (and determines the site of) syncope, entailing that the stress pattern is as in (1). In any case, the examples crucial to my argument will involve penult stress.

Given these assumptions, one argument for the cycle in Maltese comes from the fact that syncope underapplies in consonant-final verb stems which carry object suffixes (Sutcliffe 1936: 15, 155; Brame 1974: 56; Berrendonner, Le Guern and Puech 1983: 224; Odden 1993: 139; Borg 1997: 279). The pattern is illustrated by data like these, previewed earlier in the introduction:

(3)			
	(a) /ħataf-$Ø_{3.masc.sg.subj}$/	[ˈħa.taf]	'he snatched'
	(b) /ħataf-$na_{1pl.subj.}$/	[ˈħtaf.na]	'we snatched'
	(c) /ħataf-$Ø_{3.masc.sg.subj}$ -$na_{1pl.obj}$ /	[ħa.ˈtaf.na]	'he snatched us'

5 A similar point is made by Eure (1999: 16–17) in the context of a parallel OT analysis of Maltese stress. In derivational terms, after antepenult stress is assigned to /...ˈσ.CV.CV(C)/, the penult vowel will syncopate, giving surface [...ˈH.CVC]. In parallel OT, the antepenult case can therefore be understood as actually reflecting a surface-oriented preference for penult stress with heavy stressed syllables across the board.

6 This pattern of nativization co-exists with another pattern cited by Puech (2011), in which foreign words with antepenult stress are brought to have penult stress via syncope, e.g. *Aˈmerica* → *aˈmeːrka*. See also Sutcliffe (1936: 12–13), Aquilina (1959: 73), Borg (1997: 277–278), and Vella (2003: 268–269).

In (3a), where there is no overt affixation, stress falls on the penult as dictated by (1b). The second, unstressed vowel does not delete, since it falls in a syllable which is word-final as well as closed. Syncope is able to occur in (3b), where the appearance of the first-person plural suffix /-na/ as a subject marker causes stress to fall on the stem-final syllable [taf], as this is now a heavy penult. The first stem-vowel then finds itself in an unstressed open syllable /ħa/, and as a result is syncopated.

The evidence for the cycle comes from forms like (3c), where /-na/ is serving as an object marker. Here, we again find stress falling on the surface penult [taf]. However, the first vowel of the stem now fails to syncopate, even though the underlying segmental composition of the word is identical to that of (3b) where there is syncope. This inconsistency can be resolved if we assume that object-marked verbs contain an inner cyclic domain which includes the stem and any subject suffixes, but not the object suffixes. Within this domain, the first stem vowel in forms like (3c) gets stress, and this stress survives as a secondary stress on subsequent cycles, protecting the vowel from syncope. By contrast, when /-na/ is a subject maker as in (3b), it is part of the same cyclic domain as the stem, and so the first stem vowel never receives stress, thereby remaining eligible for syncope:

(4)		'we snatched'	'he snatched us'
	Input	[ħataf-na]	[[ħataf-Ø]-na]
	Cycle 1		
	Stress	ħa.ˈtaf.na	ˈħa.taf
	Syncope	ˈħtaf.na	*no change*
	Cycle 2	*n/a*	
	Stress		ˌħa.ˈtaf.na
	Syncope		*no change*
	Output	[ˈħtaf.na]	[ˌħa.ˈtaf.na]

It is important to note that the secondary stresses, like the one which this analysis must posit for the initial syllable of words like [ˌħa.ˈtaf.na], are not reported in descriptions of Maltese (e.g. Aquilina 1959: 68); Brame (1974: 44) acknowledges this, suggesting that, if needs be, the secondary stress may be deleted by a late rule ordered after syncope. (We shall see later that, notwithstanding Kiparsky 2011: 5–6, the existence of some such destressing process is not in principle incompatible with the analysis to be proposed here.)

This quite straightforward analysis faces a complication first pointed out in the theoretical literature by Odden (1990, 1993): cyclic stress, as diagnosed by the underapplication of syncope in object-marked verbs, occurs only with consonant-final verb stems. Vowel-final verbs by contrast *do* permit syncope of the first stem vowel when stress shifts rightwards under object suffixation. This fact is noted for example in Sutcliffe's (1936) grammar: 'Verbs with third weak radical such as **nesa** *to forget*, **mela** *to fill*, **beda** *to begin*, are exceptional. The first vowel drops and as the vowel **a** on receiving the accent becomes **ie**' [phonologically /ɪː/-*MW*][7, 8, 9] (p. 157; boldface in original):

(5)				
	[mlɪːni]	'he filled me'	[mlɪːna]	'he filled us'
	[mlɪːk]	'he filled you.SG'	[mlɪːkom]	'he filled you.PL'
	[mlɪːh]	'he filled him'	[mlɪːhom]	'he filled them'
	[mlɪːha]	'he filled her'		

Sutcliffe goes on to note that the same holds for verbs with indirect object markers[10] (p. 160): 'As with the simple suffixes [...] so also with the indirect suffixes, the verbs with third weak radical are exceptional, and lose their first vowel. Thus from **sewa** *to cost*' [boldface in original]:

(6)				
	[swɪːli]	1sg IDO	[swɪːlna]	1pl IDO
	[swɪːlek]	2sg IDO	[swɪːlkom]	2pl IDO
	[swɪːlu]	3sg masc. IDO	[swɪːlhom]	3pl IDO
	[swɪːlha]	3sg fem. IDO		

7 This process (the synchronic *imaala*) involves raising of stressed long /aː/ to a high vowel whose exact quality shows a good deal of dialectal variation. Here I follow Borg (1997) in transcribing the vowel as [ɪː]; as the passage from Sutcliffe (1936) indicates, the raised vowel is written <ie> in Maltese orthography.

8 For at least some present-day speakers, there are some lexical exceptions to this pattern. Michael Spagnol (p.c.) reports that object-marked forms of *nesa* and *mela* (among other verbs) resist syncope in his idiolect.

9 Because syncope for this particular verb creates an initial cluster beginning with a sonorant, these forms will be pronounced with a prothetic initial [i], unless they are immediately preceded by a vowel-final word within the same phonological phrase.

10 Indirect objects are generally marked by /-l-/ followed by the same suffix as would be used for a direct object of the same person/number/gender. One source of difference is that the third.masc.sg. object suffix has a post-vocalic allomorph /-h/ (which will appear after the final vowel of a weak-final stem) and a post-consonantal allomorph /-u/ (which will appear following the indirect-object /-l-/).

The goal of this paper is to account for this difference between consonant-final and vowel-final verb stems. Before presenting the analysis, however, I will first outline the theoretical premises upon which it is based.

11.3 Optimal Interleaving

11.3.1 OT with Candidate Chains

In Optimality Theory (Prince and Smolensky 2004 [1993]), the grammar determines output forms by producing a set of candidate outputs for a given input, and then evaluating them according to a hierarchy of ranked constraints. In the most common, parallel implementation of OT, each candidate is a direct mapping from input to (potential) output; there are no intermediate derivational stages, and the disparity between the input forms and candidate output forms is in principle unlimited (what McCarthy and Prince 1993a call FREEDOM OF ANALYSIS).

OT-CC is different in that the competing candidates are (approximately) gradual, multi-step derivations,[11] or CHAINS. For example, as a candidate where apocope feeds final devoicing, in OT-CC we would have something like <taga, tag, tak> as a candidate rather than a direct mapping /taga/→ [tak] in parallel OT. There are three principles which define what is a possible candidate chain in any given language:

(7) *Gradualness*: Given a chain $<\ldots, l_i, l_{i+1}, \ldots>$, l_{i+1} can differ from l_i by the performing of at most one basic operation.

(8) *Harmonic improvement*: Given a chain $<\ldots, l_i, l_{i+1}, \ldots>$, l_{i+1} must be more harmonic than l_i, given the constraint ranking of the language in question.

11 The 'approximately' is the result of the mechanism of CHAIN MERGER. Technically, candidates in OT-CC are not sequences of representations but rather surface forms plus a partially-ordered set of the derivational operations which were applied to reach the surface form. Merger makes these sets partially ordered by combining together candidates which reach the same output via different orders of application of the same operations, keeping only those ordering relations among the operations that are shared by all of the merged chains. While essential to the analysis of many kinds of opacity effects in OT-CC, it is possible to abstract away from merger with regards to facts analyzed in the main text of this paper. This is advantageous for expository purposes, since thinking of candidates as sequences of derivational steps is probably more intuitive for the majority of readers. (However, see notes 24, 27 and 31 below for some places where merger may play an important role with respect to the Maltese data.)

(9) *Local Optimality*: Let $\langle f_1, f_2, \dots f_n \rangle$ be a valid chain in language *L*, and let $\{g_1, g_2, \dots g_m\}$ be the set of all the forms which could be produced by applying an operation of type *T* to f_n. Then $\langle f_1, f_2, \dots f_n, g_i \rangle$ is not a valid chain of *L* unless g_i is the most harmonic member of the set $\{g_1, g_2, \dots g_m\}$. (*Informally*: starting from any given point, if there is more than one way of doing some operation, the grammar may place under consideration only the initially-best way of doing that operation.)

The gradualness requirement (7) requires that candidate derivations move from input to output in a manner that makes only one change at a time. Of course, this requirement must come with some theory of what counts as one change. Much research in OT-CC and in the related theory of Harmonic Serialism (Prince and Smolensky 2004 [1993]: 19–26, 94–97) is focused on exploring the consequences of different hypotheses about what the basic operations are.

Since the Maltese facts involve both prosodic structure and affixation, we will have to assume the existence of operations dealing with these things. Regarding prosody, I will first of all assume, with McCarthy (2010), that resyllabification happens for free at each step of the derivation, along with whatever other operation was performed. The reason for this is that unfaithful phonological mappings may be harmonically improving due the less-marked syllable structure that they make possible. If, say, a language epenthesized a vowel to get rid of syllable codas (e.g. /kan/ → [ka.nɨ]), epenthesis would yield no immediate improvement in performance on NoCoda if epenthesis and resyllabification of the erstwhile coda had to occur as separate steps (/kan/ → kan.ɨ → [ka.nɨ]).[12] The building of higher-level prosodic structure will, however, occur as distinct derivational steps. Specifically, I will assume that the following operations are available:

(10) Build one Prosodic Word node along with a single head foot. (McCarthy 2008)

(11) Build one foot. (Pruitt 2008, 2010, 2012; Kimper 2011)

Assumptions about affixation will be deferred until the next subsection, when OI and its assumptions about the interface are laid out. Finally, since we are also dealing with syncope, I will also assume that deletion of one vowel is a basic operation.

12 However, cf. Prince and Smolensky (2004 [1993]: ch. 2), Elfner (2009; this volume) and Pater (2012) on syllabification as a step in its own right.

Once the chains have been built, they compete against one another as candidates. As in parallel OT, the markedness constraints evaluate only the last form in the chain (the candidate output) and the faithfulness constraints evaluate this form's disparity from the input. The motivation for positing multi-step derivations is to account for counter-feeding and counter-bleeding opacity (Kiparsky 1973), and so it is necessary to add a new set of constraints that will refer to the intermediate stages of the derivation. These are called PRECEDENCE constraints, and have the following schema:

(12) PREC(A, B)
Assign a violation-mark for every time that:
(a) An operation of type B occurs and it is not preceded by an operation of type A.
or
(b) An operation of type B occurs and it is followed by an operation of type A.

These constraints are analogous to extrinsic ordering statements in rule-based phonology, in this case that rule A precedes rule B. These ordering statements co-exist and interact with markedness and faithfulness constraints of the familiar kind, which will generally prefer transparent interactions, as in parallel OT. Thus, ordering in OT-CC is less like the orthodox theory of rule-ordering in Chomsky and Halle (1968), where rules are extrinsically fixed in one total ordering for the language as a whole, and more like Anderson's (1969, 1974) theory of Local Ordering. In the latter theory, there is no total ordering of rules but instead only pairwise ordering statements, plus universal preferences for 'natural' orders. The main consequence of Local Ordering is that the same two rules may apply in different orders in different forms of the same language because different constraints on ordering may be applicable in different situations. (Also, two rules' functional relationship may be different in different forms, so the natural ordering principles may prefer different orders.) OT-CC is similar: since the PREC constraints as well as the markedness and faithfulness constraints are all violable, different orderings may be chosen for different forms, because some high-ranked constraint may be decisive for some forms but indifferent for others, permitting other, lower-ranked ordering preferences to make themselves felt.[13] The analysis of Maltese will exploit this aspect

13 Local Ordering also incorporates a notion of ranking between conflicting ordering requirements (Anderson 1974: 159–164, 189, 198, 217–218).

of OT-CC. (For previous discussion of OT-CC's ability to produce Local Ordering-like effects, see Wilson 2006 and Wolf 2009, 2010.)

11.3.2 OI: Morphological realization in the phonology

Optimal Interleaving (Wolf 2008) is an approach to the phonology-morphology interface which combines OT-CC with the assumption that morphological spell-out occurs in the same grammatical module as the phonology. On the morphological end, OI assumes a realizational view of morphology (Anderson 1992; Halle and Marantz 1993; Beard 1995), which means that words are produced by first creating an abstract morphosyntactic structure, whose component features are then given phonological instantiation. For instance, the derivation of the English word *cats* would proceed in something like the following manner:

(13)	[√CAT PLURAL]	→	[√CAT PLURAL]	→	[√CAT PLURAL]
			√CAT PLURAL		*√CAT PLURAL*
			/kæt-z/		[kæts]

First, on the left, the morphosyntax joins together two abstract morphemes with the meanings 'cat' and 'plural'. Next, in the middle, these abstract morphemes are paired up with MORPHS. A morph, in OI, is formally an ordered pair like <PLURAL, /-z/>, consisting of a bundle of abstract morphosyntactic features and a bundle of phonological material (an underlying form). The morphs are essentially the same as the 'vocabulary items' of Distributed Morphology (Halle and Marantz 1993). Finally, on the right, phonology (such as assimilatory devoicing of /tz/ to [ts]) applies to these underlying forms, giving us the surface pronunciation.

What OI claims is that both of the derivational steps depicted above take place in the same module of the grammar. That is, the input to the phonology consists only of the abstract morphosyntactic representation. OI's motivation for placing morphological spell-out alongside the phonology in specifically an OT-CC grammar is that it lets us use OT-CC's opacity-handling machinery to deal with several forms of opaque interaction between phonology and morphology (Wolf 2008: ch. 3–5). Most relevantly to this paper, 'cyclic' misapplication of a phonological process *P* within the base of affixation *M* can be attributed to the first clause of PREC(P, M), which demands that if *M* occurs, *P* must have occurred previously: we are required to do some phonology of type *P* on the base before *M* is added.

Inserting phonologically non-null morphs involves adding phonological structure, and thus potentially adding new phonological markedness violations. Morph insertion then would often not be harmonically-improving unless there were some constraint which specifically encouraged morph presence, and which could outrank the countervailing markedness constraints. In Wolf (2008), it is proposed that there is a Correspondence relation (McCarthy and Prince 1995) between morphemes and the morphs which express them, meaning that morph insertion is driven by a family of MAX-M constraints, which are violated if structures at the morpheme level do not have corresponding morphs. These constraints will be part of a larger collection of faithfulness constraints which demand similarity between the morphosyntactic feature-content of morphemes, and that of the morphs with which they stand in correspondence.

With these preliminaries in place, we can now proceed to the analysis of cyclic stress, and its absence, in Maltese.

11.4 Cyclic stress in consonant-final Maltese verb stems

With consonant-final verbs, we need stress to apply before the suffixation of object markers. Under the general OI assumptions described in the previous section, this would be attributed to the effects of the following constraint (recall our earlier assumption that building a Prosodic Word node necessarily includes the building of that node's head foot):

(14) PREC(build-PWd, Insert-obj)
Assign a violation-mark for every time that:
(a) An object-marker morph is inserted, and this was not preceded by earlier construction of a PWd; or
(b) An object-marker morph is inserted, and is followed by insertion of a PWd.

Before commencing the OI analysis, we will need to lay out some background assumptions about Maltese stress. Excluding final superheavies, the pre-syncope stress pattern assumed in (1) is of the Latin type: stress the penult if heavy, else stress the antepenult. We can assume that this means that a trochaic foot is built as far to the right as possible, subject to the condition that the final syllable is unfootable. This is expressed by the following ranking (identical to that employed in Prince and Smolensky's 2004 [1993]: 69–70 analysis of antepenultimate stress in Latin):

(15) Penult and ultima both light: stress on antepenult

	/LLL/	NONFINALITY(Ft)	ALL-FOOT-RIGHT
a. →	|(ˈLL)L|		1
b.	|L(ˈLL)|	1 W	L

(16) Stress on heavy penult

	/LHL/	NONFINALITY(Ft)	ALL-FOOT-RIGHT
a. →	|L(ˈH)L|		1
b.	|L(ˈHL)|	1 W	L

(17) NONFINALITY(Ft) (Prince and Smolensky 2004 [1993]: 51; Gouskova 2003: 24)
Assign one violation-mark if the head foot of a PWd is final in the PWd.

(18) ALL-FOOT-RIGHT (Prince and Smolensky 2004 [1993]: 46; McCarthy and Prince 1993b)
The right edge of every foot is aligned with the right edge of the PWd. (Violations assessed gradiently by syllables.)[14]

The notation used in (15–16) and throughout for prosodic boundaries is as follows: '.' denotes syllable boundaries; '(...)' foot boundaries, and '|...|' PWd boundaries. These tableaux are in Prince's (2002) combination format: numerals indicate the number of violation-marks incurred by each candidate; in rows for losing candidates, W indicates that a constraint prefers the winner over that loser, and L indicates that the constraint prefers that loser over the winner.

The prohibition against footing a final syllable is overridden in two cases. First, monosyllables do get stress, on their sole and therefore final syllable. With Prince and Smolensky (2004 [1993]: 51–52) we may attribute this to constraints demanding that every morphosyntactic word must be parsed into a prosodic word, together with what I will take to be an inviolable requirement that every PWd must have a head foot:

14 McCarthy (2003b), among others, is critical of the use of gradient constraints and in particular gradient alignment constraints in parallel OT. Pruitt (2008: §4.2) argues that the alternative of categorical alignment is incompatible with serial optimization, and that at least some of the pathologies associated with gradient foot alignment under parallel optimization disappear with serial optimization.

(19) Monosyllables get stress

/σ/	WdCon	NonFin(Ft)
a. → |(ˈσ)|		1
b. σ	1 W	L

(20) WdCon (cover constraint from Selkirk 1995; cf. Prince and Smolensky's 2004 [1993]: 51 Lex≈Pr)
The left and right edges of every lexical word must coincide, respectively, with the left and right edges of some prosodic word.

Second, final syllables can be footed when the final syllable is heavy enough to attract stress. Standard descriptions of stress in Maltese, summarized earlier in (1), distinguish between heavy (CVC) and super-heavy (CVCC and CV:C) syllables in this regard: the latter get stress in word-final position, but the former do not.[15] An additional case to consider involves final [CV:] syllables. These are generally absent in Maltese; however, the language does have stress on final open syllables in a few native Maltese words that end in a diphthong (Vella 2009: 66–68), as in (21a). Word-final long vowels are found in loans, especially from Italian, and these are also stressed (Aquilina 1959: 121–126, 1965: 26, 28; Fenech 1978: 18–19; Vella 2003: 270–271); as seen in (21b) below, these vowels are written with a grave accent in the standard orthography.

(21) (a) *Mulej* [mʊ.ˈlɛɪ] 'Lord'
(b) *università* 'university', *Marì* 'Mary', *tribù* 'tribe', *repò* 'refreshments', *kafè* 'coffee'

If we generalize from such examples that final [CV:], when it surfaces as long, does attract stress,[16] then it is only [CVC] syllables which are unexpectedly not heavy enough to get stress in word-final position. We can account for that fact by assuming that the last consonant of a word cannot project a mora, as Eure (1999: 14) proposes for Maltese; comparable assumptions have been made about similar phenomena in the Levantine dialects; among others see Johnson (1990: 145) and Broselow, Chen and Huffman (1997: 57). Thus, in final position [CVC] has a non-moraic coda and therefore is treated as light, whereas [CV:],

15 According to Borg (1997: 251–252) there are cases of stressed [VC#] derived from underlying [VCC#].

16 The analyses of Maltese by Odden (1990, 1993), in a version of Lexical Phonology, and Kiparsky (2011), in Stratal OT, rely on the assumption that final long vowels will be assigned stress.

[CV:C], and [CVC$^{\mu}$C] are bimoraic in final position and are treated as heavy. (I use a superscript μ on a consonant to indicate that it is moraic, and the absence of μ to indicate non-moraicity.) The violation of NONFINALITY in these cases can then be attributed to ranking the WEIGHT-TO-STRESS PRINCIPLE (Prince 1990), which requires that heavy syllables be stressed, over NONFINALITY (on the way in which this ranking yields 'extrameticality revocation' when the final syllable is heavy, see Prince and Smolensky 2004 [1993]: 64):

(22) Heavy syllable in final position attracts stress

	/CVCVCVCC/	WSP	NONFIN(Ft)
a. →	|CV.CV('CVC$^{\mu}$C)|		1
b.	|('CV.CV)CVC$^{\mu}$C|	1 W	L

(23) Final CVC syllable doesn't attract stress

	/CVCVCVC/	WSP	NONFIN(Ft)
a.	|CV.CV('CVC)|		1 W
b. →	|('CV.CV)CVC|		

(24) WEIGHT-TO-STRESS PRINCIPLE (WSP): Heavy syllables must not be metrically weak. (Prince 1990)

The issue of the length of word-final vowels will be discussed again when we come to the analysis of vowel-final verb stems, and why they lack cyclic stress.

We may now commence the OI analysis of suffixed C-final stems, using as our example the first-person plural subject- and object-suffixed forms of /ħataf/ 'snatch', as seen earlier. Assuming that morph-insertion necessarily moves inside out, the very first thing that must happen is for the stem morph to be inserted. The following tableau illustrates that this is harmonically improving under the ranking that we will assume:

(25) Insertion of stem is harmonically improving

		snatch-1pl	MAX-M	WDCON	WSP	NONFIN (Ft)	AFR	EXH (wd)	MAX-V
FFC	a.	SNATCH-1PL	2						
insert morph	b. →$_L$	ħa.taf-1PL	1	1					

Form (25a) above is the fully faithful candidate (FFC) – the form identical to the input. This is the reference point against which other candidates must be checked to see whether or not they are harmonically improving. At this point in chain construction the input is simply the two not-yet-spelled-out morphemes /SNATCH-1PL/, and so the FFC incurs two violations of MAX-M, since there are two morphemes which lack corresponding morphs. Candidate (25b) improves harmony relative to the input since it inserts the stem morph /ħataf/, thus giving a correspondent to the stem morpheme and taking away a violation of MAX-M.[17] This does however come at the expense of adding a violation of lower-ranked WDCON. With the insertion of the segments of the stem, there is now segmental (and syllabic) material which is not parsed by a PWd. If PWd construction is a separate step from morph-insertion, then insertion necessarily creates new, temporary violations of WDCON, which consequently must rank below MAX-M.

As seen in (26), we have two options for continuing on from (25b): spell out the suffix, or lay down a PWd with its head foot:

(26) Insertion of object marker, or laying down PWd, are harmonically improving

Input from (25b)	ħa.taf-1PL	MAX-M	WDCON	WSP	NONFIN (Ft)	AFR	EXH (wd)	MAX-V
FFC	a. ħa.taf-1PL	1	1					
insert morph	b. →$_L$ ħa.taf$^{\mu}$.na		1					
Build PWd	c. →$_L$ |(ˈħa)taf|-1PL	1				1	1	
	d. |(ˈħa.taf)|-1PL	1			1			
	e. |ħa(ˈtaf)|-1PL	1			1		1	

(27) EXHAUSTIVITY(word) (Itô and Mester 2003; Selkirk 1995)
One violation-mark for every syllable which is a direct dependent of the Prosodic Word.

Inserting the affixal morph, as in (26b), is harmonically-improving by virtue of removing the remaining violation of MAX-M. Alternatively, we can also improve harmony by parsing the string into a PWd, thus removing the FFC's violation of WORDCON. Given the two syllables built so far, there are three logically possible ways to build a single trochaic foot: an initial monosyllabic foot as in (26c), a binary foot as in (26d), or a final monosyllabic foot as in (26e). (I assume that a constraint requiring

17 Throughout, I treat the consonantal root, vocalism, and pattern as being inserted simultaneously; this is primarily for ease of exposition, and is not meant as making an empirical claim about whether or not these synchronically function as separate morphological units.

feet to be trochaic is undominated in Maltese.) The latter two are less harmonic than (26c) owing to their having feet which parse the word-final syllable. This violates the constraint NONFINALITY(ft). According to the Local Optimality requirement, only the most harmonic way of performing the 'build PWd' operation can be included in the set of chains under construction; throughout the paper, locally-optimal candidates like (26c) are indicated by the arrow with subscript L for 'local.' Double horizontal lines separate groups of candidates that result from the same operation, thus grouping the potential next steps into those groups of candidates which must compete with each other for Local Optimality.

A brief digression is in order about why we need to assume that a unary rather than binary foot parse is locally optimal at this point. This is so that that the second syllable of the stem remains available to be footed after we subsequently add the suffix /-na/. If we chose a binary foot instead, this would give us |(ˈħa.tafᵘ)na| after suffixation. The trouble is that we would now need /tafᵘ/ to receive the primary stress. Getting from |(ˈħa.tafᵘ)na| to |(ˌħa)(ˈtafᵘ)na| requires that we both dis-affiliate /tafᵘ/ from its previous foot and also build a new foot with /tafᵘ/ as its head. If these have to occur as separate steps, we are stuck: unfooting /tafᵘ/ is not harmonically improving because it worsens performance on ALL-FOOT-RIGHT and EXHAUSTIVITY(word) without improving performance on any higher-ranked constraint. This is seen in the tableau below; here and throughout the thumbs-down symbol is used to indicate that a candidate is not harmonically improving:

(28) Unfooting penult after suffixation worsens harmony

		|(ˈħa.tafᵘ)na|	MAX-M	WDCON	WSP	NONFIN (Ft)	AFR	EXH (wd)	MAX-V
FFC	a.	|(ˈħa.tafᵘ)na|			1		1	1	
Build PWd	b. 👎	|(ˈħa)tafᵘ.na|			1		2	2	

We could in principle bypass this conundrum by relaxing our theory of gradualness by assuming that GEN can go from |(ˈħa.tafᵘ)na| to |(ˌħa)(ˈtafᵘ)na| in one step. This, however, is undesirable insofar as it jeopardizes some of the empirical advantages of serial over parallel foot optimization. As Pruitt (2010: 502, fn. 15) and McCarthy and Pruitt (2011: §4.2) note, a definition of gradualness which permitted direct mappings like |(ˈHL)L|→|(ˈH)(ˌLL)| would make possible (at least in three vs. four-syllable words) an unattested asymmetry wherein the parsing of an initial heavy syllable (in a language with left-to-right footing) as a unary vs. as part of a binary foot is determined nonlocally by the parity of the following string. Building a unary initial foot in |(ˈH)(ˌLL)| vs. a binary initial foot in |(ˈHL)(ˌLL)| permits exhaustive parsing into (minimally) bimoraic feet. This sort of pattern is unattested (Hyde 2007, 2008), and is excluded in HS if the grammar builds feet one at a time (Pruitt 2010: 496–503): informally, gradual foot optimization means that the grammar must commit itself to either |(ˈHL)L...| or |(ˈH)LL...| before attempting to foot the rest of the string. At least in three-syllable words, allowing |(ˈHL)L|→|(ˈH)(ˌLL)| as a one-step mapping would provide a way to escape this commitment, and hence should arguably not be allowed. If the grammar is indeed barred from building a new foot over part of an existing foot in one step, then |(ˈħa)taf$^{\mu}$| must be footed as such in order for |(ˌħa)(ˈtaf$^{\mu}$)na| to be reachable.[18]

The subchain <SNATCH-1PL, ħa.taf-1PL, |(ˈħa)taf|-1PL> represents the course that we want to start on to achieve an output with 'cyclic' stress on the stem: it has placed a stress on the stem prior to any further morphology being added. Let's now investigate where further we can go from here. (The other path from (26), namely that which starts by spelling out the suffix as in (26b), will be followed up on later.) Unsurprisingly, it is harmonically improving to spell out the first-person plural morpheme. As shown below, I assume that with the PWd already in place, the morph /-na/ can be incorporated into the PWd at the point of insertion.

(29) With stem prosodified, insertion of suffix is harmonically improving

<table>
<tr><td>Input
from (26c)</td><td>|(ˈħa)taf|-1PL</td><td>MAX-
M</td><td>WDCON</td><td>WSP</td><td>NONFIN
(Ft)</td><td>AFR</td><td>EXH
(wd)</td><td>MAX
-V</td></tr>
<tr><td>FFC</td><td>a. |(ˈħa)taf|-1PL</td><td>1</td><td></td><td></td><td></td><td>1</td><td>1</td><td></td></tr>
<tr><td>insert morph</td><td>b. →$_{L}$ |(ˈħa)taf$^{\mu}$.na|</td><td></td><td></td><td>1</td><td></td><td>2</td><td>2</td><td></td></tr>
<tr><td>Delete V</td><td>c. ☹ |(ˈħatf)|-1PL</td><td>1</td><td></td><td></td><td>1</td><td></td><td></td><td>1</td></tr>
</table>

To preview what we ultimately say about syncope, the representation |(ˈħa)taf|-1PL contains a syllable /taf/ which is not parsed into a foot but instead linked directly to the PWd, which violates EXHAUSTIVITY(word). That constraint will serve ultimately as our

18 Kiparsky (2011: 5) objects to the unary foot assumed in this analysis on the grounds that Maltese has no monomoraic words, implying that the minimal foot of the language is bimoraic. This objection will not go through if we accept arguments (Piggott 1993, 2010; Garrett 1999; Moore-Cantwell this volume) that word-minimality restrictions are direct constraints on the size of words, rather than a side effect of constraints on the size of feet.

driver of syncope: syncope eliminates the vowels of unfooted syllables, and thus the syllables themselves as well.[19] However, syncope does not affect the syllable /taf/, for two reasons: it is closed as opposed to open, and it is word-final. Deletion in word-final syllables will be blocked because eliminating the final syllable would result in the head foot being final, pointlessly exchanging a violation of Exh(wd) for a violation of higher-ranked NonFinality(foot). This explanation for the failure of deletion to affect final vowels is the same as that proposed by Gouskova (2003: 232) for the corresponding generalization in Lebanese Arabic; however, owing to the serial-optimization regime being assumed here, things get slightly more complicated in the case where we spell out the suffix before attempting syncope, as we shall see momentarily.

Another thing to note is that in candidate (29b), the coda of the syllable /taf^{μ}/ has become moraic, as the addition of the suffix means that this consonant is no longer word-final. McCarthy (2007: §3.2.4.1) argues that the insertion of moras which mark the predictable weight of coda consonants occurs 'for free' at each step of an OT-CC derivation rather than being a step in its own right. This would be consistent with arguments that the insertion of such moras involves no faithfulness violation: faithfulness to the (non-)moraicity of consonants predicts the existence of contrastive moraicity and hence contrastive syllabification, which is generally held to be unattested (Bermúdez-Otero 2001; McCarthy 2003a; Campos-Astorkiza 2004; though cf. Elfner 2006, 2007). My analysis of vowel-final stems will provide another means by which the stem-final consonant would instantaneously acquire a mora upon suffixation. Specifically, I will be proposing that Maltese suffixes begin with a floating mora, to account for the lengthening of vowels before suffixes. With both V-final and C-final stems, we can assume that this floating mora docks on the stem-final segment. With V-final stems, this changes the final vowel from short to long; with C-final stems, the mora simply attaches to the previously non-moraic stem-final coda consonant, causing the stem-final CVC syllable to become heavy.[20]

19 Exhaustivity(word) will be the driver of syncope in all of the examples discussed in this paper. In /...LLL/ forms which initially get an antepenult parse |...(ˈLL)L|, syncope of the penult vowel will instead be a response to *VPl-weak, which forbids vowel place features in the dependent position of a foot, as in the analysis of metrically-conditioned syncope in McCarthy (2008).

20 Thanks to Abdelaziz Boudlal for prompting me to be more explicit about the fate of the floating mora with C-final stems.

After suffixal /-na/ has been added, it is harmonically improving to foot the now-medial syllable /taf$^{\mu}$/, since doing so no longer violates NONFINALITY(ft). Below in (30), I assume that head-foot status is immediately transferred from the previously-built foot to the new foot,[21] though this is not essential:

(30) With suffix added, putting stress on stem-final syllable is now harmonically improving

Input from (29b)	|(ˈħa)taf$^{\mu}$.na|	MAX-V (heavy)	WSP	NON FIN (Ft)	AFR	EXH (wd)	MAX-V
FFC	a. |(ˈħa)taf$^{\mu}$.na|		1		2	2	
Add foot	b. →$_{L}$ |(ˌħa)(ˈtaf$^{\mu}$)na|				3	1	
	c. |(ˌħa)(ˈtaf$^{\mu}$.na)|			1	2		
	d. ☹ |(ˌħa)taf$^{\mu}$(ˈna)|		1	1	2	1	
Delete V	e. →$_{L}$ |(ˈħa)taf$^{\mu}$n|		1		1	1	1
	f. ☹ |(ˈħat$^{\mu}$)fna|	1			1	1	1

If we continue this derivational path by building a foot on /taf$^{\mu}$/, we will have reached the output form corresponding to 'cyclic' stress: main stress is on the penult and secondary stress on the antepenult; the presence of that secondary stress means that the first syllable will not be syncopated. Deleting the first vowel would eliminate a violation of ALL-FOOT-RIGHT, but we may assume that this is blocked by a high-ranked positional faithfulness constraint (Beckman 1998) barring deletion of stressed vowels. The final vowel, though unfooted, will not delete either, since syncope (or, rather, apocope) is blocked word-finally: subsequently deleting the final vowel from (30b), yielding |(ˌħa)(ˈtaf$^{\mu}$n)|, would create a violation of NONFIN(ft) and therefore is not harmonically improving, as mentioned earlier.

Some discussion is in order, however, about vowel deletion as a possible step to take from |(ˈħa)taf$^{\mu}$.na|. Because this derivational stage is obtained by first building a unary foot on the initial syllable and then adding the suffix, the sole foot that exists so far is two syllables away from the right edge. This means that deleting either of the unfooted vowels would not be locally blocked by NONFIN(ft), since one syllable will still remain to separate the foot from the right edge. Excluding the

21 See Pruitt (2012) for arguments that it is necessary to allow transfer of primary-stress status from one foot to another to occur 'for free', simultaneously with other operations.

not-previously-discussed constraint MAX-V(heavy) seen above, the locally-optimal choice of which vowel to delete would be candidate (30f), |(ˈħatᵘ)fna|, because it eliminates the unfooted (and perforce unstressed) heavy syllable /tafᵘ/, thereby disposing of a violation of WSP. This candidate thus highlights a general concern about the analysis: since WSP is higher-ranked than MAX-V, we have to say something about why the language always copes with unstressed heavy syllables by footing them, rather than eliminating them by deleting their nuclei.

The candidate |(ˈħatᵘ)fna| is worrisome because, again given the constraints depicted above, it is more harmonic than the attested winner |(ˌħa)(ˈtafᵘ)na| by virtue of having two fewer violations of ALL-FOOT-RIGHT. Something therefore needs to rule it out, which does not come as too much of a surprise precisely because |(ˈħatᵘ)fna| has syncopated the vowel out of a closed syllable, contrary to all published formulations of Maltese syncope, which agree that it is something that happens only in (unstressed) open syllables. It seems that WSP must be outranked by some other constraint, which discourages syncope in closed syllables. It is tempting to consider the possibility that forms like *|(ˈħatᵘ)fna| are blocked due to the tri-consonantal cluster which closed-syllable syncope would give rise to. However, it is not clear that this can be the right story, because /CVCVC/ verb stems have alternants in their imperfect paradigms in which both stem vowels have been deleted through ordinary, open-syllable syncope, such as [na-ʔsm-úː-h] 'we broke him' (Sutcliffe [1936: 156]). A more promising explanation would be based on positional faithfulness (Beckman 1998) to heavy syllables, which is the option depicted in (30): deleting vowels in heavy syllables would violate a constraint MAX-V(heavy) which is ranked above WSP.

It may be independently necessary to posit such a constraint in OT-CC given a difficulty with syncope in Icelandic pointed out by Norris (2013). Icelandic syncope does not affect vowels in heavy syllables, and Norris presents a Stratal OT account of this which requires that the constraint ranking at the time feet are built differs from the ranking at the time that syncope occurs, which is possible across levels in Stratal OT but which is not countenanced in OT-CC or HS. If there were a constraint that could directly ban syncope in heavy syllables, however, it would appear that we could bypass this problem.[22]

22 For a previous proposal for positional faithfulness to heavy syllables, see Akinlabi and Urua (2000: 286).

If *|('ħat$^{\mu}$)fna| is barred on such grounds from being harmonically improving, this still leaves the apocope-derived |('ħa)taf$^{\mu}$n| as a valid next step. Where can we go from there? Deletion of the remaining unfooted vowel is ruled out by virtue of its being in a closed syllable, so the only thing further that can be done is to build a foot on the word-final CVCC syllable, which is harmonically improving given that WSP dominates NONFIN(ft):

(31)

<table>
<tr><td>Input from (30e)</td><td></td><td>|('ħa)taf$^{\mu}$n|</td><td>WSP</td><td>NON FIN (Ft)</td><td>AFR</td><td>EXH (wd)</td><td>MAX-V</td></tr>
<tr><td>FFC</td><td>a.</td><td>|('ħa)taf$^{\mu}$n|</td><td>1</td><td></td><td>1</td><td>1</td><td></td></tr>
<tr><td>Add foot</td><td>b. →$_L$</td><td>|(ˌħa)('taf$^{\mu}$n)|</td><td></td><td>1</td><td>1</td><td></td><td></td></tr>
</table>

As we will see shortly when we compare the completed chains, NONFINALITY(ft) will end up playing a role in blocking word-final vowel deletion after all. While it cannot stop the mapping |('ħa)taf$^{\mu}$.na| → |('ħa)taf$^{\mu}$n| being harmonically improving, a chain ending <...|('ħa)taf$^{\mu}$n|> will fare poorly compared to the attested winner |(ˌħa)('taf$^{\mu}$)na| (when /-na/ is an object marker) or |('ħtaf$^{\mu}$)na| (when /-na/ is a subject marker): |('ħa)taf$^{\mu}$n| violates WSP, but the attested winners do not. As seen above, the WSP violation can be eliminated in the continuation of that chain as <...|('ħa)taf$^{\mu}$n|, |(ˌħa)('taf$^{\mu}$n)|>, but that longer chain violates NONFIN(ft), whereas the attested winners do not. Thus does NONFIN(ft) contribute to ruling out deletion of word-final vowels, even in the case where it can't block such deletion from being harmonically improving.

We have now finished exploring the set of derivational paths which originate from the option of laying down stress (26c) before spelling out the suffix; these are, to speak loosely, the candidate derivational paths with 'cyclic stress'. Now we need to explore the potential further steps which could follow on from the other harmonically improving option in (26b): the form /ħa.taf$^{\mu}$.na/, which spells out the object suffix rather than building a PWd. These candidate pathways are those with 'non-cyclic stress', because they all begin from morphologically building up the whole word, stem and suffix, and only afterwards assigning stress. Starting from /ħa.taf$^{\mu}$.na/, the first thing that we can do is to build a PWd and head foot:

(32) From unprosodified stem+suffix, stressing stem-final syllable locally optimal

Input from (26b)		ħa.taf$^{\mu}$.na	MAX-M	WDCON	WSP	NONFIN (Ft)	AFR	EXH (wd)	MAX-V
FFC	a.	ħa.taf$^{\mu}$.na		1					
Build PWd	b. →$_L$	|ħa('taf$^{\mu}$)na|					1	2	
	c.	|('ħa.taf$^{\mu}$)na|			1		1	1	
	d.	|ħa('taf$^{\mu}$.na)|				1		1	
	e.	|ħa.taf$^{\mu}$('na)|			1	1		2	
	f.	|('ħa)taf$^{\mu}$.na|			1		2	2	

As seen above, the locally optimal way to build a PWd and foot parse on this string is to have the single head foot on the heavy penult /taf$^{\mu}$/. Since this is the penult and not the final syllable, it comes with one violation of ALL-FOOT-RIGHT. Three other candidates in (32) do at least as well as (32b) on AFR; however, they are less harmonic than (32b) by virtue of violating either NONFINALITY(ft) (because they foot the final syllable), WSP (because they leave the heavy penult unstressed), or both.

From here, we now find that syncopating the unfooted vowel of the first syllable will improve harmony, but parsing that syllable into a foot will not:

(33) From form above, deleting first vowel is harmonically-improving; footing it isn't

Input from (32b)		|ħa('taf$^{\mu}$)na|	MAX-M	WDCON	WSP	NONFIN (Ft)	AFR	EXH (wd)	MAX-V
FFC	a.	|ħa('taf$^{\mu}$)na|					1	2	
Delete V	b. →$_L$	|('ħtaf$^{\mu}$)na|					1	1	1
	c. 👎	|ħa('taf$^{\mu}$n)|				1		1	1
Add foot	d. 👎	|(ˌħa)('taf$^{\mu}$)na|					3	1	
	e. 👎	|ħa(ˌtaf$^{\mu}$)('na)|				1	1	1	

Footing the initial (antepenultimate) syllable removes a violation of EXH(wd), but at the cost of adding two new violations of higher-ranked AFR. Syncopating the vowel, by contrast, swaps the EXH(wd) violation for a violation of lower-ranked MAX-V, so it is harmonically-improving. With this form, |('ħtaf$^{\mu}$)na|, we have now reached the output corresponding to 'non-cyclic' stress: the sole stress is on the surface penult (expected because the penult is heavy) and all vowels in nonfinal, unstressed open syllables have been syncopated. The candidate |('ħtaf$^{\mu}$)na| is the locally-optimal way to delete because deletion of the only other unfooted vowel, the word-final one, is not harmonically improving, for the by now familiar reason that it would create a new violation of NONFIN(ft).

Having finished building our possible derivations, they will now compete against one another as candidates. There are actually two distinct competitions to consider: one in which the morph /-na/ is spelling out an object morpheme, and one in which /-na/ is spelling out a subject morpheme. In the former case, we want the derivation that ends in |(ˌħa)(ˈtaf$^{\mu}$)na| to win, and in the latter case, we want the derivation terminating in |(ˈħtaf$^{\mu}$)na| to win. If all of the chains we've built (including those which stop somewhere 'along the way') are submitted to the constraint ranking assumed so far, the chain leading to |(ˈħtaf$^{\mu}$)na| will indeed win (to save space, MAX-V(heavy) is omitted from the tableau, since no candidate violates it):

(34) Competition of chains I: 'Non-cyclic' candidate wins

<table>
<tr><th></th><th>snatch-1pl</th><th>MAX-M</th><th>WDCON</th><th>WSP</th><th>NONFIN (Ft)</th><th>AFR</th><th>EXH (wd)</th><th>MAX-V</th></tr>
<tr><td>a.</td><td><SNATCH-1PL></td><td>2 W</td><td></td><td></td><td></td><td>L</td><td>L</td><td>L</td></tr>
<tr><td>b.</td><td><SNATCH-1PL, ħa.taf-1PL></td><td>1 W</td><td>1 W</td><td></td><td></td><td>L</td><td>L</td><td>L</td></tr>
<tr><td>c.</td><td><SNATCH-1PL,
ħa.taf-1PL, ħa.taf$^{\mu}$.na></td><td></td><td>1 W</td><td></td><td></td><td>L</td><td>L</td><td>L</td></tr>
<tr><td>d.</td><td><SNATCH-1PL, ħa.taf-1PL,
ħa.taf$^{\mu}$.na, |ħa(ˈtaf$^{\mu}$)na|></td><td></td><td></td><td></td><td></td><td>1</td><td>2 W</td><td>L</td></tr>
<tr><td>e. →</td><td><SNATCH-1PL, ħa.taf$^{\mu}$-1PL,
ħa.taf$^{\mu}$.na, |ħa(ˈtaf$^{\mu}$)na|,
|(ˈħtaf$^{\mu}$)na|></td><td></td><td></td><td></td><td></td><td>1</td><td>1</td><td>1</td></tr>
<tr><td>f.</td><td><SNATCH-1PL, ħa.taf-1PL,
|(ˈħa)taf|-1PL></td><td>1 W</td><td></td><td></td><td></td><td>1</td><td>1</td><td>L</td></tr>
<tr><td>g.</td><td><SNATCH -1PL, ħa.taf-1PL,
|(ˈħa)taf|-1PL, |(ˈħa)taf$^{\mu}$.na|></td><td></td><td></td><td>1 W</td><td></td><td>2 W</td><td>2 W</td><td>L</td></tr>
<tr><td>h.</td><td><SNATCH -1PL, ħa.taf-1PL,
|(ˈħa)taf|-1PL, |(ˈħa)taf$^{\mu}$.na|,
|(ˌħa)(ˈtaf$^{\mu}$)na|></td><td></td><td></td><td></td><td></td><td>3 W</td><td>1</td><td>L</td></tr>
<tr><td>i.</td><td><SNATCH -1PL, ħa.taf-1PL,
|(ˈħa)taf|-1PL, |(ˈħa)taf$^{\mu}$.na|,
|(ˈħa)taf$^{\mu}$n|></td><td></td><td></td><td>1 W</td><td></td><td>1</td><td>1</td><td>1</td></tr>
<tr><td>j.</td><td><SNATCH -1PL, ħa.taf-1PL,
|(ˈħa)taf|-1PL, |(ˈħa)taf$^{\mu}$.na|,
|(ˈħa)taf$^{\mu}$n|, |(ˌħa)(ˈtaf$^{\mu}$n)|></td><td></td><td></td><td></td><td>1 W</td><td>1</td><td>L</td><td>1</td></tr>
</table>

Here, the 'non-cyclic' candidate (34e) bests the 'cyclic' candidate (34h), because the latter candidate has an additional non-final foot (on the initial syllable (ˌħa)) and therefore has two extra violations of ALL-FOOT-RIGHT. To get (34h) to win instead when the morpheme notated as '1PL' is an object rather than a subject marker, it must therefore be favored by some constraint which outranks AFR and which is relevant only in the object-marker case. As alluded to earlier, this 'cyclicity'-encouraging constraint will be a PREC constraint, specifically the following:

(35) PREC(build-PWd, insert-object)
Assign a violation mark for each time that:
(a) An object marker morph is inserted, and this is not preceded by the building of a PWd; or
(b) An object marker morph is inserted, and this is followed by the building of a PWd.

Tableau (36) illustrates how this constraint gives us the desired result. To save space, I omit all of the losing candidates from (34) which are knocked out by one of the other constraints ranked above AFR (however, lettering of candidates has been retained from (34) to facilitate comparison):

(36) Competition of chains II: where PREC constraint is applicable, 'cyclic' candidate wins

<table>
<tr><th></th><th></th><th>snatch-1pl</th><th>Prec
(build-Pwd,
ins-obj)</th><th>AFR</th><th>EXH
(wd)</th><th>MAX-
V</th></tr>
<tr><td></td><td>d.</td><td><SNATCH-1PL, ħa.taf-1PL,
ħa.taf$^{\mu}$.na, |ħa(ˈtaf$^{\mu}$)na|></td><td>2 W</td><td>1 L</td><td>2 W</td><td></td></tr>
<tr><td></td><td>e.</td><td><SNATCH-1PL, ħa.taf$^{\mu}$-1PL,
ħa.taf$^{\mu}$.na, |ħa(ˈtaf$^{\mu}$)na|,
|(ˈħtaf$^{\mu}$)na|></td><td>2 W</td><td>1 L</td><td>1</td><td>1 W</td></tr>
<tr><td></td><td>h. →</td><td><SNATCH -1PL, ħa.taf-1PL,
|(ˈħa)taf|-1PL, |(ˈħa)taf$^{\mu}$.na|,
|(ˌħa)(ˈtaf$^{\mu}$)na|></td><td></td><td>3</td><td>1</td><td></td></tr>
</table>

Candidate (36e) builds the PWd after (and not before) the object maker was inserted, thus violating both clauses of the PREC constraint. Candidate (36h) by contrast has PWd-construction occurring before (and not occurring after) the insertion of the object marker, so it violates neither clause. Ranking the PREC constraint above ALL-FOOT-RIGHT thus results in (36h) beating (36e) when the /-na/ morph is serving as an object marker.

Before moving on to consider the lack of cyclicity in vowel-final stems, we should briefly examine the issue of secondary stress mentioned earlier. Because secondary stress is not mentioned in descriptions of Maltese, Brame (1972, 1974) and Kiparsky (2011: 5–6) suggest that the cyclic secondary stress in a word like |(ˌħa)(ˈtafᵘ)na| can be deleted late in the derivation, after syncope has applied (to ensure that de-footing of the first syllable does not feed syncope of its vowel). Nothing in principle would prevent us from adding the same idea to the OT-CC analysis presented here. Under the ranking already assumed, a 'cyclic' derivation in which we ultimately delete the cyclic stress would be harmonically improving, as shown in the following harmonic-improvement tableau:[23, 24]

(37) Harmonic-improvement tableau for 'cyclic' stress followed by deletion of previous-cycle stress

	|ħa.taf|-1PL	MAX-M	WDCON	WSP	AFR	EXH (wd)	MAX-V
Input	a. SNATCH-1PL	2					
Insert root	b. ħataf-1PL	1	1				
Build PWd	c. |(ˈħa)taf|-1PL	1			1	1	
Insert suffix	d. |(ˈħa)tafᵘ.na|			1	2	2	
Build foot	e. |(ˌħa)(ˈtafᵘ)na|				3	1	
Delete foot	f. |ħa(ˈtafᵘ)na|				1	2	
Delete V	g. |(ˈħtafᵘ)na|				1	1	1

The only thing which would be necessary to add would be to ensure that, in chain comparison, the chain which stops at (37f) beats the one which continues on to (37g) – that is, to ensure that defooting cannot be followed by syncope. If PREC(MAX-V, 'MAX-FOOT') outranks EXH(wd),

23 This analysis requires that we allow GEN to undo an earlier operation, in this case removing a foot it had previously built. McCarthy (2007) assumes that this is not allowed, though he notes that such derivations can be harmonically improving, as this one is. That derivations of this sort should be allowed has previously been argued for by Zhang and van de Weijer (2008).

24 While the chain which ends in (37f) does end in the same surface form as chain (34d) (and likewise the chain ending in (37g) in the same surface form as (34e)), these pairs of chains will not be merged, since the sets of operations they perform en route to that surface form are not the same: (37f) has a step of building a foot (as opposed to building a PWd and head foot), but (34d) has no such step.

we can satisfy this desideratum.[25] Lastly, it should also be noted that while the analysis I am proposing is *compatible* with the assumption that the secondary-stress foot is deleted, it is not committed to that assumption either – if we wanted to retain the secondary stress, it would suffice to assume that ALL-FT-RIGHT is outranked by a faithfulness constraint which prohibited the removal of a foot.

11.5 The lack of cyclicity in vowel-final stems

The analysis just given for consonant-final stems will, if left as is, lead us to expect that vowel-final stems will also show cyclic stress in object-marked forms, since that is what PREC(build-PWd, insert-object) will prefer. Since vowel-final stems do not have cyclic stress with object suffixes, that constraint's demands will have to be overridden by a factor connected with something that happens only in vowel-final stems. Building on the insights in Odden (1990, 1993),[26] the analysis I will give below will assume that this something is the lengthening of stem-final vowels before suffixes.

Outside of the mainly borrowed words with final (stressed) long vowels mentioned earlier in (21), Maltese patterns with many Arabic dialects in that morph-final vowel length is predictable: these vowels are short word-finally but long before consonant-initial suffixes. (The feminine

25 If this de-footing analysis should prove unfeasible, at least two other general responses to the secondary-stress issue are imaginable. First, we could adopt proposals to the effect that feet may exist without corresponding to an overt stress (Crowhurst 1996; Hyde 2002; Buckley 2009; Vaysman 2009; among others). Second, it might be possible to claim that the syncope of unstressed vowels in Maltese serves to partially obscure whatever phonetic cues would differentiate secondary stress from unstressedness, such that the cyclic secondary stresses are present on the surface, but are not directly recognizable as such. I am grateful to Maria Gouskova for pointing out this possibility.

26 Odden's (1990, 1993) analysis posits that all morphology precedes all phonology, but that phonology does operate cyclically on morphological constituents. He proposes that pre-suffixal lengthening applies pre-cyclically, so on the stem cycle, the final vowel is already long if there is a suffix on the next cycle. This results in the stem-final vowel getting stress on the first cycle. The problem for this approach is that it predicts that phonologically-conditioned allomorphy of an affix cannot be sensitive to derived phonological properties of the affix's base; despite the counter-arguments given in Odden (1990, 1993) against certain proposed examples, this seems to be well-attested (see e.g. Kiparsky [1996: 24–25], Wolf [2009]).

ending /-a/ in Maltese is an exception (Borg 1997: 267); it is consistently short, even before a following suffix.) As with any property which varies allophonically, there is potential indeterminacy regarding the *underlying* length of such vowels. Are they underlyingly long, shortening when word-final, or are they underlyingly short, lengthening before suffixes? Both assumptions have widely been made in the literature on the phonology of Arabic dialects; see McCarthy (2005), who favors the shortening analysis, for extensive references. Maltese is no exception: Borg (1997: 266–267) assumes shortening, and Odden (1990, 1993) assumes lengthening.

Here, I will be taking the latter option and assuming that the stem-final vowels of Maltese third-weak-radical verbs are short in the underlying representation. In an OT context, this immediately poses a worry about Richness of the Base: we cannot assume any systematic language-specific restrictions on underlying forms, so we must ensure that inputs with underlying final long vowels are mapped onto some licit surface form of the language. I will suggest that words like those in (21) already furnish the evidence we need about what happens to underlyingly long word-final vowels in Maltese: they surface faithfully as long, and receive stress.[27]

27 An anonymous reviewer objects that word-final monophthongal long vowels are limited to borrowed vocabulary, which might imply that there is still a ROTB problem with respect to the Semitic vocabulary. If this objection goes through, we would have to take into account the possibility of native stems with underlyingly final long vowels, and also to countenance the existence of final shortening. That might not be a problem, though. If having a final short vowel is the source of the failure of cyclic stress in V-final stems (as I propose), then the worst that would happen is that stems with a final long vowel would revert to having cyclic stress (i.e., not syncopating the first stem vowel). Since there are idiolects in which some exceptional V-final stems do resist syncope (see n. 7 above), this is fine: the underlying contrast in final vowel length would then map onto a surface contrast in showing or not showing underapplication of syncope. In that case, the basic generalization under analysis here would be that V-final stems can differ in whether they show underapplication of syncope, whereas C-final stems cannot.

Here, in a bit more formal detail, is the argument as to why we might expect long-vowel-final stems to show the 'cyclic' pattern. Consider the object-marked derivatives of a stem of the form /mela:/. Native words ending in diphthongs (like (21a)) make it reasonable to infer that a final long vowel, if left intact, would get stress. Therefore, there are two convergent derivations leading to surface [ˈmla:.na], one which assigns stress before suffixation, and one which assigns stress after:

<FILL-1PL, me.la:-1PL, |me(ˈla:)|-1PL, |me(ˈla:)na|, |(ˈmla:)na|>
<FILL-1PL, me.la:-1PL, me.la:.na, |me(ˈla:)na|, |(ˈmla:)na|>

If stem-final vowels in third weak radical verbs are underlyingly short, then they must systematically lengthen before suffixes. What mechanism is responsible for this? As mentioned earlier, I will assume that all verbal suffixes have come to be analyzed by speakers as containing a floating mora, which docks on and lengthens a preceding stem-final vowel.[28] The account of the absence of cyclic effects will then run like this: ranked above PREC(build-PWd, insert-object) is another PREC constraint which demands that stress be preceded by lengthening. With consonant-final stems, there is no opportunity for lengthening and so the constraint will be equally violated by all candidates (except those where no stress is assigned at all, which will be knocked out by higher-ranked WDCON and/or MAX-M). With vowel-final verbs, on the other hand, we can satisfy the new PREC constraint by applying stress only after lengthening. Since lengthening is caused by suffixation, doing stress after lengthening entails deferring stress until after suffixation – that is, not having cyclic stress.

Being convergent, these chains will be merged prior to chain comparison (see note 10 above); that is, combined into one candidate which retains only those pairwise ordering relations among operations which is shared by all of the convergent chains. In this case, the merged candidate will assert no ordering relation between PWd-construction and object suffixation, since the pairwise order of these differs in the two convergent chains. The merged candidate leading to |(ˈmla:)na|—the 'non-cyclic' form—thus will violate PREC(build-PWd, ins-obj); it will also violate PREC(IDENT(long), build-PWd), since it has no IDENT(long)-violating step.

Amongst the competing candidates will be this one (assuming, again, per the reviewer's objection, that shortening of final vowels is harmonically improving):

<FILL-1PL, me.la:-1PL, me.la-1PL, |(ˈme)la|-1PL, |(ˈme)la:.na|, |(ˌme)(ˈla:)na|>

This chain, which corresponds to the 'cyclic' outcome (i.e., the first stem vowel doesn't syncopate), also violates PREC(IDENT(long), build-PWd), since the re-lengthening of the stem-final vowel occurs after stress assignment; however, it satisfies PREC(build-PWd, insert-obj), since it assigns stress prior to object suffixation. Thus, we expect the chain ending in |(ˌme)(ˈla:)na| to beat the merged candidate for surface |(ˈmla:)na|.

28 Joe Pater points out that this leaves open the possibility that there could be suffixes that fail to trigger lengthening. The obvious place to look for this in Maltese would be amongst suffixes borrowed from Italian or English, as opposed to those inherited from Arabic. One such possibility is the English plural suffix /-s/, which some have suggested is being rapidly integrated into the language, and which is reportedly inert with respect to stress, at least when added to a consonant-final noun (Vella 2009: 67).

I will illustrate the analysis by showing the derivation of [(i)mlɪːna] 'he filled us'. In the following tableaux illustrating the construction and comparison of derivational paths, I will ignore for the sake of brevity two processes which affect this word: (a) the prothesis of [i] before the syncope-created sonorant-initial cluster (see n. 29), and (b) the raising of stressed long /aː/ to /ɪː/ (see n. 6). As with the consonant-final stems, the first harmonically-improving step which can be taken is to insert the stem morph:

(38) Insertion of stem is harmonically improving

	fill-1pl	MAX-M	WDCON	WSP	NONFIN (Ft)	AFR	EXH (wd)	MAX-V
FFC	a. FILL-1PL	2						
insert morph	b. → L me.la-1PL	1	1					

After the stem is inserted, the derivation splits (again as before): it is harmonically improving to lay down a PWd and its head foot (with the initial syllable being the locally optimal location for stress), and it is also harmonically improving to insert the suffix morph. For ease of presentation, I will assume that the insertion of the segments of the pronominal suffix and the associated lengthening mutation of the stem-final vowel occur simultaneously, in one step. This is not crucial to the argument, though: the same results will obtain if we assume that insertion of the object morph results in an intermediate stage /me.la.$^{(\mu)}$na/, where the floating mora in the suffix's UR is not yet docked on the preceding vowel.[29]

29 The reasoning is as follows. For a stem like /mela/, it will never be locally optimal to place stress on the second syllable until after lengthening (and therefore, after suffixation) has occurred. If we tried assigning stress before suffixation, the locally optimal parse would be |(ˈme)la| (see (39c-e).) If we assigned a foot parse to the string /me.la.na/ (after suffixation but before docking the floating mora), the locally optimal parse would be |(ˈme.la)na|, with antepenult stress. By contrast, after lengthening, the locally-optimal choice is to stress the heavy penult (see (40)). After reaching |me(ˈlaː)na|, it is not harmonically improving to build any more feet (see (41)). Ergo, to get a stress anywhere other than on the second syllable, we would have to assign stress before lengthening, and thus violate PREC(ID(long), build-PWd)).

(39) Insertion of object marker, or laying down PWd, are harmonically improving

Input from (38b)	me.la-1PL	Max-M	WdCon	WSP	NonFin (Ft)	AFR	Exh (wd)	Max -V
FFC	a. me.la-1PL	1	1					
insert morph	b. $\rightarrow_L$ me.laː.na		1					
Build PWd	c. $\rightarrow_L$ |(ˈme)la|-1PL	1				1	1	
	d. |me(ˈla)|-1PL	1			1		1	
	e. |(ˈme.la)|-1PL	1			1			

On the derivational branch initiated by inserting the suffix morph (with concomitant morpho-phonological lengthening of the stem-final vowel), the one thing that it is harmonically improving to do next is to build a PWd whose head coincides with the stem-final syllable:

(40) From unprosodified stem+suffix, stressing stem-final syllable locally optimal

Input from (39b)	me.laː.na	Max-M	WdCon	WSP	NonFin (Ft)	AFR	Exh (wd)	Max-V
FFC	a. me.laː.na		1					
Build PWd	b. $\rightarrow_L$ |me(ˈlaː)na|					1	2	
	c. |(ˈme.laː)na|			1		1	1	
	d |me(ˈlaː.na)|				1		1	
	e. |me.laː(ˈna)|			1	1		2	
	f. |(ˈme)laː.na|			1		2	2	

This is closely analogous to what we saw earlier with CVC-final stems. Adding a suffix puts the stem-final syllable in penult position and also makes it heavy, either by allowing its coda to become moraic (with C-final stems) or by inducing lengthening of the stem-final vowel (with V-final stems). In either case, the resultant /σHL/ sequence will subsequently be assigned penult main stress.

After this stress has been assigned, it is harmonically improving to syncopate the first vowel so as to eliminate a violation of Exhaustivity(word). As before, this is locally-optimal relative to the alternative option of syncopating the word-final vowel, which is harmonically worsening because it would create a new violation of higher-ranked NonFin(foot). Also as before, it is not harmonically improving to eliminate the Exhaustivity(word) violation of the initial syllable by footing it; just as with consonant-final stems (see (33)), this is ruled out by ALL-FOOT-RIGHT being ranked above EXHAUSTIVITY(word):[30]

(41) From form above, deleting first vowel is harmonically-improving; footing it isn't

Input from (40b)	|me(ˈla:)na|	MAX-M	WDCON	WSP	NONFIN (Ft)	AFR	EXH (wd)	MAX-V
FFC	a. |me(ˈla:)na|					1	2	
Delete V	b. →$_L$ |(ˈmla:)na|					1	1	1
	c. 👎 | me(ˈla:n)|				1		1	1
Add foot	d. 👎 |(ˌme)(ˈla:)na|					3	1	
	e. 👎 |me(ˌla:)(ˈna)|				1	1	1	

30 Since syncope at this step can create a sonorant-initial onset cluster /ml/, EXHAUSTIVITY(word) must outrank the constraint against such clusters. Maltese adds a prothetic /i/ before such word-initial clusters, unless the preceding word ends in a vowel and is in the same phonological phrase (Sutcliffe 1936: 16; Brame 1972: 33–34; Borg 1997: 273–274). For prothesis to improve harmony, it must not create a new violation of EXHAUSTIVITY(word), i.e. the syllable headed by the prothetic vowel must fall outside of the PWd. Along the lines of de Lacy (2004) and Flack (2009: 292–294), we may assume that the PWd left- and right-edge alignment constraints abbreviated as WDCON require only syllables whose nuclei belong to the morphological word to be parsed into the PWd. Since epenthetic vowels lack a morphological affiliation, the prothetic vowel can then be extraprosodic, and thus not incur a violation of EXHAUSTIVITY(word). This solution obviously would not work for vowel epenthesis into word-medial CRC clusters; however, claimed cases of this seem amenable to re-analysis. First, there is epenthesis to the left of the indirect-object suffix /-l-/ (see n. 9) when it occurs between two consonants. This could, if necessary, be analyzed as a choice between listed allomorphs {l, il} rather than as phonological epenthesis. Second, Hume (1994: ch. 6) argues that CRVCV → CVRCV metathesis is really syncope followed by epenthesis (see also Kiparsky [2003]). However, in order to predict the quality of the vowel inserted before the sonorant, the analysis in Hume (1994) assumes that 'syncope' leaves behind a floating vocalic melody which then associates to an epenthetic V-slot. The empirical argument for this comes from cases where the metathesized vowel is /i/. Maltese has a process of lowering /i/ to [a] when it is adjacent to a guttural (mentioned earlier in n. 3), but metathesized vowels do not undergo lowering in either their pre- nor post-metathesis position. Therefore, Hume argues, the 'metathesized' vowel melody must remain temporarily floating so as to wait out the application of lowering. It might be possible to account for this while retaining the assumption of literal metathesis by assuming the following: (a) lowering cannot apply to the output of metathesis due to an ordering requirement like PREC(IDENT(high), LINEARITY), and (b) if lowering is analyzed as some sort of autosegmental feature spreading

Now let us go back to consider the other branch of the derivation from (39): the one with 'cyclic' stress (39c), where we construct a foot parse without having yet inserted the suffix morph. If we first assign a PWd and foot parse to the stem (with initial stress), it is then harmonically-improving to spell out the suffix (which again per our assumptions comes with simultaneous lengthening of the stem-final vowel). Deleting the unfooted second syllable will not improve harmony at this point, again due to the blocking effect of NONFIN(ft):

(42) With stem prosodified, insertion of suffix is harmonically improving

Input from (39c)	|(ˈme)la|-1PL	MAX-M	WDCON	WSP	NONFIN (Ft)	AFR	EXH (wd)	MAX-V
FFC	a. |(ˈme)la|-1PL	1				1	1	
insert morph	b. →$_L$ |(ˈme)laː.na|			1		2	2	
Delete V	c. ♀ |(ˈmel)|-1PL	1			1			1

After this, it will be harmonically improving to foot the (now long and penultimate) stem-final syllable, as well as to apocopate the final vowel. Deleting the stem-final vowel is ruled out by MAX-V(heavy):

(43) From above, putting stress on stem-final syllable is harmonically improving

Input from (42b)	|(ˈme)laː.na|	MAX-V (heavy)	WSP	NON FIN (Ft)	AFR	EXH (wd)	MAX-V
FFC	a. |(ˈme)laː.na|		1		2	2	
Add foot	b. →$_L$ |(ˌme)(ˈlaː)na|				3	1	
	c. |(ˌme)(ˈlaː.na)|			1	2		
	d. ♀ |(ˌme)laː(ˈna)|		1	1	2	1	
Delete V	e. →$_L$ |(ˈme)laːn|		1		1	1	1
	f. ♀ |(ˈmel)na|	1			1	1	1

From |(ˌme)(ˈlaː)na|, there is nothing further to do: once again, deletion of the final vowel will be blocked by NONFIN(foot). From |(ˈme)laːn|, however, it will be harmonically improving to foot the final syllable:

(44)

Input from (43e)	|(ˈme)laːn|	WSP	NON FIN (Ft)	AFR	EXH (wd)	MAX-V
FFC	a. |(ˈme)laːn|	1		1	1	
Add foot	b. →$_L$ |(ˌme)(ˈlaːn)|		1	1		

With our chains now completed they compete as candidates. In the tableau below, marks from PREC(build-PWd, insert-obj) are shown under the assumption that /-na/ is serving as an object marker, since this is the situation where cyclic stress would have been (incorrectly) expected if we left the analysis from Section 11.4 unmodified. Also, as in (36), I omit for space reasons those candidates knocked out by markedness or faithfulness constraints ranked above All-Foot-Right. The candidates in (45) are lettered so as to link them with the analogous candidates in the consonant-final case seen in (34) and (36), again for ease of comparison.

(45) Competition of derivational paths: candidate without 'cyclic' stress wins

<table>
<tr><th></th><th>fill-1pl</th><th>Prec (Ident(long), build-PWd)</th><th>Prec (build-PWd, ins-obj)</th><th>AFR</th><th>Exh (wd)</th><th>Max-V</th></tr>
<tr><td>d.</td><td><fill-1pl, me.la-1pl, mela:na, |me('la:)na|></td><td></td><td>2</td><td>1</td><td>2 W</td><td>L</td></tr>
<tr><td>e. →</td><td><fill-1pl, me.la-1pl, mela:na, |me('la:)na|, |('mla:)na|></td><td></td><td>2</td><td>1</td><td>1</td><td>1</td></tr>
<tr><td>h.</td><td><fill-1pl, mela-1pl, |('me)la|-1pl, |('me)la:.na|, |(ˌme)('la:)na|></td><td>2 W</td><td>L</td><td>3 W</td><td>1</td><td>L</td></tr>
</table>

Because the vowel-final status of the stem results in lengthening occurring at the point that we insert the direct object marker, Prec(Ident(long), build-PWd) serves to knock out the 'cyclic stress' candidate (45h), since it assigns stress before spelling out the suffix (and hence before lengthening). This leaves the 'non-cyclic stress' candidates (45d) and (45e), with Exh(word) crucially deciding in favor of (45e), which syncopates the unfooted initial syllable, over (45d), which does not.[31]

from the guttural to the vowel, it would be impossible to lower and then metathesize, because the vowel is tethered to its original position by a [+low] feature which is shared with the guttural consonant. In any case, in light of Hume's (1994) observations, these cases would remain enough like literal metathesis to plausibly require some sort of special treatment.

31 Regarding the mechanism for the suppression of cyclic stress assumed here, there is a legitimate locality worry noted by Kiparsky (2011: 6): there could be languages like real Maltese but where the lengthening that disrupts cyclicity is quite distant from the inner 'cyclic' domain in the word's morphological bracketing. However, the more extreme worry he raises about cyclic stress being suppressed if there is lengthening in another word of the same utterance is probably mistaken. Since there is little chance of lengthening in one word influencing the placement of stress in another word, the two operations would not be *crucially* ordered with respect to one another, and in OT-CC non-crucial orders are filtered out via the mechanism of chain merger (see n. 11 above).

While Prec(Ident(long), build-PWd) serves to discourage 'cyclic' stress in vowel-final stems, it will have no such effect with consonant-final stems. With C-final stems, there is no lengthening, so every candidate which actually does build a prosodic word will equally violate Prec(Ident(long), build-PWd): there simply is no lengthening for PWd-construction to be preceded by. This violation could be avoided by not building a PWd at all, but those candidates will be knocked out by their violations of Max-M and/or WdCon. All other candidates tie on Prec(Ident(long), build-PWd), which allows lower-ranked Prec(build-PWd, ins-obj) and its preference for 'cyclic' stress to emerge as decisive.

There are also several places in Maltese where Prec(Ident(long), build-PWd) proves to be violable with vowel-final stems. The most trivial case is when there are no suffixes at all, as in the third-person masculine singular perfective without object suffixes. Bare stems like /mela/ do indeed receive stress, despite having no opportunity to undergo lengthening owing to the lack of suffixes. This is because WdCon, which demands the presence of a PWd and hence of stress, outranks Prec(Ident(long), build-PWd).

Less trivially, there are two morphological environments in which vowel-final stems do show evidence of undergoing cyclic stress. One occurs in the imperfect conjugation. In imperfects, the stem is preceded by a /CV-/ prefix which marks the person of the subject, and is followed by a suffix /-u/ if the subject is plural; any object markers occur to the right of this plural suffix. This plural suffix is argued by Brame (1972, 1974) to be cyclic: stress applies in the /CV+*stem*/ constituent it attaches to before the /-u/ is added. The argument for the cyclic status of /-u/ applies even when it attaches to a V-final stem. In forms like the one below (Sutcliffe 1936: 118), we can diagnose the presence of cyclic stress from the fact that the first stem vowel, but not the prefix vowel, is syncopated:

(46)	/jV-ʔara-u/	→	[jaʔˈraw]	'they read (imperf)'
	3.subj-read-pl.subj			
	cf. /jV-ʔara/	→	[ˈjaʔ.ra]	'he reads'

As we can see, when /-u/ is attached to a V-final stem, it devocalizes to become an off-glide of the stem-final vowel. The resulting diphthong is bimoraic and so it attracts stress. If we waited to assign stress until after diphthong-formation, the expected parse would be |jaʔa(ˈraw)|. This would leave us without an explanation of why the second but not the first of the unfooted vowels is syncopated. (Indeed, under the ranking we have assumed, syncope of the second vowel would not improve harmony: doing so would result in |jaʔ(ˈraw)|, with a new unfooted heavy syllable. Since WSP outranks EXH(wd), the motivator of syncope, this would be harmonically worsening.) Likewise, if we assigned stress after suffixation but before diphthong-formation, we would expect antepenult stress: |ja(ˈʔa.ra)u|. This is even worse, because it puts a stress on the very vowel we need to syncopate. By contrast, if we apply stress before adding the suffix, the expected parse is antepenult |(ˈja.ʔa)ra|, which we subsequently expect to undergo syncope of the second vowel as it falls in the weak position of a foot (see n. 19). This is just what we see above in the unsuffixed form of the word. What this suggests is that there is another constraint PREC(build-PWd, insert-imperf-pl) which outranks PREC(IDENT(long), build-PWd). That is, assigning a 'cycle' of stress before adding /-u/ is more important than deferring stress-assignment until after lengthening.

The other place where V-final stems show cyclic behavior is when they are negated. Negation is marked by circumfixation of /ma ... -ʃ/, though this arguably is really two separate morphs, since there are specific morphological conditions under which each part will appear without the other (Borg and Azzopardi-Alexander 1997: 88–91).[32] What is important for our purposes is that /-ʃ/, like other suffixes, causes lengthening of a preceding vowel. When /-ʃ/ is attached to a vowel-final stem, the final vowel lengthens and gets stress (and, if /aː/, undergoes raising), but the first stem vowel is not deleted (Misfud [1994: 249], [1995: 301]):

(47) Cyclic stress with /ma ... -ʃ/

bena	'he built'	[ma beniːʃ]	'he did not build'	
dara	'he got used'	[ma daraːʃ]	'he did not get used'	(also Misfud [1995: 119])
nesa	'he forgot'	[ma nesiːʃ]	'he did not forget'	(also Misfud [1995: 119])

This suggests that the first stem vowel is protected by virtue of having been assigned stress 'cyclically', prior to the suffixation of /-ʃ/. That means assigning stress before lengthening, so as with /-u/, we can infer

32 Specifically, negated imperatives take only /-ʃ/ (p. 88), while /ma/ by itself is used when the verb occurs with certain negative pronouns like *ħadd* 'nobody', *xejn* 'nothing', *imkien* 'nowhere', or with the negative adverb *qatt* 'never' (pp. 90–91).

that PREC(IDENT(long), build-PWd) is dominated by a constraint which requires us to build a PWd before adding the negative suffix.[33, 34]

11.6 Conclusion

The distribution of syncope in Maltese shows that neither cyclic stress nor its absence is an immutable property of the language. Normally there is cyclic stress before object markers, but this fails to obtain with vowel-final stems. As we just saw, though, vowel-final stems can and do undergo cyclic stress before other types of suffixes. In this paper I have shown that such facts obtain easily in a theory like OT-CC (and specifically the OI variant of it) where the ordering of processes takes the form of violable pairwise ordering statements. The pressure to assign stress before object suffixation is overridden by a pressure to defer stress until after (pre-suffixal) lengthening. OI is able to model such patterns because the ordering used for any given form is chosen by the constraint hierarchy itself, rather than being hard-wired into the overall modular stucture of the grammar, as in Stratal OT (Kiparsky 2000). OT-CC thus echoes Local Ordering (Anderson 1969, 1974) in that orderings are decided on a form-by-form basis, rather than being pre-established and fixed for the language as a whole.

In a Stratal OT model, if C-final stems undergo stress assignment before receiving object suffixation, we would be committed to the assumption that V-final stems do as well. In order to account for the

33 For reasons too involved to go into here, the constraint in question cannot be a PREC constraint, but instead an alignment constraint (McCarthy and Prince 1993b) requiring /-ʃ/ to appear at the right edge of a PWd. If ranked above MAX-M, this constraint will prevent it from being harmonically-improving to insert /-ʃ/ unless a PWd is already in place.

34 For reasons of space I have not dealt at all here with another potential source of violations of PREC(IDENT(long), build-PWd), namely when the vowel of a V-final suffix lengthens before another suffix, but we would need to apply a 'cycle' of stress to a constituent inside of the suffix which induces the lengthening. This is a concern but not a very big one, since the range of suffix combinations possible in the language produces relatively few such situations where cyclic stress is actually required. Another source of potential cases of stress applying before lengthening involves lengthening/vocalization induced by the abstract consonant /ʕ/ (Brame 1972); we may simply have to extrinsically order /ʕ/-vocalization after stress (in OT-CC, via another PREC constraint), as Brame's analysis in fact assumes.

evident lack of inner-cycle stress on the first vowel of [[mela]-na], we would have to assume that V-final stems get stress on their second vowel when stress is assigned on the stem cycle. Kiparsky (2011) proposes just such an analysis, with the assignment of final stress to V-final stems following from the assumption that the third radical glide remains synchronically underlying (e.g. /melaj/); since V-final stems end in an underlying diphthong, we would expect them to get final stress, in line with (21). The trouble with this analysis is that in the stem-level phonology, we cannot know if an object suffix is destined to be added at the word level or not. Therefore, the unaffixed form of a third-weak-radical stem will receive final stress at the stem level, and then undergo stress retraction as a result of final glide deletion at the word level: /melaj/$\rightarrow_{\text{Stem}}$ me.ˈlaj$\rightarrow_{\text{Word}}$ ˈme.la. Thus, the Stratal OT analysis of Maltese requires that the same underlying string – even without the addition of affixes in the course of the derivation – receive different metrical parses at successive levels. In Wolf (2012), I argue that this needed capability of Stratal OT gives rise to pernicious consequences regarding the surface distribution of stress-conditioned segmental phonology; for example, it lets us model unattested languages in which consonant-final words have aspiration on all voiceless stops in the onsets of stressed syllables, whereas vowel-final words have aspiration on stops in the onsets of *un*stressed syllables. I also show there that HS and OT-CC are immune from this problem because, unlike Stratal OT, they use only a single constraint ranking for all stages of the derivation, which tightly limits how much the metrical parses at different stages of a derivation can differ from one another.

The Maltese example does not seem to be alone in posing problems for traditional implementations of the cycle like Stratal OT. In Wolf (2009), I present an OI analysis of an ordering paradox on Tigrinya (Buckley 1994) involving the order of vowel epenthesis with respect to plural and possessive suffixation. Buckley (to appear) discusses an ordering paradox in Kashaya wherein certain processes must be assigned, but where the choice of whether to perform these processes or not is sensitive to the shape of affixes added at the word level. In Bedouin Hijazi Arabic (Al-Mozainy 1981; McCarthy 2007; Wolf 2010), there is an ordering paradox involving a postlexical process of raising which counterfeeds a lexical process of syncope. These and other examples seem to point to a common conclusion that the order of phonological and morphological processes must be decided *by* the grammar, via constructing and comparing entire candidate derivations with different orders, rather than written into the procedural architecture of the grammar, as in more traditional conceptions.

References

Akinlabi, Akinbiyi, and Eno Urua (2000) Tone in Ibibio verbal reduplication. In H. Ekkehard Wolff and Orin D. Gensler (eds), *Proceedings of the Second World Conference of African Linguistics, Leipzig 1997*, 279–291 Köln: Rüdiger Köppe Verlag.

Al-Mozainy, Hamza Q. (1981) *Vowel Alternations in a Bedouin Hijazi Arabic Dialect: Abstractness and Stress*. Doctoral dissertation, University of Texas, Austin.

Anderson, Stephen R. (1969) *West Scandinavian Vowel Systems and the Ordering of Phonological Rules*. Doctoral dissertation, MIT, Cambridge, MA.

Anderson, Stephen R. (1974) *The Organization of Phonology*. San Diego, CA: Academic Press.

Anderson, Stephen R. (1992) *A-Morphous Morphology*. Cambridge: Cambridge University Press.

Aquilina, Joseph (1959) *The Structure of Maltese: A Study in Mixed Grammar and Vocabulary*. Valletta: Royal University of Malta.

Aquilina, Joseph (1965) *Teach Yourself Maltese*. London: English Universities Press.

Berrendonner, Alain, Michel Le Guern, and Gilbert Puech (1983) *Principes de grammaire polylectale*. Lyon: Presses Universitaires de Lyon.

Beard, Robert (1995) *Lexeme-Morpheme Base Morphology*. Albany, NY: SUNY Press.

Beckman, Jill N. (1998) *Positional Faithfulness*. Doctoral dissertation, University of Massachusetts, Amherst.

Bermúdez-Otero, Ricardo (2001) Underlyingly nonmoraic consonants, faithfulness, and sympathy. Ms., University of Manchester. Available online at http://bermudez-otero.com/DEP-mora.pdf.

Borg, Albert, and Marie Azzopardi-Alexander (1997) *Maltese*. London: Routledge.

Borg, Albert (1997) Maltese phonology. In Alan S. Kaye (ed.), *Phonologies of Asia and Africa*, vol. 1, 245–285. Winona Lake, IN: Eisenbrauns.

Brame, Michael K. (1972) On the abstractness of phonology: Maltese ʕ. In Michael K. Brame (ed.), *Contributions to Generative Phonology*, 22–61. Austin, TX: University of Texas Press.

Brame, Michael K. (1974) The cycle in phonology: Stress in Palestinian, Maltese, and Spanish. *Linguistic Inquiry* **5**: 39–60.

Broselow, Ellen, Su-I Chen, and Marie Huffman (1997) Syllable weight: Convergence of phonology and phonetics. *Phonology* 14: 47–82.

Buckley, Eugene (1994) Tigrinya vowel features and vowel coalescence. *Penn Working Papers in Linguistics* 1: 1–28.

Buckley, Eugene (2009) Locality in metrical typology. *Phonology* 26: 389–435.

Buckley, Eugene (to appear). Global effects in Kashaya prosodic structure. In Vera Gribanova and Stephanie Shih (eds), *The Morphosyntax-Phonology Connection: Locality and Directionality at the Interface*. Oxford: Oxford University Press.

Available online at "http://www.ling.upenn.edu/~gene/papers/Buckley2014_global.pdf%22%20%5Ct%20%22_blank"http://www.ling.upenn.edu/~gene/papers/Buckley2014_global.pdf.

Campos-Astorkiza, Rebeka (2004) Faith in moras: A revised approach to prosodic faithfulness. In Keir Moulton and Matthew Wolf (eds), *Proceedings of the 34th Annual Meeting of the North East Linguistic Society*, 163–174. Amherst: GLSA.

Chomsky, Noam, and Morris Halle (1968) *The Sound Pattern of English*. New York: Harper and Row.

Chomsky, Noam, Morris Halle, and Fred Lukoff (1956) On accent and juncture in English. In Morris Halle, Horace Lunt, Hugh McLean and Cornelis van Schooneveld (eds), *For Roman Jakobson: Essays on the Occasion of his Sixtieth Birthday*, 65–80. The Hague: Mouton.

Crowhurst, Megan J. (1996) An optimal alternative to conflation. *Phonology* 13: 409–424.

de Lacy, Paul (2004) Maximal words and the Māori passive. In John J. McCarthy (ed.), *Optimality Theory in Phonology: A Reader*, 295–512. Oxford: Blackwell.

Elfner, Emily (2007) Moraic faithfulness: Evidence from Blackfoot and English. Talk presented at 15th Manchester Phonology Meeting, University of Manchester, UK. Handout available online at http://emilyelfner.sites.olt.ubc.ca/files/2014/10/elfner_2007_morafaith_mfm_handout.pdf

Elfner, Emily (2009) Syllabification and stress-epenthesis interactions in Harmonic Serialism. Ms., University of Massachusetts Amherst. Available on Rutgers Optimality Archive, ROA-1047.

Elfner, Emily (This volume) Stress-epenthesis interactions in Harmonic Serialism.

Eure, John (1999) *Maltese Arabic Syllabic Phenomena in Optimality Theory*. Senior essay in Linguistics, Yale University, New Haven, CT.

Fenech, Edward (1978) *Contemporary Journalistic Maltese: An Analytical and Comparative Study*. Leiden: E. J. Brill.

Flack, Kathryn (2009) Constraints on onsets and codas of words and phrases. *Phonology* 26: 269–302.

Garrett, Edward (1999) Minimal words aren't minimal feet. *UCLA Working Papers in Linguistics* 1: 68–105. Available on Rutgers Optimality Archive, ROA-1031.

Gouskova, Maria (2003) *Deriving Economy: Syncope in Optimality Theory*. Doctoral dissertation, University of Massachusetts Amherst.

Hume, Elisabeth V. (1994) *Front Vowels, Coronal Consonants, and their Interaction in Nonlinear Phonology*. New York: Garland.

Hyde, Brett (2002) A restrictive theory of metrical stress. *Phonology* 19: 313–359.

Hyde, Brett (2007) Non-finality and weight-sensitivity. *Phonology* 24: 287–334.

Hyde, Brett (2008) The odd-parity parsing problem. Ms., Washington University, St. Louis. Available on Rutgers Optimality Archive, ROA-971.

Itô, Junko, and R. Armin Mester (2003) Weak layering and word binarity. In Takeru Honma, Masao Okazaki, Toshiyuki Tabata, and Shin-Ichi Tanaka (eds), A *New Century of Phonology and Phonological Theory: A Festschrift for Professor Shosuke Haraguchi on the Occasion of his Sixtieth Birthday*, 26–65. Tokyo: Kaitakusha.

Johnson, C. Douglas (1990) Levantine cyclogenesis. In Mushira Eid and John J. McCarthy (eds), *Perspectives on Arabic Linguistics II*, 143–166. Amsterdam: John Benjamins.

Kimper, Wendell (2011) Locality and globality in phonological variation. *Natural Language and Linguistic Theory* 29: 423–465.

Kiparsky, Paul (1973) Phonological representations. In Osamu Fujimura (ed.). *Three Dimensions of Linguistic Theory,* 1–135. Tokyo: TEC.

Kiparsky, Paul (1996) Allomorphy or morphophonology? In Rajendra Singh (ed.), *Trubetzkoy's Orphan,* 13–31. Amsterdam: John Benjamins.

Kiparsky, Paul (2000) Opacity and cyclicity. *The Linguistic Review* 17: 351–367.

Kiparsky, Paul (2003) Syllables and moras in Arabic. In Caroline Féry and Ruben van de Vijver (eds), *The Syllable in Optimality Theory,* 161–182. Cambridge: Cambridge University Press.

Kiparsky, Paul (2011) Chains or strata?: The case of Maltese. Ms., Stanford University. Available on Lingbuzz lingbuzz/001379.

McCarthy, John J. (2003a) Sympathy, cumulativity, and the Duke-of-York gambit In Caroline Féry and Ruben van de Vijver (eds), *The Syllable in Optimality Theory,* 23–76. Cambridge: Cambridge University Press.

McCarthy, John J. (2003b) OT constraints are categorical. *Phonology* 20: 75–138.

McCarthy, John J. (2005) The length of stem-final vowels in colloquial Arabic. In Mohammad T. Alhawary and Elabbas Benmamoun (eds), *Perspectives on Arabic Linguistics XVII–XVIII,* 1–26. Amsterdam: John Benjamins.

McCarthy, John J. (2007) *Hidden Generalizations: Phonological Opacity in Optimality Theory.* London: Equinox.

McCarthy, John J. (2008) The serial interaction of stress and syncope. *Natural Language and Linguistic Theory* 26: 499–546.

McCarthy, John J. (2010) Studying Gen. *Journal of the Phonetic Society of Japan* 13 (2): 3–12. Available on Rutgers Optimality Archive, ROA-1049.

McCarthy, John J., and Alan Prince (1993a). *Prosodic Morphology I: Constraint Interaction and Satisfaction.* Technical Report 3, Rutgers University Center for Cognitive Science, New Brunswick, NJ. Available on Rutgers Optimality Archive, ROA 482.

McCarthy, John J., and Alan Prince (1993b) Generalized Alignment. *Yearbook of Morphology 1993*: 79–153. Available on Rutgers Optimality Archive, ROA-7.

McCarthy, John J., and Alan Prince (1995) Faithfulness and reduplicative identity. In Jill N. Beckman, Suzanne Urbanczyk, and Laura Walsh Dickey (eds), *University of Massachusetts Occasional Papers in Linguistics 18*: *Papers in Optimality Theory,* 249–384. Amherst: GLSA. Available on Rutgers Optimality Archive, ROA-60.

McCarthy, John J., and Kathryn Pruitt (2011) Sources of phonological structure. In Hans Broekhuis and Ralf Vogel (eds), *Linguistic Derivations and Filtering: Minimalism and Optimality Theory.* London: Equinox.

Misfud, M. (1994) The weak-final conjugation in the Semitic component of Maltese. In Joseph M. Brincat (ed.), *Languages of the Mediterranean: Substrata – The Islands – Malta: Proceedings of the Conference Held in Malta: 26–29 September 1991* 244–265. Msida: Institute of Linguistics, The University of Malta.

Misfud, Manwel (1995) *Loan Verbs in Maltese: A Descriptive and Comparative Study.* Leiden: E. J. Brill.

Norris, Mark (2013) The architecture of derivational OT: Evidence from Icelandic syncope. In Yelena Fainleib, Nicholas LaCara, and Yangsook Park (eds), *Proceedings of the 41st Annual Meeting of the North East Linguistic Society,* 55–68. Amherst: GLSA.

Odden, David (1990) Phonology and its interaction with syntax and morphology. *Studies in the Linguistic Sciences* 20 (2): 69–108.

Odden, David (1993) Interaction between modules in Lexical Phonology. In Sharon Hargus and Ellen M. Kaisse (eds), *Phonetics and Phonology 4: Studies in Lexical Phonology*, 111–144. San Diego, CA: Academic Press.

Pater, Joe (2012) Serial Harmonic Grammar and Berber syllabification. In Toni Borowsky, Shigeto Kawahara, Taka Shinya, and Mariko Sugahara (eds), *Prosody Matters: Essays in Honour of Elisabeth O. Selkirk*. London: Equinox. Available on Rutgers Optimality Archive, ROA-1085.

Piggott, Glyne (1993) Satisfying word minimality. *McGill Working Papers in Linguistics* 8 (2): 194–233.

Piggott, Glyne (2010) Cyclic spell-out and the typology of word minimality. Ms., McGill University, Montréal. Available online at http://www.mcgill.ca/files/linguistics/WordMinimality.pdf

Prince, Alan (1990) Quantitative consequences of rhythmic organization. In Michael Ziolkowski, Manuela Noske, and Karen Deaton (eds), *Proceedings of the 26th Regional Meeting of the Chicago Linguistic Society: Volume 2: Parasession on the Syllable in Phonetics and Phonology*, 355–398. Chicago, IL: CLS.

Prince, Alan (2003) Arguing Optimality. In Angela C. Carpenter, Paul de Lacy, and Andries W. Coetzee (eds), *University of Massachusetts Occasional Papers in Linguistics 26: Papers in Optimality Theory II*, 269–304. Amherst, MA: GLSA. Available on Rutgers Optimality Archive, ROA-562.

Prince, Alan, and Paul Smolensky (2004 [1993]) *Optimality Theory: Constraint Interaction in Generative Grammar*. Oxford: Blackwell. Available on Rutgers Optimality Archive, ROA-537.

Pruitt, Kathryn (2008) Iterative foot optimization and locality in stress systems. Ms., University of Massachusetts Amherst. Available on Rutgers Optimality Archive, ROA-999.

Pruitt, Kathryn (2010) Serialism and locality in constraint-based metrical parsing. *Phonology* **27**: 481–526.

Pruitt, Kathryn (2012) *Stress in Harmonic Serialism*. Doctoral dissertation, University of Massachusetts Amherst.

Puech, Gilbert (2011) Syllabic structure and stress in Maltese. In Sandro Caruana, Ray Fabri, and Thomas Stolz (eds), *Variation and Change: The Dynamics of Maltese in Space, Time, and Society*, 277–316. Berlin: Akademie Verlag.

Pullum, Geoffrey K. (1976) The Duke of York gambit. *Journal of Linguistics* 12: 83–102.

Selkirk, Elisabeth O. (1995) The prosodic structure of function words. In Jill N. Beckman, Suzanne Urbanczyk, and Laura Walsh Dickey (eds), *University of Massachusetts Occasional Papers in Linguistics 18: Papers in Optimality Theory*, 439–470. Amherst, MA: GLSA.

Sutcliffe, Edmund F. (1936) *A Grammar of the Maltese Language: With Chrestomathy and Vocabulary*. London: Humphrey Milford/Oxford University Press.

Vaysman, Olga (2009) *Segmental Alternations and Metrical Theory*. Doctoral dissertation, MIT, Cambridge, MA.

Vella, Alexandra (2003) Language contact and Maltese intonation: Some parallels with other language varieties. In Kurt Braunmüller and Gisella Ferraresi (eds),

Aspects of Multilingualism in European Language History, 261–283. Amsterdam: John Benjamins.

Vella, Alexandra (2009) Maltese intonation and focus structure. In Ray Fabri (ed.), *Maltese Linguistics: A Snapshot: In Memory of Joseph A. Cremona (1922–2003)*, 63–92. Bochum: Brockmeyer.

Wilson, Colin (2006) Counterfeeding from the past. Ms., University of California, Los Angeles. Available online at http://quote.ucsd.edu/blogs/sadphig/files/2009/01/counterfeedingfromthepast.pdf

Wolf, Matthew (2008) *Optimal Interleaving: Serial Phonology-Morphology Interaction in a Constraint-Based Model*. Doctoral dissertation, University of Massachusetts Amherst. Available on Rutgers Optimality Archive, ROA-996.

Wolf, Matthew (2009) Local ordering in phonology/morphology interleaving: Evidence for OT-CC. 83rd Linguistic Society of America Annual Meeting, San Francisco, CA. Handout available online at http://wolf.phonologist.org/LSA%20 2009%20handout.pdf.

Wolf, Matthew (2010) On the existence of counter-feeding from the past. 84th Linguistic Society of America Annual Meeting, Baltimore, MD. Handout available online at http://wolf.phonologist.org/OnTheExistenceOfCFFTP-corrected.pdf

Wolf, Matthew (2012) Inversion of stress-conditioned phonology in Stratal OT. Ms., Yale University, New Haven, CT. [Available online at "http://ling.auf.net/lingBuzz/001547%22%20%5Ct%20%22_blank" http://ling.auf.net/lingBuzz/001547]

Zhang, Jisheng, and Jeroen M. van de Weijer (2008) OT-CC: A minimal approach to Shaoxing trisyllabic tone sandhi. Ms., China East Normal University and University of Leiden. Available on Rutgers Optimality Archive, ROA-986.

Chapter 12

Learning serial constraint-based grammars[1]

Robert Staubs and Joe Pater

12.1 Introduction

In this paper we describe a method for learning grammars in the general framework of Harmonic Serialism (see McCarthy this volume for references and an introduction). We have two main goals. The first is to address the hidden structure problem that serial derivations introduce. The second is to address the problem of learning variation, which has yet to be confronted in this framework (see also Staubs *et al.* 2010 and Tessier and Jesney 2014 on the learning of Harmonic Serialism). In the remainder of this section, we illustrate the grammar model, and our approach to the learning of hidden structure, with a simple example of stress-epenthesis interaction. Our main contribution comes in the next section, where we introduce a method for calculating probabilities over the unbounded non-monotonic derivations that are characteristic of a stochastic serial theory. The third and last section consists of an application to data from French 'schwa' deletion that display variation.

In Harmonic Serialism, the path from the grammar's initial input to its final output is a series of derivational steps. In each step, a set of operations first applies to create a candidate set of outputs, from which one is chosen by a set of constraints. This output becomes the input for

1 This research was supported by the National Science Foundation under grant S121000000211 to the first author and grant BCS-0813829 to the University of Massachusetts Amherst. We are grateful to an anonymous reviewer for comments on a draft of the manuscript, and to members of our grant group for discussion.

the next step of the derivation. The derivation terminates, or converges, when the chosen output is identical to the input of that step. From the perspective of learning, these derivations are an instance of what Tesar and Smolensky (2000) call 'hidden structure'. Hidden structure refers to properties of the learning data that are not supplied to the learner, but must instead be inferred as part of the learning process. Our approach to learning this case of hidden structure is a generalization of Eisenstat's (2009) method for learning phonological Underlying Representations (URs). We will explain and illustrate Eisenstat's proposal as we discuss our toy stress-epenthesis case, since it also includes a UR learning problem.

The words in our mini-language are [tikát] and [píkat], stressed on the second and first syllable respectively. These surface forms can be produced by the derivations shown in (1). This is the standard rule-based analysis of opaque stress-epenthesis interaction (Brame 1974). As Elfner (this volume) shows, Harmonic Serialism can also produce derivations that follow these steps.

(1)	/tkat/	/pikat/	UR
	tkát	píkat	Stress
	tikát	–	Epenthesis
	[tikát]	[píkat]	SR

The data given to the learner are the overt forms [tikát] and [píkat]. We do not include any hidden structure in the Surface Representations (SRs), such as metrical structure, so the overt forms and SRs are equivalent.

One type of hidden structure that the learner must discover is the UR for each SR. Here we restrict the search space to /tkat/ and /tikat/, and /pkat/ and /pikat/. In Eisenstat's (2009) model, the probabilities of the overt forms are the summed probabilities of the UR-SR mappings corresponding to each. Thus for our example the probabilities of [tikát] and [píkat] are calculated as in (2).

(2) Probabilities of overt forms as summed probabilities of UR-SR mappings
p ([tikát]) = *p* (/tkat/ → [tikát]) + *p* (/tikat/ → [tikát])
p ([píkat]) = *p* (/pkat/ → [píkat]) + *p* (/pikat/ → [píkat])

The learning objective is to maximize the probability or likelihood that the grammar assigns to the observed overt forms. We will discuss how this is done shortly.

In the serial setting, the steps between the underlying form and the surface form are further instances of hidden structure. Here we assume that each step of the derivation adds the epenthetic vowel [i], or a stress to a syllable of a word that lacks it. When the derivation starts with a UR that lacks [i] and stress, like /tkat/, there are thus two paths that lead to an SR like [tikát] with stress on the final syllable, with either epenthesis or stress placement applying first. The probability of an overt form is now the summed probability of the derivations that lead to it, as shown in (3).

(3) Probabilities of overt forms as summed probabilities of derivations
p ([tikát]) = *p* (/tkat/ → tikat → [tikát]) + *p* (/tkat/ → tkát → [tikát]) + *p* (/tikat/ → [tikát])
p ([píkat]) = *p* (/pkat/ → pikat → [píkat]) + *p* (/pikat/ → [píkat])

The probability of each of the derivations is the product of the probabilities of each of its steps. The probability of each of the steps is the probability that the grammar grants to that outcome relative to all other possible outcomes of that step, that is, relative to all of the other members of the current candidate set.

Like Eisenstat (2009) we use Goldwater and Johnson's (2003) Maximum Entropy Grammar (MaxEnt) to define the probability distribution over a candidate set. Our learning proposal is in principle compatible with other stochastic variants of Optimality Theory (OT) and Harmonic Grammar (HG) (see Coetzee and Pater 2011 for an overview and Kimper 2011a on a serial implementation), but MaxEnt has a conveniently direct definition of probability. MaxEnt is a probabilistic version of HG in that it computes the Harmony of each candidate as the weighted sum of constraint violations (see Pater this volume for an introduction to and overview of HG, and Boersma and Pater this volume on Noisy HG, another stochastic implementation). In MaxEnt, the probability of a candidate is proportional to the exponential of its Harmony. MaxEnt in a serial framework is thus a probabilistic version of serial HG; for other work in serial HG see Elfner (this volume), Kimper (2011b), Mullin (2011) and Pater (2012, this volume).

For this example, which includes UR learning, we also follow Eisenstat (2009) and Pater, *et al.* (2012) in using constraints on URs, first introduced to OT by Zuraw (2000) and Boersma (2001), and first applied to the learning of underlying forms by Apoussidou (2007). These constraints demand particular mappings from meaning (or morpho-syntactic features) to phonological URs (our formalism differs somewhat from earlier proposals; see also Pater *et al.* 2012). Here we have the set of four UR constraints in (4a), which assign violations as in (4b). We give [píkat] the meaning 'paper', and [tikát] the meaning 'table'.

(4) UR constraints

a. 'paper' → /pkat/ 'paper' → /pikat/ 'table' → /tkat/ 'table' → /tikat/

b. Assign a penalty of −1 if meaning 'X' does not map to UR /Y/.

Our derivations start with an input meaning, and a candidate set of the two URs that satisfy each of the two constraints for each meaning. At this initial step of the derivation, the only constraints that apply are the UR constraints, and these constraints do not apply to any of the further steps of the derivation. The assumption that the phonological constraints do not apply at the step of UR insertion is not crucial; we make it only to produce derivations that look more like standard rule-based ones (see Wolf 2008 for relevant discussion).

The tableau in (5) illustrates the first step of a derivation from 'table'. The weights of the constraints are shown underneath the constraint names; these are the weights that our learner found in a simulation with overt [tikát] and [píkat] (paired with their meanings). The Harmony scores for each of the candidate URs is shown to the right of the row. Beside each of the candidate URs is its probability under the MaxEnt definition given above, rounded to three decimal points ('1' = 1.000). With these weights, /tikat/ is vanishingly improbable as the first step.

(5) First step of a derivation from 'table'

	'table'	'table' → /tkat/ 7.94	'table' → /tikat/ 0	
a. 1	/tkat/		−1	0
b. 0	/tikat/	−1		−7.94

We will follow the derivational path for the highly likely /tkat/.

The candidate sets in subsequent steps consist of the unchanged input, as well as all results of applying one of two operations. Stress placement adds stress to any syllable, if a word does not already have a stress. Epenthesis adds the vowel [i] between any two adjacent consonants. The candidate sets are evaluated by the constraints Stress-L and Stress-R, which demand stress on the leftmost and rightmost syllable respectively, and assign a violation if that syllable is not stressed, as well as by *Cluster, which penalizes adjacent consonants.

The tableau for the second step of the derivation shows the activity of these operations and constraints. The unchanged (6a.) violates all three constraints, and thus has very low probability. Adding a vowel, as in (6b.), resolves the *CLUSTER violation, but still incurs violations of the stress constraints. Since [tkát] is monosyllabic, it satisfies both stress constraints, at the cost of retaining the *CLUSTER violation. In (6) the sum of the weights of the stress constraints is greater than the weight of the *CLUSTER constraint, so [tkát] has higher probability than [tikat].[2] This shows how the constraints can affect the order of operations: with these weights, stress is extremely likely to precede epenthesis.

(6) Second step of a derivation from 'table'

	/tkat/	STRESS-R 12.01	STRESS-L 18.57	*CLUSTER 24.42	
a. 0	tkat	−1	−1	−1	−55.01
b. 0.002	tikat	−1	−1		−30.58
c. 0.998	tkát			−1	−24.43

We again follow the path resulting from the most probable output. In the next step, the only available operation is epenthesis, since we are assuming stress placement only applies to unstressed words. Adding the vowel as in (7b.) adds a violation of STRESS-L relative to the unchanged form in (7a.), because the leftmost syllable is now unstressed. The unchanged (7a.), on the other hand, retains the violation of *CLUSTER. With *CLUSTER having a higher weight than STRESS-L, [tikát] gets higher probability than [tkát].

(7) Third step of a derivation from 'table'

	tkát	STRESS-R 12.01	STRESS-L 18.57	*CLUSTER 24.42	
a. 0.001	tkát			−1	−24.42
b. 0.999	tikát		−1		−18.57

With [tikát] as an input, there are no further operations that can apply, so [tikát] will be the sole output candidate, and the derivation will thus converge.

2 It is not clear whether all cases of this type could be analyzed using a gang effect between constraints wanting stress in particular locations: it's likely that early stress placement would need to be compelled by a constraint penalizing stressless words, as in Elfner (this volume).

Given ‘table’ as the initial input, the probability of this derivation is nearly 1, so the observed [tikát] with second syllable stress has probability approaching its maximum. These weights also give observed [píkat] with initial stress probability approaching 1. The difference between them comes from the difference in the weights of their respective UR constraints. As shown in (8), ‘paper’ is extremely likely to map in the first step to bisyllabic /pikat/.

(8) First step of a derivation from ‘paper’

	‘paper’	‘paper’ → /pkat/ 0	‘paper’ → /pikat/ 7.94	
a. 0	/pkat/		−1	−7.94
b. 1	/pikat/	−1		0

The next step in the derivation, from /pikat/, is shown in (9).

(9) Second step of a derivation from ‘paper’

	/pikat/	Stress-R 12.01	Stress-L 18.57	*Cluster 24.42	
a. 0.001	pikát	−1			−12.01
b. 0.999	píkat		−1		−18.57
c. 0	pikat	−1	−1		−30.58

The most probable outcome is the observed [píkat] because Stress-L has a greater weight than Stress-R. The next step of the derivation converges on this form, since neither operation can apply.

We have now explained and illustrated how the grammar produces probabilities over the overt forms as the sum of probabilities of the derivations leading to them, with the probabilities of the derivations themselves arising as joint probabilities of their steps. We are therefore now ready to explain how the learner finds the constraint weights that yield the probabilities over the candidate sets. As we mentioned above, the learning objective is to maximize the likelihood of the observed forms. In a weighted constraint approach, the variables that the learner manipulates are the constraint weights. Set up in this way, this optimization problem can be solved by a range of algorithms. The particular optimization algorithm we used was the L-BFGS-B method (Byrd *et al.* 1995) as implemented in R (R Development Core Team 2010). L-BFGS-B is a quasi-Newton method implementing box constraints on solutions. We use these box constraints to set a zero minimum for constraint weights.

The only remaining issue is that the solution space is unbounded above, because with finite weights the probability that the grammar assigns to the observed forms can only approach 1. To resolve this, we take the standard approach of adding to the objective function a regularization term that penalizes the constraint weights in proportion to their distance from zero. The weights shown above were found with a Gaussian (or L2) prior with variance 100,000, which is sufficiently weak that it does little more than stop the weights from going to infinity.

In the general case, the calculation of the probabilities over overt forms that the grammar generates is more complex than in our toy example. The simplicity of the calculation above was due to the monotonic nature of the derivations: structure was only added, and never taken away or changed. The complexity of the general case can be appreciated by considering the result of adding an operation that deletes the vowel [i]. The set of derivations leading to [tikát] is now infinite – a few paths are illustrated in (10).

(10) A sample of an infinite set of derivations with the same result
  a. /tkat/ → tikat → tikát → [tikát]
  b. /tkat/ → tikat → tikát → tkát → tikát → [tikát]
  c. /tkat/ → tikat → tikát → tkát → tikát → tkát → tikát → [tikát]

The path in (10a.) is the one that our grammar produced with probability approaching 1. The path in (10b.) results from deleting [i] after placing stress, and reinserting it, and (10c.) iterates this loop. Deletion is preferred by STRESS-L, and insertion by *CLUSTER, so even if we remove harmonically bounded candidates (see Section 12.2), there is always some probability of choosing the result of these operations rather than the unchanged form, thus postponing convergence. The length of the derivations is therefore unbounded, and the set of derivations producing the same result is infinite. As Kimper (2011a: 458) points out, in a stochastic serial theory the probability of non-convergence in such cases becomes arbitrarily close to zero as derivation length increases. The issue for our learning model, which we address in the next section, is how to calculate probabilities over infinite sets of paths like those in (10) that lead from a given initial input to the same final output.

12.2 Computing probabilities over unbounded derivations

12.2.1 Unbounded derivations

To discuss our approach to unbounded derivations, we will use a schematic case involving obstruent voicing. The input /bada/ has two voiced obstruents. Figure 12.1 depicts the possible derivations starting from this input as a directed graph. Here the operations assumed in GEN are simply VOICE and DEVOICE – that is, the voicing of an obstruent can be changed in either direction. The one exception is that [pada] does not map to [bada], because [pada] harmonically bounds [bada] with our constraint set – we return to this point in Section 12.2.2.

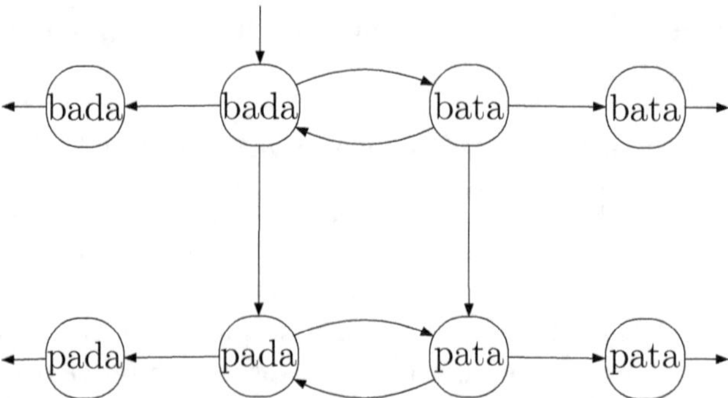

Figure 12.1*:* Graph of space of derivations for voicing example

From the input /bada/, an output [pata] is reachable in a number of ways, as shown in the sample of derivational paths in (11).

(11) Derivational paths from /bada/ to [pata]
/bada/ → pada → pata → [pata]
/bada/ → bata → pata → [pata]
/bada/ → bata → bada → bata → pata → [pata]
/bada/ → bata → pata → pada → pata → [pata]
...

A given derivation may meander through the graph in any way that obeys the directions of its arcs. It is only when the derivation 'loops' back on a single particular node that it enters a convergence state and the derivation as a whole converges. Thus, in general, the length of derivations is unbounded, and the number of them producing a particular output for a given input is infinite.

If the number of derivations producing an input/output mapping is infinite, we must incorporate this knowledge into our computation of derivational probabilities. This incorporation will in effect sum over the infinite number of possible derivations, yielding a probability distribution over convergence states. We return to the details of this computation after first discussing the construction of graphs like Figure 12.1.

12.2.2 From operations to graphs

In order to explicitly calculate derivational probabilities, we will require an explicit representation of the possible steps of a serial derivation in the given domain. In particular, given a set of operations GEN and a set of constraints CON we must be able to construct a graph like Figure 12.1 depicting the possible transitions between two steps.

Such graphs allow the finite representation of an infinite number of derivations. This finiteness is crucial for the approach adopted here. Finiteness follows immediately from non-probabilistic serial constraint-based theories: any particular Harmonic Serialist derivation has only a finite potential for improvement under a given ranking of the constraints in CON and must always improve harmony, step by step. Taken together, these facts imply that all Harmonic Serialist derivations are of bounded length (McCarthy, 2008: 274). Thus the possible steps included in a derivational graph should include at least all steps (eventually) derivable from the input(s) through operations in GEN whose results are not harmonically bounded in their candidate sets according to the constraint set CON (where a harmonically bounded candidate has a proper superset of violation marks of another member of the same candidate set). This criterion yields the set of derivations fulfilling the criterion of (potential) harmonic improvement.

With a non-probabilistic model, the set of all steps to non-harmonically bounded candidates provides the full space of possible derivations. This is true also of any probabilistic framework that guarantees that harmonically bounded candidates do not receive any probability mass. Of the OT and HG theories of stochastic grammar surveyed in Coetzee and Pater (2011), only MaxEnt grants probability to harmonically bounded forms (see Jäger and Rosenbach 2006, Jesney 2011 on this aspect of MaxEnt). Since we adopt MaxEnt to define probabilities over candidate sets, we must instead impose an additional delimiting heuristic on the space of derivations in order to avoid creating infinite graphs. For example, a finite graph is guaranteed to exist if only non-harmonically bounded candidates are permitted, as in the graph in Figure 12.1.[3]

With the criteria described here we can create graphs describing the derivational spaces relevant to arbitrary sets GEN and CON. With these mechanisms in place we can proceed to the calculation of probabilities over the final outputs of the grammar.

12.2.3 Derivations as Markov chains

Given a particular weighting of constraints, the space of derivations does not simply look like Figure 12.1. Arcs are not merely present or absent; instead, arcs are annotated by their various probabilities of occurrence. Figure 12.1 can thus be written as in Figure 12.2.

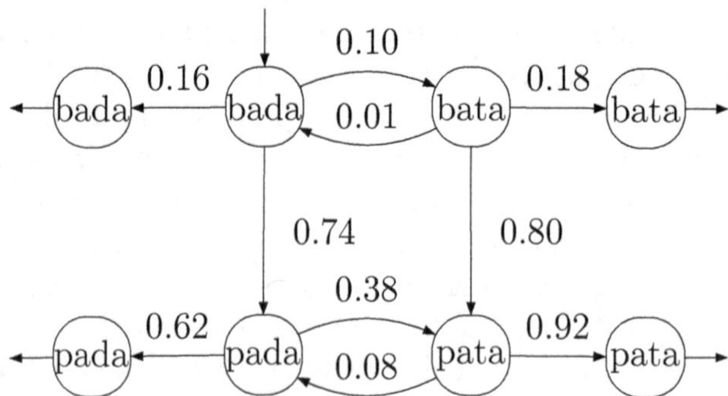

Figure 12.2: Markov chain for voicing example

Here the probabilities are those given by a Maximum Entropy model with the following constraints: IDENT(voice) ('do not change voicing,' $w = 1.5$), *VOICE ('do not have voicing in the output,' $w = 3.0$), and *VtV ('do not have voiceless obstruents between vowels in the output,' $w = 2.0$).

Written in this form, the derivational space is a particular type of probabilistic graph known as a Markov chain. Derivational Markov chains have a particularly useful property: they are *absorbing*. From any

3 Although the exclusion of harmonically bounded candidates is guaranteed to result in finite graphs, their inclusion does not always break finiteness. In both of the simulations reported in Sections 12.1 and 12.3, harmonically bounded candidates were included in the candidate sets. The representational spaces remained finite because epenthesis was in both cases limited to applying between consonants.

given step of a derivational chain, there must be at least one convergence state reachable from that step. Once a derivation enters one of these convergence states, the derivation has 'converged' and may not leave that state. In the terminology of Markov chains, these states are *absorbing states*. States that are not absorbing are called *transient*. A Markov chain of this sort permits us to perform some helpful manipulations for calculating derivational probabilities.

To proceed we restate the information contained within the graph itself in the form of a matrix **P**. Each row (and column) of the matrix corresponds to a particular state (i.e. derivational step). The values in the matrix are transitional probabilities. Thus the value in a particular row and column is the probability of moving between the state represented by that row and the state represented by that column.

$$\mathbf{P} = \begin{array}{c} \\ \text{bada} \\ \text{bata} \\ \text{pada} \\ \text{pata} \\ [\text{bada}] \\ [\text{bata}] \\ [\text{pada}] \\ [\text{pata}] \end{array} \begin{array}{c} \begin{array}{cccccccc} \text{bada} & \text{bata} & \text{pada} & \text{pata} & [\text{bada}] & [\text{bata}] & [\text{pada}] & [\text{pata}] \end{array} \\ \begin{pmatrix} 0 & 0.10 & 0.74 & 0 & 0.16 & 0 & 0 & 0 \\ 0.01 & 0 & 0 & 0.80 & 0 & 0.18 & 0 & 0 \\ 0 & 0 & 0 & 0.38 & 0 & 0 & 0.62 & 0 \\ 0 & 0 & 0.08 & 0 & 0 & 0 & 0 & 0.92 \\ 0 & 0 & 0 & 0 & 1 & 0 & 0 & 0 \\ 0 & 0 & 0 & 0 & 0 & 1 & 0 & 0 \\ 0 & 0 & 0 & 0 & 0 & 0 & 1 & 0 \\ 0 & 0 & 0 & 0 & 0 & 0 & 0 & 1 \end{pmatrix} \end{array}$$

Figure 12.3: Transition matrix for voicing case. Forms in square brackets represent convergent states

We can arrange this matrix so that transient states are listed first by both columns and rows. We can then partition the transition matrix of an absorbing Markov chain in the following way. The upper-left portion (corresponding to transient-transient mappings) we call **Q**. The upper-right portion (transient-absorbing mappings) we call **R**. As the derivation never leaves absorbing states, the lower-left (absorbing-transient) is a zero matrix **0**. Finally, as absorbing states must self-loop, the lower-right (absorbing-absorbing) is an identity matrix **1** with the same number of rows and columns as there are absorbing states.

$$\mathbf{P} = \begin{array}{c} \\ \text{transient} \\ \text{absorbing} \end{array} \begin{array}{c} \begin{array}{cc} \text{transient} & \text{absorbing} \end{array} \\ \begin{pmatrix} \mathbf{Q} & \mathbf{R} \\ \mathbf{0} & \mathbf{1} \end{pmatrix} \end{array}$$

Figure 12.4: Partitioning a transition matrix

By raising this matrix to a given power we obtain a matrix that describes the probability of going between any two states in a given number of steps. The particular structure of our chains allows us to find the limiting case in which we obtain the probability after *any* number of steps. With proper manipulation we can obtain a matrix **B** in which the rows represent the transient states and the columns represent absorbing states. A particular element of this matrix thus contains the probability of eventually ending in a particular state from a particular starting state. Stated in more concrete terms, this is the derivational probability of a particular output if starting from a particular input. This matrix can be calculated as shown in Figure 12.5 (see e.g. Grinstead and Snell 2006: 417–421). The identity matrix **1** here has a row and a column for every transient state.

$$\mathrm{B} = (\mathbf{1} - \mathrm{Q})^{-1}\,\mathrm{R}$$

Figure 12.5: Calculation of derivational probability matrix

Applied to the voicing example, the resulting derivational probability matrix is as shown in Figure 12.6.

$$\mathrm{B} = \begin{array}{c} \\ \text{bada} \\ \text{pata} \\ \text{pada} \\ \text{pata} \end{array} \begin{array}{c} \begin{array}{cccc} [\text{bada}] & [\text{bata}] & [\text{pada}] & [\text{pata}] \end{array} \\ \begin{pmatrix} 0.16 & 0.02 & 0.48 & 0.34 \\ 0.00 & 0.18 & 0.05 & 0.77 \\ 0 & 0 & 0.64 & 0.36 \\ 0 & 0 & 0.05 & 0.95 \end{pmatrix} \end{array}$$

Figure 12.6: Derivational probability matrix

The probability of going from an initial input /bada/ to any of the possible output forms is given in the first row of **B**. Note that this result differs from the individual step probabilities; we will return to this point in the next section.

12.2.4 Error minimization

In the preceding subsections we detailed a way of calculating the probabilities over final outputs given an initial input, as well as a particular GEN, CON, and constraint weighting. We must also consider how weights may be learned given an observed distribution over mappings from initial inputs to final outputs. In Section 12.1 we stated the learning objective as maximizing the probability of the observed final outputs, which each had probability 1. Here we state the objective as error minimization, which is a more general formulation that is also suitable for real numbered probabilities over final outputs.

A standard measure of model error is Kullback-Leibler (KL) divergence between the observed distribution p and the fitted distribution q (Kullback and Leibler 1951). The KL divergence between two distributions is the average log difference between them, with the average taken in terms of p. It is minimized (at zero) if and only if p and q are equal. Thus by picking weights w to minimize the KL divergence between the two distributions, we are choosing weights which best match the observed distribution.

$$w^* = \operatorname*{argmin}_w D(p||q_w) = \operatorname*{argmin}_w \sum_{i \in \{x|p(x)>0\}} \left(p(i) \log \frac{p(i)}{q_w(i)} \right)$$

Figure 12.7: KL minimization

In the present context, the observed and fitted distributions are probability distributions over the possible final outputs for each of a set of initial inputs; to get single distributions for optimization, we average over all inputs in calculating p and q, giving equal weight to the distribution for each input. The observed distributions for each input are given to the learner. Given the construction of the probability matrix **B** in the preceding section, we may easily calculate the output distributions for each input, and hence values of q. By numerically minimizing KL, then, we are finding weights that come as close as possible to producing the observed probability distributions.

12.3 An application to variation

In the previous section, an abstract example was used to illustrate the calculation of probabilities over unbounded derivations. In that example, a pair of obstruents in a word like /bada/ changed their voicing back and forth from voiced to voiceless, travelling through the representational space of {[bada], [bata], [pada], [pata]}. Here we apply the learning model to a case of variation in natural language that can be characterized in a similar way. French 'schwa' is usually an orthographic 'e' and a phonetic back rounded vowel [œ]: the term schwa is used to refer to those instances of this vowel that alternate with zero. Given a phrase like *je me prépare*, where *je* and *me* both contain schwa, we will consider a representational space consisting of the four possibilities resulting from each vowel being pronounced or not. This space is illustrated in (12), using the French orthographic convention of notating a deleted schwa with an apostrophe.

(12) Representational space for *je me prépare* 'I prepare myself'
  a. Je me prépare
  b. J' me prépare
  c. Je m' prépare
  d. J' m' prépare

Our operations are schwa insertion and deletion, which can take either (12a.) or (12d.) back and forth from either (12b.) or (12c.). As in the example in Section 12.1, vowel insertion only inserts a vowel between two consonants, thus keeping the representational space finite.

We model some impressionistic data on the relative probability of the two pronunciations in (12b.) and (12c.) from Delattre (1949b: 46), who notes when a schwa is deleted, it is usually the second one, as in (12c.), but that the first schwa is also sometimes lost, as in (12b.). He estimates the occurrence of (12b.) as '... *disons – moins d'une fois sur dix*' ('let's say less than once in ten'). We also include the contrastive case of *je te répondrais*, which Delattre states is usually pronounced with initial deletion as *j'te répondrais*, and sometimes as *je t' répondrais*, '*peut-être une fois sur vingt*' ('maybe once in twenty').

We use four constraints to derive this pattern. Delattre (1949ab) and Côté (2000) attribute the relative ill-formedness of *je t' répondrais* to stops being difficult to produce and/or perceive without a following vowel. For the forms we are considering, in which [t] is the only stop, we use a constraint *t' that penalizes [t] without a following schwa. Along the same lines, we might attribute the general preference for retention of the initial schwa to the fact that this allows both of the consonants to be adjacent to a vowel. For our data, it is sufficient to have a constraint *j' that penalizes [ʒ] without a following schwa. As the constraint to motivate deletion we use simple *Schwa, which penalizes each pronounced schwa. Finally, we also include a constraint that penalizes deletion of both schwas, as in (12d.). Following Grammont (1894), we use a constraint against a sequence of three consonants, *CCC. In all of this we simplify greatly relative to the complexity of actual French; see Eychenne (2006) for a recent overview of the generative literature.

We provided our learner with the distributions over the output patterns shown in the 'Obs.' (for Observed) column of (13). The relative probabilities of forms in rows (13b.) *vs.* (13c.) correspond to Delattre's estimates. *J' me prépare* is 'less than one in ten' with respect to *Je m' prépare* and *Je t' répondrais* is 'one in twenty' with respect to *J' te répondrais.* French schwa deletion is generally described as optional, so the forms in row (13a.) must have some probability, but Delattre (1949ab) does not give any information about the likelihood of these types of pronunciation. We used values that our grammar model could represent.[4] In particular, because the low probability of *je t' répondrais* requires a relatively high weighted *t' constraint, *je te répondrais* must get higher probability than *je me prépare.*

(13) Distributions in learning data (Obs.), and produced by the grammar (Fit.)

		Obs.	Fit.		Obs.	Fit.
a.	Je me prépare	0.13	0.125	Je te répondrais	0.6	0.605
b.	J' me prépare	0.07	0.075	J' te répondrais	0.38	0.374
c.	Je m' prépare	0.8	0.8	Je t' répondrais	0.02	0.02
d.	J' m' prépare	0	0	J' t' répondrais	0	0

In this simulation, we abstracted from UR learning by assuming schwaful *je, me* and *te* (/ʒœ/, /mœ/, /tœ/) as underlying representations. The constraint weights were found using the same parameters as in Section 12.1, that is, L-BFGS-B with regularization variance 100,000 and starting weights of zero. The 'Fit.' (for Fitted) column in (13) shows the probabilities that our learner's grammar generates with these weights, calculated using the method described in Section 12.2. The weights themselves are given beneath the constraint names in the tableaux below.

The tableau in (14) shows the first step from underlying *je te* (the unchanging *répondrais* is omitted from the tableaux, which should be borne in mind especially in assessing *CCC violations). The probabilities generated in this first step are quite different from those in the table in (13). The probability of 0.425 for [ʒœtœ] in (14) is much lower than the total probability in (13), and the probability of the other outcomes is

4 This was done by repeatedly fitting the grammar to hypothesized observed distributions. The final observed values were chosen as values that came close to, but weren't exactly the same, as what the fitted values would be. We offer this simulation as only a demonstration that our learner can cope with unbounded derivations; comparisons of this particular learner and/or constraint set to others would of course not proceed in this way.

somewhat higher. The probability of [ʒœtœ] in (14a.) is the probability that the derivation converges on this first step. For the other outcomes, we can consider what happens on subsequent steps.

(14) First step from /ʒœtœ/

	ʒœtœ	*CCC 4.7	*t' 3.6	*SCHWA 1.98	*j' 1.84	
a. 0.425	ʒœtœ			−2		−3.97
b. 0.49	ʒtœ			−1	−1	−3.82
c. 0.08	ʒœt		−1	−1		−5.58

From (14b.), the second step is as shown in (15). We see that a significant portion of the probability (0.463) goes to [ʒœtœ] (15a.), from which the next step would be again as in (14). Since the probability of convergence would then be 0.425, we have just added 0.096 (= 0.49 × 0.463 × 0.425) to the probability of [ʒœtœ] as a final output, bringing us somewhat closer (0.521) to the total probability of 0.605. The probability of convergence in the second step is given in (15b.). Note that we abstract from the optional devoicing of [ʒ] when it is adjacent to [t], which would occur in a subsequent step.

(15) A second step from /ʒœtœ/

	ʒtœ	*CCC 4.7	*t' 3.6	*SCHWA 1.98	*j' 1.84	
a. 0.463	ʒœtœ			−2		−3.97
b. 0.535	ʒtœ			−1	−1	−3.82
c. 0.001	ʒt	−1	−1		−1	−10.13

For the second step from [ʒœt] (14c.), even more of the probability is granted to [ʒœtœ] (16a.).

(16) Another second step from /ʒœtœ/

	ʒœt	*CCC 4.7	*t' 3.6	*SCHWA 1.98	*j' 1.84	
a. 0.832	ʒœtœ			−2		−3.97
b. 0.166	ʒœt		−1	−1		−5.58
c. 0.002	ʒt	−1	−1		−1	−10.13

We have just seen, then, why the probability of this grammar mapping initial /ʒœtœ/ to final [ʒœtœ] is higher than the probability of the single step of [ʒœtœ] mapping to itself. The probability of SR [ʒœtœ] comes not only from the probability of converging to it on the first step, but also from the probability of the other outputs of the first step, [ʒtœ] and [ʒœt], mapping back to [ʒœtœ] in the second step and then converging on [ʒœtœ] in the third, as well as from the probability of all of the infinitely many other derivational loops that wind up eventually converging on [ʒœtœ].

The above three tableaux show the probabilities from inputs at three of the four points in our representational space for *je te répondrais*. The last case is the input [ʒt], which can arise in the third step from /ʒœtœ/ through either tableau (15) or (16).

(17) A third step from /ʒœte/

	ʒt	*CCC 4.7	*t' 3.6	*SCHWA 1.98	*j' 1.84	
a. 0.851	ʒtœ			−1	−1	−3.82
b. 0.147	ʒœt		−1	−1		−5.58
c. 0.002	ʒt	−1	−1		−1	−10.13

This tableau shows that not only is it improbable for [ʒt] to be produced as an output for either [ʒœt] or [ʒtœ], it is also improbable for the derivation to then converge in the next step. The case of [ʒt] illustrates how the probability of an ultimate output can be much smaller (in this case 0.000001) than the probability of any of the individual steps that produce it.

12.4 Conclusions

The introduction of this method for learning serial constraint-based grammars has several immediate positive consequences, and also opens up a number of avenues for future research. The first immediate consequence is that as we have just demonstrated in Section 12.3 for French 'schwa' deletion, it is now possible to find a grammar in the serial framework that generates a given probability distribution over surface representations (if that probability distribution can be represented with a given set of constraints and operations, and *modulo* any local minima – see the next paragraph). This has the same practical benefit for linguistic analysis in the Serial Variation framework proposed by Kimper (2011a) as that of learning algorithms for stochastic theories in parallel OT and HG (and slightly further afield, of regression models for variable rules theories as introduced in Cedergren and Sankoff 1974). A second immediate consequence is that it is now possible to compare the ability of a serial theory to capture particular attested patterns of variation to that of parallel theories, so that we can now potentially bring fine-grained probabilistic data to the question of whether a serial or parallel theory is better supported empirically. Finally, this work demonstrates that any difficulties imposed by the hidden structure of serial derivations are not necessarily insuperable. Despite an apparent increase in the information learned – namely, an extension to non-surface forms – we obtain encouraging learning results.

As hinted at in the last paragraph, one avenue for future research is the determination of the extent to which serial derivations introduce or remove local minima in learning problems. Once these sorts of differences between parallel and serial theories are identified, then one might next ask which learning theory better characterizes human behavior, or study ways in which the local minima might be avoided. It may of course be the case that serial derivations introduce intractable local minima that do not correspond to human learning difficulties, but we currently have no reason to suspect that this is true (or false).

Another important avenue for further research is the development of gradual learning algorithms in this framework, which could be used in modeling the course of human language acquisition.

A final, additional, area for future research is the extension of these methods to other serial constraint-based grammatical frameworks. The formal problem of learning these sorts of grammars – stochastic or categorical – in, for example, Stratal Optimality Theory (Kiparsky 2000) or the derivational version of Targeted Constraints theory in Wilson (2013) seems very much the same as the problem approached here. With appropriately adapted construction of the relevant Markov chains, our strategy might well yield useful learning frameworks for these theories as well.

References

Apoussidou, Diana (2007) *The learnability of metrical phonology.* Doctoral Dissertation, University of Amsterdam. Available at http://roa.rutgers.edu/.

Boersma, Paul (2001) Phonology-semantics interaction in OT, and its acquisition. In Robert Kirchner, Wolf Wikeley and Joe Pater (eds) *Papers in Experimental and Theoretical Linguistics* 6, 24–35. Edmonton: University of Alberta.

Boersma, Paul and Joe Pater (This volume) Convergence properties of a gradual learning algorithm for Harmonic Grammar.

Brame, Michael (1974) The cycle in phonology: Stress in Palestinian, Maltese and Spanish. *Linguistic Inquiry* 5 (1): 39–60.

Byrd, Richard H., Lu, Peihuang, Nocedal, Jorge and Zhu, Ciyou (1995) A limited memory algorithm for bound constrained optimization. *SIAM J. Scientific Computing* 16, 1190–1208.

Cedergren, Henrietta J. and David Sankoff (1974) Variable rules: Performance as a statistical reflection of competence. *Language* 50: 333–355.

Coetzee, Andries and Pater, Joe (2011) The place of variation in phonological theory. In John Goldsmith, Jason Riggle, and Alan Yu (eds), *The Handbook of Phonological Theory* (2nd edn.), 401–431. Malden, MA: Blackwell.

Côté, Marie-Hélène (2000) Consonant cluster phonotactics: A perceptual approach. Ph.D. dissertation, MIT.

Delattre, Pierre (1949a) Le jeu de <e> instable de monosyllabe initial en français. *The French Review* 22: 455–459.

Delattre, Pierre (1949b) Le jeu de <e> instable de monosyllabe initial en français (suite). *The French Review* 23: 43–47.

Eisenstat, Sarah (2009) Learning underlying forms with MaxEnt. MA thesis, Brown University.

Elfner, Emily (This volume) Stress-Epenthesis Interactions in Harmonic Serialism.

Eychenne, Julien (2006) *Aspects de la phonologie du schwa dans le français contemporain optimalité, visibilité prosodique, gradience.* PhD. dissertation, Université de Toulouse-Le Mirail.

Goldwater, Sharon and Johnson, Mark (2003) Learning OT constraint rankings using a maximum entropy model. In Jennifer Spenader, Anders Eriksson and Östen Dahl (eds), *Proceedings of the Stockholm Workshop on 'Variation within Optimality Theory',* 111–120. Stockholm: Stockholm University.

Grammont, Maurice (1894) La loi des trois consonnes. *Mémoires de la societé de linguistique de Paris* 8: 53–90.

Grinstead, Charles M. and Laurie J. Snell (2006) *Introduction to Probability,* 2nd edition. Providence, RI: American Mathematical Society.

Jäger, Gerhard and Anette Rosenbach. (2006) The winner takes it all – almost: Cumulativity in grammatical variation. *Linguistics* 44: 937–971.

Jesney, Karen (2011) *Cumulative Constraint Interaction in Phonological Acquisition and Typology*. PhD dissertation, University of Massachusetts Amherst.

Kimper, Wendell A. (2011a) Locality and globality in phonological variation. *Natural Language and Linguistic Theory* 29: 423–465.

Kimper, Wendell A. (2011b) *Competing Triggers: Transparency and Opacity in Vowel Harmony.* PhD dissertation, University of Massachusetts Amherst.

Kiparsky, Paul (2000) Opacity and cyclicity. *The Linguistic Review* 17: 351–367.

Kullback, Solomon and Richard Leibler (1951) On information and sufficiency. *Annals of Mathematical Statistics* 22 (1): 79–86.

McCarthy, John J. (2008) The gradual path to cluster simplification. *Phonology* 25: 271–319. Available at: http://works.bepress.com/john_j_mccarthy/30

McCarthy, John J. (This volume) The theory and practice of Harmonic Serialism.

Mullin, Kevin (2011) Strength in Harmony Systems: Trigger and Directional Asymmetries. Ms., University of Massachusetts Amherst. [Available at http://people.umass.edu/kmullin/Mullin2011StrengthHarmonySystems.pdf]

Pater, Joe (This volume) Universal Grammar with Weighted Constraints.

Pater, Joe (2012) Serial harmonic grammar and Berber syllabifiction. In Toni Borowsky, Shigeto Kawahara, Takahito Shinya and Mariko Sugahara (eds), *Prosody Matters: Essays in Honor of Elisabeth O. Selkirk*, 43–72. London: Equinox Press. [Available at http://roa.rutgers.edu/]

Pater, Joe, Robert Staubs, Karen Jesney and Brian Smith (2012) Learning probabilities over underlying representations. In the *Proceedings of the Twelfth Meeting of the ACL-SIGMORPHON: Computational Research in Phonetics, Phonology, and Morphology*. 62–71.

R Development Core Team (2010) *R: A language and environment for statistical computing*. R Foundation for Statistical Computing, Vienna, Austria. URL http://www.R-project.org.

Staubs, Robert, Michael Becker, Christopher Potts, Patrick Pratt, John J. McCarthy and Joe Pater (2010) OT-Help 2.0. Software package. Amherst, MA: University of Massachusetts Amherst.

Tesar, Bruce and Smolensky, Paul (2000) *Learnability in Optimality Theory*. Cambridge, MA: MIT Press.

Tessier, Anne-Michelle and Karen Jesney (2014) Learning in Harmonic Serialism and the necessity of a richer base. *Phonology* 31: 155–178.

Wilson, Colin (2013) A Targeted Spreading Imperative for Nasal Place Assimilation. In Yelena Fainleib, Nicholas LaCara and Yangsook Park (eds), *Proceedings of the 41st Annual Meeting of the North East Linguistics Society*, 261–272.

Wolf, Matthew (2008) *Optimal Interleaving: Serial Phonology-Morphology Interaction in a Constraint-Based Model*. Ph.D. dissertation, University of Massachusetts Amherst.

Kie Zuraw. 2000. *Exceptions and regularities in phonology*. Ph.D. dissertation, UCLA.

Chapter 13

Convergence properties of a Gradual Learning Algorithm for Harmonic Grammar[1]

Paul Boersma and Joe Pater

13.1 Introduction

The central question in research on language learnability in generative linguistics (e.g. Wexler and Culicover 1980, Dresher and Kaye 1990, Gibson and Wexler 1994, Fodor 1998, Tesar and Smolensky 1998, 2002, Yang 2002) is the following: is a given learning algorithm guaranteed to converge on a grammar that is correct for any language defined by a given theory of grammar? Research on language acquisition, on the other hand, studies the path taken to adult competence by human learners. Though clearly related, these two lines of research are often pursued completely independently of one another. In this paper, we investigate the convergence properties of a gradual learning algorithm for Harmonic Grammar (HG; Legendre, Miyata and Smolensky 1990/2006) and compare its behavior with similar learning algorithms developed for Optimality Theory (OT; Prince and Smolensky 1993/2004). An

1 Thanks to Diana Apoussidou, Rajesh Bhatt, Michael Collins, Chris Davis, Karen Jesney, Robert Staubs, Kie Zuraw and participants in Linguistics 794, Fall 2007 UMass for comments on earlier drafts. Thanks also to Gaja Jarosz, Mark Johnson, Giorgio Magri, Jason Naradowsky, Patrick Pratt, and Colin Wilson for useful discussion, and to Bruce Tesar for supplying a list with the overt forms for the languages needed for the RIP simulations. This research was partially supported by grants 016.024.018 and 277-70-008 from the Netherlands Organization for Scientific Research (NWO) to Boersma and by grant BCS-0813829 from the National Science Foundation to the University of Massachusetts Amherst.

attractive feature of a gradual learning algorithm is its ability to model human learning paths (see e.g. Boersma and Levelt 2000 and Curtin and Zuraw 2002 on Boersma's 1998 GLA for stochastic OT, and Jäger 2007, Farris-Trimble 2008, Jesney 2011 and Jesney and Tessier 2011 on the present GLA for HG). This paper strengthens the relationship between research on acquisition and learnability by showing that a gradual learning algorithm can also meet goals of learnability theory. By examining a learning algorithm that has its roots in neural modelling and statistical learning, this paper also strengthens the connections between these areas of research and the generative study of acquisition and learnability.

In Section 13.2, we illustrate the HG model of grammar, and the learning algorithm, with a simple case involving stress placement. In Section 13.3, we adapt the perceptron convergence proof to show that if the learner is given fully structured language data, this learning algorithm is guaranteed to find a correct set of constraint weights for any non-varying language, i.e. any language in which each possible input to the grammar generates the same output form on every occasion. The uniqueness of optima in the languages being learned means that, like most other generative learnability research, including Tesar and Smolensky's (1998, 2000) foundational work in OT, we operate under the idealization that languages do not display variation. In Section 13.4, we test an implementation of the algorithm, and show that it quickly converges on a correct set of weights both for the original version of HG assumed in the convergence proof, and for Exponential HG, in which a restriction to positive values for weights is enforced through their exponentiation.

Section 13.5 extends the proof to a version of HG with probabilities on the weights, or noise, as in stochastic OT (Boersma 1998, Boersma and Hayes 2001). While we continue to operate with the idealization of non-varying target languages, the addition of noise to the grammar increases its realism as a model of language acquisition, since the learner itself displays variation between grammatical 'stages'. The convergence of noisy HG learners on correct grammars for languages with unique optima is of further interest because the GLA for stochastic OT has been shown to fail on a case of this type (Pater 2008). In simulations, noisy HG and noisy Exponential HG learners converge on correct weights, while stochastic OT learners occasionally fail to converge.

The proofs and simulations in Sections 13.3–13.5 operate under the idealization that learners have access to the full structure of the learning data. Section 13.6 presents simulations involving learning with hidden structure that use Tesar and Smolensky's (2000) Robust Interpretive

Parsing. When applied to Tesar and Smolensky's test cases on the learning of stress, these simulations show that HG learners converge on correct grammars more frequently than OT learners with similar on-line learning algorithms (Error-Driven Constraint Demotion; Tesar and Smolensky 1998, 2000, Gradual Learning Algorithm; Boersma 1998, Boersma and Hayes 2001). We also find that convergence success rates are improved by noise on the constraint values. We use a simple example to show how symmetric update (promotion and demotion), noise, and weighted constraints can decrease the probability of a learner ending up in a trap caused by hidden structure.

13.2 Harmonic Grammar and its Gradual Learning Algorithm

We assume a model of grammar identical to that of OT except that it works with constraint weights, as in OT's predecessor Harmonic Grammar (see Smolensky and Legendre 2006 and Pater 2009a for overviews of research in HG).

The tableaux in (1) illustrate HG evaluation of stress placement for a pair of words differing in syllable shape. Simplifying greatly, we assume that syllable structure is given in the underlying form, and only the placement of stress has to be determined. The first tableau (1a), then, is for an underlying /.ba.tan.ma./, which has a penultimate syllable with a coda /n/. We assume that this syllable is heavy, as opposed to the two CV syllables, which are light. Tableau (1a) determines where the stress lies in the phonological surface form, in which we simplify away the foot structure (see Section 13.7 for an account *with* foot structure). If the second syllable is stressed, as in the candidate surface form [.ba.tán.ma.], the constraint WEIGHT-TO-STRESS (W-TO-S; 'every heavy syllable is stressed') is satisfied, so there is no violation indicated under W-TO-S for this candidate. Placing stress instead on the final syllable, as in the candidate [ba.tan.má.], violates W-TO-S once, which we indicate with a score of −1 (following Legendre, Sorace and Smolensky 2006 in regarding a violation as a negative satisfaction). This finally stressed alternative does satisfy the constraint ALIGN-R ('assign a violation to every syllable separating the stress from the right edge of the word'; McCarthy and Prince 1993), which the other candidate violates once, as indicated in the tableau.

In tableaux (1), the constraint weights (strengths) are shown beneath the constraint names. To the right of each row we see each candidate's *harmony*, which is calculated by multiplying its violation score on each constraint by that constraint's weight, and then summing the results. The optimal candidate is the one with the highest harmony, and is indicated by a pointing finger. The higher weight of W-TO-S (0.9) makes the score of the candidate that violates it (i.e. [.ba.tan.má.]) lower than one that violates ALIGN-R once (i.e. [.ba.tán.ma.]). Thus, the higher weight of W-TO-S as compared to ALIGN-R makes stress on the penultimate syllable optimal, just as its higher rank would do in OT.

(1) Weighted constraint evaluation

a. /.ba.tan.ma./	W-TO-S 0.9	ALIGN-R 0.6	
☞ [.ba.tán.ma.]		−1	−0.6
[.ba.tan.má.]	−1		−0.9
[.bá.tan.ma.]	−1	−2	−2.1

b. /.ban.ta.ma./	W-TO-S 0.9	ALIGN-R 0.6	
☞ [.ban.ta.má.]	−1		−0.9
[.bán.ta.ma.]		−2	−1.2
[.ban.tá.ma.]	−1	−1	−1.5

Tableau (1b) shows that the outcome is different when the heavy syllable is further from the right. The candidate surface form [.bán.ta.ma.] does satisfy the highest-valued constraint, W-TO-S, but its stressed syllable is two syllables away from the right edge; this candidate therefore incurs two violations of ALIGN-R, so that its harmony is $-2 \times 0.6 = -1.2$. This is a lower harmony than that of the alternative with final stress and an unstressed heavy first syllable, [.ban.ta.má.], which has a harmony of $-1 \times 0.9 = -0.9$ and is therefore optimal. This example illustrates the difference between HG's numerical measure of harmony and OT's ranking: with W-TO-S ranked above ALIGN-R, [.bán.ta.ma.] would be optimal in OT. For discussion of the typological consequences of the differences in outcomes with these constraints, see Legendre, Sorace and Smolensky (2006), Pater (2009a, this volume), and Jesney (this volume).

Tableaux (1) not only illustrate a *grammar* (with weights 0.9 and 0.6), but also a subset of a *language*, namely the set of underlying–surface pairs /.ba.tan.ma./–[.ba.tán.ma.] and /.ban.ta.ma./–[.ban.ta.má.]. As in other theories of grammar, the language that can be handled by the set of constraint weights is infinite, i.e. the grammar also applies to forms that are longer than the three syllables in (1): for forms with an arbitrary number of syllables of which exactly one is heavy, the grammar in (1) will place the stress on the penultimate syllable if it is heavy, otherwise on the final syllable. The goal of a human learner acquiring this language is to arrive at such constraint weights that her outputs are precisely the surface forms [.ba.tán.ma.] and [.ban.ta.má.] given the underlying forms /.ba.tan.ma./ and /.ban.ta.ma./ (the correct outputs for longer forms then follow automatically).

The learning algorithm discussed in this paper can now be illustrated informally. Suppose that at a certain point during the acquisition of her language a human learner has the grammar (constraint weights) in (2). She will produce the underlying form /.ba.tan.ma./ correctly as [.ba.tán.ma.], but the underlying form /.ban.ta.ma./ incorrectly as *[.bán.ta.ma.], as indicated in the tableau by a pair of asterisks around the pointing finger. The correct surface form [.ban.ta.má.] is indicated by a checkmark.

(2) The learner

a.	a. /.ba.tan.ma./	W-TO-S 1.2	ALIGN-R 0.3	
☞	[.ba.tán.ma.]		−1	−0.3
	[.ba.tan.má.]	−1		−1.2
	[.bá.tan.ma.]	−1	−2	−1.8

b.	/.ban.ta.ma./	W-TO-S 1.2	ALIGN-R 0.3	
√	[.ban.ta.má.]	−1		−1.2
☞	[.bán.ta.ma.]		−2	−0.6
	[.ban.tá.ma.]	−1	−1	−1.5

Suppose now that the learner, having the grammar in (2), hears an adult produce the surface form [.ban.ta.má.]. If we assume that the learner can correctly infer the underlying form /.ban.ta.ma./ from this, we can conclude that the learner must notice that her grammar is in error, because she herself would produce a different surface form, namely [.bán.ta.ma.], from the given underlying form. The learner will then regard the second candidate in (2b) as her *incorrect optimum* (*☞*), and the first candidate as the *correct output* (√). Realizing that something must be wrong with the weight of the constraints W-TO-S and/or ALIGN-R (because these are the constraints that prefer either of the two forms), she can take action by improving the grammar, i.e. by changing one or both of these weights in an appropriate direction. Such an action is an example of error-driven learning: the learner is given a single correct pair of grammar inputs and outputs (e.g. a single correct underlying–surface pair) at a time, and adjusts the constraint weights only if this pair is not optimal under the learner's current grammar hypothesis. Error-driven learning is known from perceptrons (Rosenblatt 1958) and their connectionist relatives, from theories of language acquisition (Wexler and Gulicover 1980: 127), from Error-Driven Constraint Demotion (EDCD) in OT (Tesar 1995), and from parameter setting algorithms (Gibson and Wexler 1994, Fodor 1998, Yang 2002).

The question is what precise action the learner should take. If she follows the learning algorithm discussed in this paper, she will both lower the weight of W-TO-S and raise the weight of ALIGN-R, making it more likely that she will turn an underlying /.ban.ta.ma./ into the correct output [.ban.ta.má.] on a future occasion. In general, the Gradual Learning Algorithm for HG (HG-GLA) will raise the weight of each constraint that favors the correct output over the incorrect optimum by an amount ε (the *learning rate* or *plasticity*) multiplied by the number of violations by which that constraint favors the correct output, and it will lower the weight of each constraint that favors the incorrect optimum over the correct output by – times the number of violations by which that constraint favors the incorrect optimum. So if $\varepsilon = 0.1$, then for the correct output [.ban.ta.má.] in (2b), the learner will lower the weight of W-TO-S by ε, changing it to 1.1, and she will raise the weight of ALIGN-R by 2ε, changing it to 0.5. This will not yet make the correct output the winner in her tableau for /.ban.ta.ma./, but it will do so after another instance of [.ban.ta.má.] has arrived, which will change the weights to 1.0 and 0.7, respectively.

To make our discussion of learning more formal, we will represent the difference between the correct output and the incorrect optimum as an *error vector* (see also Pater this volume, where this is called a comparative vector). For tableau (2b) we get the error vector in (3). An error vector is obtained by subtracting the violation scores of the learner's incorrect optimal candidate from those of the desired correct output form. Error vectors are the HG equivalent of OT's winner-loser pairs (Tesar and Smolensky 1998, 2000; Prince 2003): positive scores indicate the degree to which a constraint prefers the correct output, and negative ones the degree of dispreference. For example, ALIGN-R prefers the correct [.ban.ta.má.] in (3) by 2 violations, and W-TO-S disprefers it by 1 violation.

(3) Error vector

/.ban.ta.ma./	W-TO-S	ALIGN-R
[.ban.ta.má.] – [.bán.ta.ma.]	−1	+2

From this error vector we can calculate how to update the weights. The error-driven learning algorithm we investigate updates each constraint's weight by multiplying its value on the error vector by a positive integer that is constant across constraints (the learning rate, or plasticity), and adding the result to its pre-update value (see Section 13.3.1 for a formal statement). This update rule has been broadly applied in neural modeling and statistical learning under the names *perceptron update rule* (Rosenblatt 1958), *delta rule* (Widrow and Hoff, 1960), and *stochastic gradient ascent.* In generative linguistics, it is very similar to the update rule used in the gradual learning algorithm for OT in Boersma (1998) and Boersma and Hayes (2001) (see Section 13.4.2 below on the difference), and it is identical to the update rules used in the connectionist implementation of HG in Soderstrom, Mathis and Smolensky (2006: eqs. 14, 18, 21, 35d), the application of stochastic gradient ascent to a stochastic version of HG in Jäger (2007), and the application of the perceptron convergence procedure to categorical HG in Pater (2008). We will refer to the error-driven learning algorithm using this update rule with HG as HG-GLA (Harmonic Grammar's Gradual Learning Algorithm).

In the learning proofs and simulations that follow, we make the standard assumption that learners are exposed to a finite set of data sampled from the infinite set of pairs of grammar inputs and outputs (here, underlying and surface forms) that a language defines. The goal for a learning algorithm is to find a correct grammar with sufficient exposure to such learning data. Another standard assumption is that data that inform the learner that her grammar is still incorrect, such as [.ban.ta.má.] in the example above, are not 'maliciously withheld' (Tesar and Smolensky 1998, 2000) from the learner (see Section 13.3.3 for details). With this assumption about the availability of 'informative data' we can derive learning paths for the example in (2). Figure 13.1 (left) shows the state in (2), with W-TO-S having a weight of 1.2 and ALIGN-R having a weight of 0.3, as a dot marked with '1'. The dashed lines embrace the region where the learner should end up, namely the region where the weight of W-TO-S is greater than the weight of ALIGN-R (so that the grammar will generate [.ba.tán.ma] correctly) but less than twice the weight of ALIGN-R (so that the grammar will produce [.ban.ta.má.] correctly).

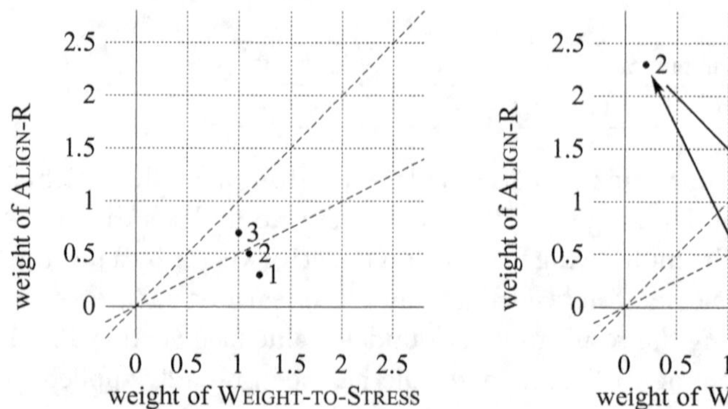

Figure 13.1: Two learning paths from the same initial state, at the left with a plasticity of 0.1, at the right with a plasticity of 1.0.

Let us follow the two learning paths shown in Figure 13.1. In the left figure, the plasticity is 0.1, as above, and the informative adult [.ban.ta.má.] moves the weights to 1.1 and 0.5, as above; this point is labeled labeled as '2'. The next [.ban.ta.má.] datum makes the learner move the weights to 1.0 and 0.7 (labeled as '3'), thus arriving in the target region. When her grammar is in the target region, she cannot make any errors any longer (her own optimum will always be identical to the incoming correct adult form), and learning stops forever. In other words, the learner has *converged* upon a correct grammar of the target language, in only two learning steps.

In the right-hand figure, the plasticity is 1.0, so that the first incoming adult [.ban.ta.má.] moves the weights to 0.2 and 2.3, labeled as '2'. The learner has thereby stridden past the target region. Fortunately, an incoming adult [.ba.tán.ma.] will now become become informative, as the learner's optimum will be the incorrect *[.ba.tan.má.]. From (2a) we can infer that the learner will change her weights to 1.2 and 1.3 (equal steps for both constraints, because the relevant error vector computed from (2a) has +1 for W-to-S and –1 for Align-R), labeled as '3' in the figure. Next, another incoming [.ba.tán.ma.] will move the weights to 2.2 and 0.3 (labeled as '4'), and a subsequent incoming [.ban.ta.má.] will move the weights to 1.2 and 2.3 ('5'), after which an incoming [.ba.tán.ma.] will move the weights to 2.2 and 1.3 ('6'), which is in the target region and is therefore the final state of the learner's grammar. In this case, convergence was reached after five learning steps. So we see that the learner eventually arrives in a correct grammar both for a small plasticity of 0.1 and for a large plasticity of 1.0.

It remains to be explained what happens if, as a rare coincidence, the learner arrives exactly on the boundary of the target region. In such a case, at least two candidates (one correct, the other incorrect) must be equally harmonic in the learner's grammar. The assumption about such a *tie* is that the learner will choose one of the tied candidates as her optimum at random. As the data come in repeatedly, the learner is sure to choose the incorrect candidate as her optimum at some point in time, and once she does so, she will notice her error and move the weights in a new learning step, moving away from the boundary.

Finally, there is the question of negative weights. It can be seen that if the plasticity is greater than both constraint weights, the first learning step in the above example will already move the weight of one of the constraints to a negative value. The learning algorithm will be seen to have no trouble with this situation.

With sufficient exposure to informative data, it turns out that an algorithm using the update rule mentioned above will always find a correct set of weights, independently of the initial weights of the constraints, independently of the plasticity, and independently of whether any weights ever become negative. This is shown in our adaptation of the perceptron convergence proof in Sections 13.3 and 13.4, and is illustrated in our simulations of convergence times that follow the proofs. Specifically, these proofs show that for any non-varying language (i.e. for any language in which every underlying form has a unique correct output) and for any set of constraints, an HG-GLA learner fed with a random sequence of correct underlying–surface pairs will need only a finite number of errors (learning steps) before it finds a correct grammar for the language.

13.3 Convergence of HG-GLA for Non-noisy Learners

In this section, we first provide a formal definition of the learning algorithm described in Section 13.2 (HG-GLA), restricted to learners without evaluation noise. We then provide a convergence proof for such non-noisy learners, and finally show by computer simulation that the algorithm converges very fast in practice.

13.3.1 Formalization of the Language Data and the Learner

As mentioned in Section 13.2, no learning algorithm can be guaranteed to converge on a grammar that is appropriate for a target language unless it receives sufficient exposure to informative data. Our proof therefore operates on the basis of a finite *dataset*, which contains a sufficient amount of informative data and from which the learning data that are supplied to the learner are sampled. This dataset, then, consists of a finite number M of 'correct' *input-output* pairs. That is, the dataset has M different *input* (e.g. underlying) forms i_m (m = 1...M), each of which is coupled with a unique 'correct' *output* (e.g. surface) form o_m. Note that we use the term 'input' here in the OT sense of 'input to the grammar', not in the entirely different sense of 'language input to the learner' known from the language acquisition literature; here, the language input to the learner consists of both the inputs and the outputs of the grammar.

We assume that the language user, especially the learner, maintains a finite number K of constraints C_k (k = 1..K) with weights w_k; these weights are used for input-output processing, as in (1), and can change as a result of learning, as in Figure 13.1. We have five assumptions about the learner's processing, and three about learning.

The first assumption about processing, shared with existing proofs for EDCD and MaxEnt, is that for each input i_m of the dataset the learner can compute (or is given) a finite number N_m of output candidates o_{mn} (n = 1.. N_m), as in tableaux (1); this finiteness is required for the definition of the margin of separability in Section 13.3.2 and D_{max} in Section 13.3.3.[2] The second assumption is that the learner can compute (or is given) the extent to which each input-output pair (i_m, o_{mn}) satisfies each constraint C_k (k = 1..K). We denote the extent of constraint satisfaction by s_{mnk}; for instance, the value of s_{132} in tableaux (1) is –2 (first tableau, third row, second column). The third assumption is that the learner can compute a harmony value for each input-output pair (i_m, o_{mn}), as a weighted sum of satisfaction scores:

(4)
$$H_{mn} \equiv \sum_{k=1}^{K} w_k s_{mnk}$$

In (1), for instance, the value of H_{23} is –1.5. The fourth assumption is that the learner, when asked to evaluate an input form i_m, can identify

2 Under some circumstances, a finite set of possibly optimal candidates might be derived from an infinite candidate set (e.g., Riggle 2004 on 'contenders' in OT), and the proof could still work.

the (one or more) *optimal* forms o_{mr}, i.e. the forms that have a maximum harmony among the output candidates:

(5)
$$\forall n = 1..N_m : \quad \sum_{k=1}^{K} w_k s_{mrk} \geq \sum_{k=1}^{K} w_k s_{mnk}$$

The fifth assumption about processing is that if there are multiple optimal forms, the learner is able to randomly select one of them as the *winning candidate* of her evaluation.

The objective of the learning procedure is to find constraint strengths w_k such that the optimal outputs for all inputs i_m in the learner's grammar are identical to the correct outputs o_m of the dataset. The first assumption about learning is that for every input-output pair (i_p, o_p) that arrives in the language data, the learner computes her own winning output candidate for the given input i_p. Thus, this winning candidate is o_{pr}, for some r in the range $1..N_p$; it has constraint satisfactions s_{prk} and harmony H_{pr}. The second assumption is that the output form o_p given in the language data is among the learner's set of output candidates for i_p, i.e., the learner can identify the given output form as o_{pq}, where $1 \leq q \leq N_p$, and thereby establish the satisfactions s_{pqk} and harmony H_{pq} of this form in her current grammar. The third assumption about learning is that the learner compares this form o_{pq} with her own winning candidate o_{pr}. If it turns out that the two forms are different (and therefore indicate that the learning objective has not been reached yet), the learner labels the situation as an *error*: the form o_{pq} is now considered the *correct output* and o_{pr} the *incorrect optimum*. The learner then takes action by changing the weights of the constraints from w_k to w_k':

(6)
$$w_k' = w_k + \varepsilon \cdot \left(s_{pqk} - s_{prk} \right)$$

where ε is the plasticity, or learning rate, a positive real number. The formula is identical to that used by Jäger (2007) for Maximum Entropy grammars and a notational variant of the formulas used by Soderstrom, Mathis and Smolensky (2006: eqs. 14, 18, 21, 35d) for their connectionist implementation of HG. The update rule implies that the learner raises the weights of all the constraints satisfied better with the correct output o_{pq} than with the incorrect optimum o_{pr}, and lowers the weights of all the constraints satisfied better with the incorrect optimum than with the correct output.

13.3.2 The Unit Sample Target Grammar

In this section we introduce two concepts that we need for applying the perceptron convergence proof to HG-GLA in Section 13.3.3: the *unit sample target grammar*, and one of its properties, the *margin of separation of harmony*.

No learning algorithm can learn a language that its underlying grammar model cannot represent. Thus, the learning algorithm defined in Section 13.3.1 will fail to learn any language that cannot be described by a Harmonic Grammar. What we really want to prove, therefore, is that HG-GLA is successful for any set of unique optima that can be generated by a Harmonic Grammar (it is possible, for instance, to construct languages in HG in which two candidates with different constraint violation patterns still tie for optimality; HG-GLA will typically not find a correct set of weights for such cases; our proof is therefore limited to target languages in which only a single candidate is correct for each tableau).

By safe assumption, therefore, there exists at least one weighting of the constraints that makes the set of correct forms in the dataset uniquely optimal. Given the nature of the weights (namely, real numbers) and the nature of the interaction of the finite set of K constraints, there probably exist many such weightings (e.g. any point between the two lines in Figure 13.1), but here we select just one, and call it the *sample target grammar*. The goal of the learner, then, is to arrive either at this sample target grammar or at any other grammar that describes the dataset correctly.

We do not know much about the constraints' weights in the sample target grammar, but we do know that not all weights are zero, because if they were, all candidates would be optimal, and the sample target grammar would exhibit variation in its outputs (unless all inputs have only a single output candidate, but in that case the learner already has a correct grammar from the start). Since not all weights are zero, we can normalize the sample target grammar by dividing every constraint weight by the square root of the sum of the squares of the constraint weights. Thus, if the sample target grammar has six constraints with weights 17, 5, –3, 14, –5, and 9, the sum of their squares is $17^2+5^2+(-3)^2+14^2+(-5)^2+9^2 = 625$, whose square root is 25. When we divide all constraint weights by 25, we get the weights (0.68, 0.20, –0.12, 0.56, –0.20, 0.36). This grammar generates exactly the same language as it did before the division. We call such a form of the grammar a *unit grammar*: the sum of the squares of its weights is 1. For the unit sample target grammar, we write these weights as u_k (k=1..K).

In the sample grammar, the optimal outputs are unique (because they are the correct language forms, and the language is non-varying). Therefore, the harmony (in the sample grammar) of any optimal form is greater than that of any other candidate for the same input. Suppose that for each input i_m the optimal output is o_{mn_m} (n_m is the index of this output in the learner's tableau for i_m). Then the harmony comparison reads:

(7)
$$\forall m = 1..M, n = 1..N_m, n \neq n_m : \sum_{k=1}^{K} u_k s_{mn_m k} > \sum_{k=1}^{K} u_k s_{mnk}$$

We now assume that for every input a second-best candidate can also be found (it does not have to be unique); it is the winner of the partial tableau that results if the best form is left out. The harmony differences between the second-best forms and the best forms are δ_m (m=1..M); for input i_m, the harmony (in the unit target grammar) of all non-optimal candidates is at least δ_m less than the harmony of the optimal candidate. Since the number M of inputs is finite, there must exist a value δ that is the minimum of all M values δ_m. We therefore know that the harmony of any non-optimal candidate for any input is separated by at least δ from the harmony of the optimal candidate for that same input (in the unit target grammar):

(8)
$$\forall m = 1..M, n = 1..N_m, n \neq n_m : \sum_{k=1}^{K} u_k s_{mn_m k} \geq \sum_{k=1}^{K} u_k s_{mnk} + \delta$$

Adapting terminology from the perceptron literature, we call δ the *margin of separation of harmony*; this value is independent of the language data fed to the learner, and it is greater than 0 for any Harmonic Grammar that generates a nonvariable language with a finite number of possible inputs.

13.3.3 The Perceptron Convergence Proof Applied To HG-GLA

The perceptron is a model of linear classification developed by Rosenblatt (1958). Its learning rule is convergent: if two sets of data are linearly separable, the perceptron update rule is guaranteed to find a set of weights that will correctly separate them (Rosenblatt 1962, Block 1962, Novikoff 1962, Minsky and Papert 1969; see Bishop 1995 for a textbook introduction). Our convergence proof for HG-GLA is closely analogous to the perceptron convergence proof by Block and Novikoff and many later formulations, including in linguistics (Collins 2002 applies it to part-of-speech tagging). In proving convergence for the update rule (7) in a constraint-weighting model, we are preceded by Fischer (2005), who provides a proof of its convergence for Maximum Entropy grammars; unfortunately, that proof does not immediately generalize to Noisy HG learners, whereas the perceptron proof does, as we show in Section 13.4. Within HG, Soderstrom, Mathis and Smolensky (2006) did use the update rule (7), but provided no convergence proof.

The simplest kind of perceptron proof would work for a single HG tableau with initial zero weights for all constraints; our proof here has to work from arbitrary initial weights (an element found in various existing perceptron proofs) and for any number of tableaux (a straightforward extension that also appears in e.g. Collins' proof). We will now show that the perceptron proof can be applied to HG-GLA because in Section 13.3.2 we managed to define the unit sample target grammar and the margin of separation of harmony.

The convergence proof for HG-GLA involves showing (1) that given any finite set of language data that can be generated by an HG grammar, the learner is guaranteed to make only a finite number of errors, (2) that after having made the last error, the learner has reached a correct grammar, and (3) that the last error will arrive within finite time.

The first goal of our proof, then, is to show that for any given set of language data, we can define an upper bound on the number of errors that the learner will make.

The initial state of the learner (i.e. the state after zero errors) is characterized by a set of initial constraint weights $w_k(0)$. These weights could be all zero, or all 10.0, or weights corresponding to any of the initial rankings proposed in the OT literature (e.g. all markedness constraints high, all faithfulness constraints low, as proposed by Demuth 1995, Levelt 1995, Ohala 1996, Smolensky 1996; see Jesney and Tessier 2011 in HG).

As the language data come in, the learner receives a sequence of 'correct' input-output pairs. However, since only the pairs that cause errors will make changes to the weights, we here consider only the pairs that cause an error. It therefore makes sense to define $w_k(t)$ as the weight of constraint C_k after t errors; the set of $w_k(t)$ for all k and all t therefore defines the entire history of the learner's grammar. Let the tth error-causing input-output pair be $(i_{p(t)}, o_{p(t)})$, the incorrect winning candidate $o_{p(t)r(t)}$, and the correct output form $o_{p(t)q(t)} = o_{p(t)}$. The update rule for this data pair is then

(9) $$w_k(t) = w_k(t-1) + \varepsilon \cdot \left(s_{p(t)q(t)k} - s_{p(t)r(t)k}\right)$$

We can now start off on some mathematical manoeuvers that together allow us to establish an upper bound on the number of errors. First, we take the inner product of all terms in (10) with the weights of the unit sample target grammar:

(10) $$\sum_{k=1}^{K} u_k w_k(t) = \sum_{k=1}^{K} u_k w_k(t-1) + \varepsilon \cdot \left(\sum_{k=1}^{K} u_k s_{p(t)q(t)k} - \sum_{k=1}^{K} u_k s_{p(t)r(t)k} \right)$$

With the help of (8), we can establish a lower bound on the last term, so that (10) becomes

(11) $$\sum_{k=1}^{K} u_k w_k(t) \geq \sum_{k=1}^{K} u_k w_k(t-1) + \varepsilon\delta$$

By induction, starting with $t = 0$, we derive

(12) $$\sum_{k=1}^{K} u_k w_k(t) \geq \sum_{k=1}^{K} u_k w_k(0) + \varepsilon t \delta$$

The first term on the right is the inner product of the weight vector u_k of the unit sample target grammar and the initial weight vector $w_k(0)$ of the learner's grammar. We abbreviate it as $W_0 \cdot$ U, so that (12) can be written as

(13) $$\sum_{k=1}^{K} u_k w_k(t) \geq \varepsilon t \delta + W_0 \cdot U$$

According to the Cauchy-Schwarz inequality, the absolute value of the inner product of two vectors is always less than or equal to the product of the norms of the two vectors:

(14) $$\sqrt{\sum_{k=1}^{K} u_k^2} \sqrt{\sum_{k=1}^{K} w_k^2(t)} \geq \left| \sum_{k=1}^{K} u_k w_k(t) \right|$$

Since the first factor on the left in (14) is 1 (since u_k defines a unit grammar, its norm is 1), we can combine (13) and (14) into

(15) $$\sqrt{\sum_{k=1}^{K} w_k^2(t)} \geq \varepsilon t \delta + W_0 \cdot U$$

Equation (15) is valid only for values of t that exceed $-W_0 \cdot U/(\varepsilon\delta)$, which can be a positive time if e.g. the learner's initial grammar has negative weights while the sample target grammar has positive weights. Loosely speaking, (15) says that the overall constraint weights move away from zero with time, i.e. with the number of mistakes t (at least for $t \geq -W_0 \cdot U/(\varepsilon\delta)$). The term on the left in (15) is the norm of w_k. Equation (15) expresses the fact that the norm is bounded from below by a line that rises with t. Equivalently, we can say that the square of the norm is bounded from below by a concave parabola (see Figure 13.2):

(16) $$\sum_{k=1}^{K} w_k^2(t) \geq \left(\varepsilon t \delta + W_0 \cdot U\right)^2$$

at least for $t \geq -W_0 \cdot U/(\varepsilon\delta)$.

Beside a lower bound we can also determine an *upper* bound on this squared norm. We start by writing it in terms of what it was before the latest learning step:

(17) $$\sum_{k=1}^{K} w_k^2(t) = \sum_{k=1}^{K} \left(w_k(t-1) + \varepsilon \cdot \left(s_{p(t)q(t)k} - s_{p(t)r(t)k}\right)\right)^2 =$$
$$= \sum_{k=1}^{K} w_k^2(t-1) + 2\varepsilon \sum_{k=1}^{K} w_k(t-1)\left(s_{p(t)q(t)k} - s_{p(t)r(t)k}\right) + \varepsilon^2 \sum_{k=1}^{K} \left(s_{p(t)q(t)k} - s_{p(t)r(t)k}\right)^2$$

According to (4), the factor after 2ε involves a harmony difference:

(18) $$\sum_{k=1}^{K} w_k(t-1)\left(s_{p(t)q(t)k} - s_{p(t)r(t)k}\right) = H_{p(t)q(t)}(t-1) - H_{p(t)r(t)}(t-1)$$

This is the difference between the harmony of the correct form in the learning data (a non-optimal, or perhaps tied, candidate in the learner's current grammar) and the incorrect optimum (an optimal candidate in the learner's current grammar before the learning step), and must therefore be less than or equal to zero. We can now bound the squared norm in (17) from above by throwing away the 2ε term from (17):

(19) $$\sum_{k=1}^{K} w_k^2(t) \leq \sum_{k=1}^{K} w_k^2(t-1) + \varepsilon^2 \sum_{k=1}^{K} \left(s_{p(t)q(t)k} - s_{p(t)r(t)k}\right)^2$$

We can proceed by noting that the sum in the last term is again bounded from above: it can never be more than the sum of squares of maximum constraint satisfaction differences between the optimal candidates and their nonoptimal counterparts. If the maximum number of constraint violations in the unit grammar for any (potentially incorrectly optimal) input-output pair is V_{max}, and the maximum number of positive constraint satisfactions for any input-output pair is S_{max}, the sum in the last term can never be more than K times the square of the sum of V_{max} and S_{max}. For instance, if no cell in any of the M tableaux contains more than five violations or three positive satisfactions, the last term in (19) can never be more than $8^2\varepsilon^2K$. Generally, (19) becomes

(20) $$\sum_{k=1}^{K} w_k^2(t) \leq \sum_{k=1}^{K} w_k^2(t-1) + \varepsilon^2 K D_{max}^2$$

where D_{max} is defined as $V_{max} + S_{max}$. This value is independent of the language data fed to the learner, and is guaranteed to exist, since the number of output candidates for each input is finite.

By induction on t, starting again at $t = 0$, we get

(21) $$\sum_{k=1}^{K} w_k^2(t) \leq \varepsilon^2 K D_{max}^2 t + \sum_{k=1}^{K} w_k^2(0)$$

The last term on the right in (21) is the squared norm of the learner's initial weight vector. We abbreviate it as $|W_0|^2$, so that (21) becomes

(22) $$\sum_{k=1}^{K} w_k^2(t) \leq \varepsilon^2 K D_{max}^2 t + |W_0|^2$$

This formula establishes that the squared norm of the learner's weight vector lies below a line that rises with the number of errors t (see Figure 13.2). But in (16) we have seen that the squared norm also lies *above* a function that increases quadratically with t. Since a rising parabola cannot stay below a rising line forever with increasing t (see Figure 13.2), the two bounding conditions together necessarily indicate that t itself must be limited. More formally, we can combine (16) and (22) to yield

$$(23)\quad \left(\varepsilon\delta t + W_0 \cdot U\right)^2 \le \varepsilon^2 K D_{max}^2 t + \left|W_0\right|^2$$

for $t \geq -W_0 \cdot U/(\varepsilon\delta)$. Equation (23) is a quadratic equation in t that can be solved:

$$(24)\quad \varepsilon^2\delta^2 t^2 - \left(\varepsilon^2 K D_{max}^2 - 2\varepsilon\delta W_0 \cdot U\right)t + \left(W_0 \cdot U\right)^2 \le \left|W_0\right|^2$$

$$(25)\quad \left(t - \left(\frac{KD_{max}^2}{2\delta^2} - \frac{W_0 \cdot U}{\varepsilon\delta}\right)\right)^2 \le \left(\frac{KD_{max}^2}{2\delta^2} - \frac{W_0 \cdot U}{\varepsilon\delta}\right)^2 + \frac{\left|W_0\right|^2 - (W_0 \cdot U)^2}{\varepsilon^2\delta^2}$$

The first thing to establish is that the right-hand side of this equation cannot be negative: if we write $t = 0$ in (14), and take into account that the norm of u_k is one, we see that $|W_0 \cdot U|$ must be less than or equal to $|W_0|$, so that the second term on the right in (25) must be positive or zero. We are therefore allowed to rewrite (25) as

$$(26)\quad \left|t - \left(\frac{KD_{max}^2}{2\delta^2} - \frac{W_0 \cdot U}{\varepsilon\delta}\right)\right| \le \sqrt{\left(\frac{KD_{max}^2}{2\delta^2} - \frac{W_0 \cdot U}{\varepsilon\delta}\right)^2 + \frac{\left|W_0\right|^2 (W_0 \cdot U)^2}{\varepsilon^2\delta^2}}$$

Since (26) is valid for all $t \geq -W_0 \cdot U/(\varepsilon\delta)$, it is certainly valid for all $t \geq KD^2_{max}/(2\delta^2) - W_0 \cdot U/(\varepsilon\delta)$. For those values of t, (26) is equivalent to (27):

$$(27)\quad t \le \frac{KD_{max}^2}{2\delta^2} - \frac{W_0 \cdot U}{\varepsilon\delta} + \sqrt{\left(\frac{KD_{max}^2}{2\delta^2} - \frac{W_0 \cdot U}{\varepsilon\delta}\right)^2 + \frac{\left|W_0\right|^2 (W_0 \cdot U)^2}{\varepsilon^2\delta^2}}$$

For smaller values of t, i.e. $t < KD^2_{max}/(2\delta^2) - W_0 \cdot U/(\varepsilon\delta)$, (26) is not equivalent to (27), but (27) holds a fortiori. Therefore, (27) expresses a true upper bound on t.

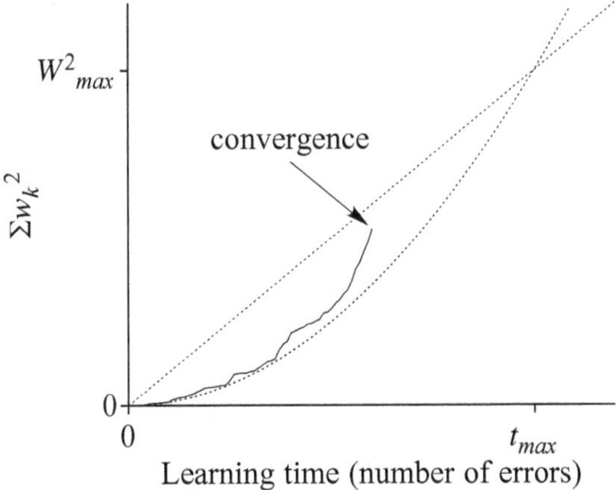

Figure 13.2: A possible learning path (in terms of $\Sigma_k w_k^2$) if all weights start at zero. The path stays between the parabola of equation (16) and the line of equation (22). Convergence is reached well before $t_{max} = KD^2_{max}/\delta^2$. The value of W^2_{max} is $\varepsilon^2 K^2 D^4_{max}/\delta^2$.

The interpretation of (27) is that t can never be greater than a value t_{max} (also visible in Figure 13.2) that is defined as the right-hand side of (27). This t_{max} is determined completely by properties of the hypothetical unit target grammar (D_{max}, K, U, δ), properties of the initial state ($W_k(0)$), and fixed *a priori* properties of the learning algorithm (ε); crucially, t_{max} does not depend on any properties of the language data as they are stochastically presented to the learner. This means that given an initial state and a target grammar, we know that the learning algorithm can never make more than a fixed number of t_{max} mistakes. The interpretation is that learning will stop after at most t_{max} mistakes. In other words, the learner will end up in a never-changing grammar after at most t_{max} mistakes.

What will this final stable grammar look like? Well, it is guaranteed to be a correct grammar for the target dataset. For if it were not, the grammar would have to be one that is able to produce an incorrect output for at least one input. That input is certain to arrive at the learner at some point in time (under a common assumption addressed in the next paragraph); when it does, it will cause a mistake and must force learning, which is impossible because learning has stopped. Hence, the final grammar is a correct grammar of the target data set.

The last question is whether the final grammar will ever be reached. After all, t is not the time but the number of mistakes. Between any two consecutive mistakes there may be any number of input-output pairs that cause no mistake, i.e. input-output pairs for which the learner's current grammar produces the same output as that given by the language data. In fact, the likelihood that an input-output pair generates a mistake probably decreases as the learner's grammar approaches one that is appropriate for the target dataset; as a result, the average expected time between two consecutive mistakes tends to rise as acquisition proceeds. A necessary condition for convergence, therefore, is the same as the one mentioned by Tesar and Smolensky (1998: 246) for the Constraint Demotion convergence proof, namely that informative input-output pairs "are not maliciously withheld from the learner". For the finite dataset under discussion, the continual availability of all forms to the learner is guaranteed if the data are presented in an infinite sequence of sufficiently randomly selected input-output pairs. Similarly, any incorrect form that in the learner's grammar ties for harmony with a correct form should not be maliciously withheld from being chosen as the incorrect winning candidate; the random selection between ties mentioned in §3.1 guarantees this (random selection was not addressed in the EDCD convergence proof by Tesar 1995, and therefore led to misconvergences for that algorithm; as here, the solution for EDCD lies in the random selection between ties, as shown by Boersma 2009).

Since the number of possible inputs M, the number of output candidates per input N_m, and the number of constraints K are finite, D_{max} in (28), which is the maximum number of constraint violations-minus-satisfactions in at least one correct grammar, will be finite as well. Therefore, t_{max} will have a finite value if a margin of separation δ exists in at least one grammar of the target language; the hypothetical sample target grammar of §3.2 is just such a grammar, again as a result of the finiteness of all the dimensions involved. This rounds up our proof: HG-GLA learners will converge in finite time, regardless of the weights in their initial grammars.

3.4 Convergence Times: Non-noisy Learners

It is one thing to show that a learning algorithm converges within a finite time, but quite another thing to show that it converges within some time that could reasonably be a characterization of a human learner's youth. Depending on the characteristics of a learning problem, the upper bound on the number of errors expressed by t in (27) can be quite large. We address this issue by showing with computer simulations that the learning algorithm converges in a reasonable amount of time, for a number of randomly generated languages.

To test whether a non-noisy HG learner converges not just in theory but also in practice, and to test how fast it converges, we generate a random target grammar and a random dataset, as follows. We create the grammar as a set of K constraints (with K randomly drawn from the range 2...20), with weights randomly drawn from a uniform distribution between 2.0 and 18.0. We then define M inputs i_m (M randomly drawn from 1...20) each with N_m output candidates (N_m randomly drawn from 2...20 for each input separately), with each candidate violating each constraint between 0 and D_{max} times (D_{max} randomly drawn from 1...5 for the whole language), as determined by an evenly distributed random choice from among these $D_{max}+1$ possibilities. That is, we have M tableaux with up to 20 candidates each, each with random integer satisfaction scores from 0 to $-D_{max}$ on each of K constraints. We then use the constraint weights to select the optimal output for each tableau. If an output happens not be unique (which can happen if the best two candidates for that input have the same violation pattern), we discard the whole dataset and start again with new inputs, weights, and violation patterns (and a new K, M, and D_{max}). If necessary, this is repeated until all outputs are unique, so that we know that at least one target grammar exists (it is the grammar we just created). In this way, we create a dataset consisting of M "correct" input-output pairs.

Once the target grammar exists, we create a nonnoisy HG learner with the same inputs, output candidates, and violation patterns as the target grammar, but with all constraints weighted at $w_k(0) = 10.0$ ($k = 1..K$). We feed this learner "correct" input-output pairs of the target language, randomly and evenly selected from among the M possible "correct" input-output pairs in the dataset. The learner's grammar evaluates each input with the HG evaluation mechanism, and if the optimal output thus determined is different from the "correct" output given by the target language, the learner takes an HG learning step with a plasticity of $\varepsilon = 1.0$.[3] After every 10 pieces of learning data, we check whether the learner has arrived in a grammar in which all outputs are correct and

3 In a computer implementation of a nonnoisy learner, the plasticity has to have an integer value, so as to avoid floating-point rounding errors, which is necessary to allow the learner to go through stages with tied candidates, i.e. stages where multiple candidates within a tableau have exactly equal harmonies. Such situations lead to random variation, and hence to more learning (Section 13.3.1).

uniquely optimal. If the learner has reached such a grammar at e.g. the 18th check, we conclude that the learner has converged between the 171st and 180th learning datum.

We perform this procedure not just for a single virtual learner, but for 100,000 virtual learners. They all converge, and do not take much time. Figure 3 shows a histogram of the convergence times.

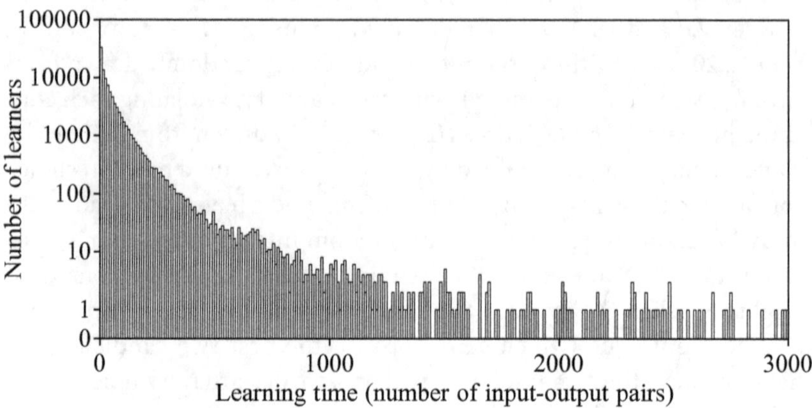

Figure 13.3: Histogram of convergence times of 100,000 non-noisy HG learners. The vertical scale is logarithmic.

The slowest learner requires 24,530 pieces of data, but most learners require less than 1000 pieces of data to converge.

HG learners should not only be able to learn from datasets generated by HG grammars, but also from datasets generated by OT grammars, because any finite dataset generated by OT can also be generated by HG. We therefore test another 100,000 non-noisy HG learners on datasets generated by target OT grammars. We create such a target grammar as a random total ranking of constraints, and compute the M optimal outputs in the usual OT way. Again, all 100,000 HG learners converge, although it takes them on average slightly more time than they need for the target HG grammars, perhaps because the weights have to be drawn further apart.

We conclude that in practice non-noisy HG-GLA learners converge quite fast.

13.3.5 Tests of Learners with a Positivity Restriction: Exponential HG

If HG is to function as an OT-like theory of language typology, then it appears necessary to ban negative weights (Keller 2000, 2006, Prince

2002, Pater 2009, Potts, Pater, Jesney, Bhatt and Becker 2010). To see why, consider the tableau in (28).

(28) Harmonic bounding subverted

	i_1	C_1 1.0	C_2 −2.0	
	o_{11}	−1		−1.0
	o_{12}		−1	+2.0
☞	o_{13}	−1	−2	+3.0

In OT, output candidate o_{13} cannot be optimal under any ranking of the constraints, because its violations are a proper superset of those of one of the other candidates (in fact, of both of them); candidate o_{13} is therefore said to be *harmonically bounded* (Samek-Lodovici and Prince 1999, 2005): it can become optimal under no ranking of the constraints. Prince (2002) points out that such harmonic bounding relations are preserved in HG if constraints are restricted to have positive weights. Technically, if such a positively-weighted HG permits constraints to assess both positive satisfaction and negative violation scores, we can say that a candidate A harmonically bounds another candidate B if and only if A and B are members of the same candidate set, and the error vector produced by subtracting B's constraint scores from those of A contains at least one positive value, and no negative ones (Becker and Pater 2007).

One way to preserve OT-like harmonic bounding in HG, therefore, is not to use negative constraint weights. To determine whether, and with what set of weights a set of input-output pairs can be rendered jointly optimal in positively-weighted HG, one could use the *simplex* algorithm of linear programming, as shown by Potts *et al.* (2010); this method functions as the HG equivalent of the recursive version of Constraint Demotion, whose inconsistency detection properties are useful for calculating predicted typologies for OT (see Hayes, Tesar and Zuraw 2003 on typology calculations using recursive Constraint Demotion). The simplex algorithm is unlikely, however, to form the basis of a realistic model of human learning. Fortunately, the learning algorithm discussed in this paper can be adapted for positively-weighted HG, as we now show.

The HG-GLA learning algorithm can be used with a model of grammar that automatically enforces the positivity restriction directly in the harmony function. In *Exponential HG,* which has been available in the

Praat program (Boersma and Weenink 1992–2013) since February 2006, the harmony of a representation is not given by (4), but by (29).

(29)
$$H_{mn} \equiv \sum_{k=1}^{K} s_{mnk} e^{w_k}$$

Although the 'weights' w_k can have positive and negative values, the weighting of the constraints in the harmony function is e^{w_k}, which is always positive. For the learning algorithm, the update rule is the same as in (6). Hence, the learning algorithm may shift the weight (w_k) of a constraint to a negative value; according to (29), however, such a constraint will still always contribute positively to harmony if it assigns rewards, or negatively if it assigns violations.

Every finite OT-generated dataset should be representable in Exponential HG (Prince and Smolensky 1993 [2004: 236]). We therefore test 100,000 learners in the fashion described in Section 13.3.4 on datasets generated by non-varying OT grammars. Some care has to be taken to set the plasticity to a reasonable value: since a value of 1.0 (as in Section 13.3.4) would increase the contribution of a constraint by a giant factor of approximately 2.7183 at every learning step, we take a moderate plasticity of 0.1, which leads to a factor of 1.105.[4]

It turns out that all 100,000 learners find a correct weighting. This convergence could be expected on the basis of the observation by Magri (2007) that an equivalent combination of grammar model and learning algorithm (the usual linear HG, with a 'Multiplicative GLA') has a convergence proof (based on the 'multiplicative perceptron'); we leave the full adaptation of this proof to grammar learning as a topic for further research. The learners are not only correct, but fast as well: Figure 13.4 shows a histogram of the convergence times. The slowest learner needed 1,160 input-output pairs to converge.

Exponential HG is thus a good candidate for adding OT-like behavior to HG; it has OT-like harmonic bounding as well as a working learning algorithm. It can be mentioned here that another manner of banning negative weights, namely simply clipping all weights at zero from below (Keller 2000), also achieves 100% convergence in our simulations.

4 In a computer implementation, with its floating-point rounding errors, the trick of the previous footnote to handle ties no longer works. Instead, one can use a tiny evaluation noise in w_k, of e.g. 0.001, to force random choices between candidates with analytically 'equal' harmonies.

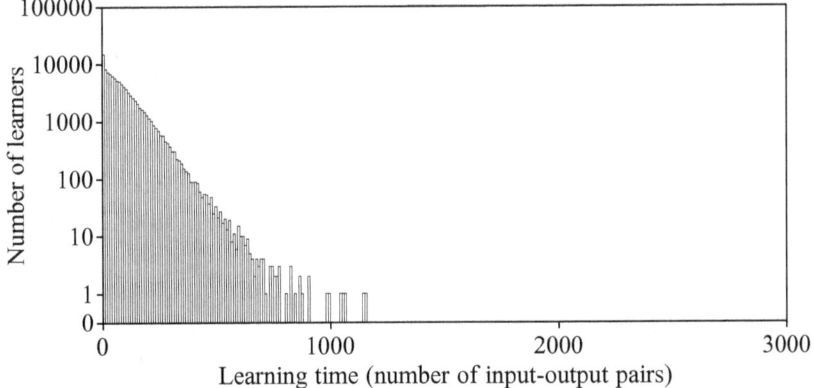

Figure 13.4: Convergence times of 100,000 Exponential HG learners on OT-generated datasets.

13.4 Convergence of HG-GLA for Noisy Learners

Up to this point we have assumed a version of HG that is categorical in the sense that each time the same grammar is used to evaluate a candidate set, it returns the same candidate as optimal (or set of candidates, in the case of a tie). In both OT and HG, variation in the choice of optima across instances of evaluation has been widely used in accounting for natural language phenomena termed 'free variation' (see Anttila 2007 and Coetzee and Pater 2011 for overviews). There are two extant stochastic versions of HG that yield this sort of variation. One stochastic variant of HG is the log-linear model of probabilistic grammar usually referred to as *Maximum Entropy grammar* (Johnson 2002, Goldwater and Johnson 2003, Fischer 2005, Wilson 2006, Jäger 2007). In this model, every output candidate has a relative probability that rises exponentially with its harmony. Another approach to stochastic HG is to incorporate the evaluation noise of stochastic OT (Boersma 1997; see also Ackley, Hinton and Sejnowski 1985 and Hinton and Sejnowski 1986 on noise on connection weights in neural networks, and Goldrick and Daland 2009 on a relationship of evaluation noise in HG to connectionist models of speech processing).

In *Noisy HG*, every time the grammar is used to evaluate a candidate set, the weighting values are perturbed by noise: each constraint's value is temporarily altered by adding a random positive or negative number sampled from a normal distribution with zero mean. This will lead to different weightings across instances of evaluation, which will produce variation. The proportion of times a candidate in any given candidate set is chosen is thus determined by the weights of the constraints and by the width of the noise distribution. The effect of noisy evaluation is illustrated in (30), which shows two evaluations for the input i_2 in the same noisy grammar. The basic weights (corresponding to the *ranking values* of Stochastic OT) of the constraints are 1.1 and 1.0 for C_1 and C_2 respectively. The weights after the addition of noise (corresponding to the *disharmonies* of Stochastic OT) are shown in parentheses. Depending on these stochastic weights, either output o_{22} or output o_{21} will be chosen, as these two tableaux show. Because o_{22} violates the constraint with a lower basic weight, it will be optimal in a greater proportion of evaluations than o_{21}.

(30) Variation in choices of optima in Noisy HG: one grammar, two outputs

i_2	C_1 1.1 (1.1)	C_2 1.0 (0.9)	
o_{21}	−1		−1.1
☞ o_{22}		−1	−0.9

i_2	C_1 1.1 (1.0)	C_2 1.0 (1.1)	
☞ o_{21}	−1		−1.0
o_{22}		−1	−1.1

This kind of additive noise was first combined with HG in Praat in January 2006 (with the correct HG-GLA implemented in April 2007 on the basis of Pater 2008). Jesney (2007), Boersma and Escudero (2008), Pater (2009) and Coetzee and Pater (2011) apply Noisy HG to the modelling of variation. As well as accounting for variation in a target language, noise in the learner's grammar plays a role in modeling gradual acquisition. In combination with a learning algorithm that adjusts the grammar gradually (e.g. OT-GLA or HG-GLA), noisy evaluation produces realistic sigmoid learning curves (Boersma 1998: 284; Boersma and Levelt 2000). Gradual learning with a categorical version of HG would produce an acquisition path with sudden changes from one optimum to another (e.g. sudden acquisition of codas, without a period of variation between deletion and accurate production).

Jäger (2007) shows that Maximum Entropy grammar with the HG-GLA update rule (that is, stochastic gradient ascent) yields the desired realistic learning curves. We here pursue the Noisy HG alternative because these two theories of stochastic HG have interesting differences in the patterns they produce. Maximum Entropy grammar, even a version that limits weights to positive values, always grants some probability to harmonically bounded candidates (Jäger and Rosenbach 2006). With the positivity restriction, these candidates can still never be granted the highest probability in their candidate sets. The empirical prediction is that there should be candidates that occur only in variation, but never as unique optima. In Noisy HG, however, such candidates have zero probability if the weights (including the evaluation noise) are limited to positive values (Jesney 2007; see here Section 13.3.5; for a boundary case, see Pater this volume Section 1.5). In this theory, the set of possible unique optima, and the set of candidates that can occur in variation, are identical. These differences in the theories of grammar are highly deserving of further investigation; here we aim to show that noisy HG is a viable alternative to Maximum Entropy grammar, and to stochastic OT, in terms of its convergence properties in combination with HG-GLA.

Section 13.3 showed that HG-GLA finds correct constraint weights for non-varying languages if the learner's processing involves no evaluation noise. The present section shows that HG-GLA also finds correct weights for non-varying languages if the learner's processing does include evaluation noise. We show this first by extending the proof of section 3 to this case. We then provide learning simulations that show that convergence is fast in practice, and compare the HG results with those obtained for OT-GLA with Stochastic OT.

13.4.1 A Convergence Proof for HG-GLA: Noisy Learners

Instead of by (5), the harmony of an output candidate o_{mn} given an input i_m is now given by

(31)
$$H_{mn} \equiv \sum_{k=1}^{K} (w_k + \mathbf{N}_k) s_{mnk}$$

where $\mathbf{N}_k$ is a Gaussian random variable. The lower bound on the sum of the squares of the constraint weights in (16) is still valid. The computation of the upper bound, however, changes, because (18) is no longer true. Instead, it is

(32)
$$\sum_{k=1}^{K} (w_k(t-1) + \mathbf{N}_k(t-1)) \left(s_{p(t)q(t)k} - s_{p(t)r(t)k} \right) = H_{p(t)q(t)}(t-1) - H_{p(t)r(t)}(t-1)$$

Thereby, (19) becomes

(33) $$\sum_{k=1}^{K} w_k^2(t) \le \sum_{k=1}^{K} w_k^2(t-1) + \varepsilon^2 \sum_{k=1}^{K} \left(s_{p(t)q(t)k} - s_{p(t)r(t)k}\right)^2 - 2\varepsilon \sum_{k=1}^{K} \mathbf{N}_k(t-1)\left(s_{p(t)q(t)k} - s_{p(t)r(t)k}\right)$$

Now suppose that the evaluation noise has a maximum absolute value during acquisition of N_{max}; for instance, if acquisition lasts 100,000 input-output pairs, N_{max} tends to be in the vicinity of $6N$, where N is the standard deviation of the evaluation noise. A lower bound on the value within the last sum in (33) is then $-N_{max}D_{max}$, so that an upper bound of the last sum is $2\varepsilon KN_{max}D_{max}$. Instead of (20) we thus get

(34) $$\sum_{k=1}^{K} w_k^2(t) \le \sum_{k=1}^{K} w_k^2(t-1) + \varepsilon^2 KD_{max}^2 + 2\varepsilon KN_{max}D_{max}$$

In the end, convergence is guaranteed by

(35) $$t \le \frac{\varepsilon KD_{max}^2 + 2KN_{max}D_{max}}{2\varepsilon\delta^2} - \frac{W_0 \cdot U}{\varepsilon\delta} + \sqrt{\left(\frac{\varepsilon KD_{max}^2 + 2KN_{max}D_{max}}{2\varepsilon\delta^2} - \frac{W_0 \cdot U}{\varepsilon\delta}\right)^2 + \frac{|W_0|^2 (W_0 \cdot U)^2}{\varepsilon^2\delta^2}}$$

The value for t_{max} that one can derive from (35) is, other than in (27), no longer independent of the language data stochastically presented to the learner. This is because N_{max} is a property of the learning process rather than a property of the unit target grammar, the initial state, or the learning algorithm. However, the probability that N_{max} exceeds e.g. the non-learning-dependent value of $30N$ even during the whole lifetime of a human learner is vanishingly small (it is in the order of 10^{-190} if the learner hears 100,000,000 utterances). For all imaginable practical purposes, then, t_{max} will exist. Formally speaking, the algorithm converges *almost surely*, i.e. it *converges with probability one.*

13.4.2 Convergence Times for HG-GLA and OT-GLA: Noisy Learners

We now test 100,000 HG-GLA learners on their ability to converge on correct grammars for non-varying languages, when the learners use evaluation noise in learning. The procedure is the same as that in

Section 13.3.4, except that the learners incorporate a constraint weight noise with a standard deviation of $N = 2.0$ during the evaluation of every incoming input-output pair. That is, the learner takes the observed input, computes a weighting of its constraints by temporarily adding evaluation noise, and computes the optimal output. If this optimal output is different from the 'correct' output that the learner has observed in the given input-output pair, an 'error' has occurred and the learner changes some constraint weights. We set the plasticity to a value of 0.1, so that it will take the learner multiple errors to overcome the noise. Again we check the learner's grammar after every 10 learning data (input-output pairs). We judge that the learner has converged to the target *non-varying* language if we find that the learner's grammar, *if the evaluation noise is set to zero*, produces correct singly optimal outputs for all M possible inputs. Figure 13.5 shows a histogram of the convergence times.

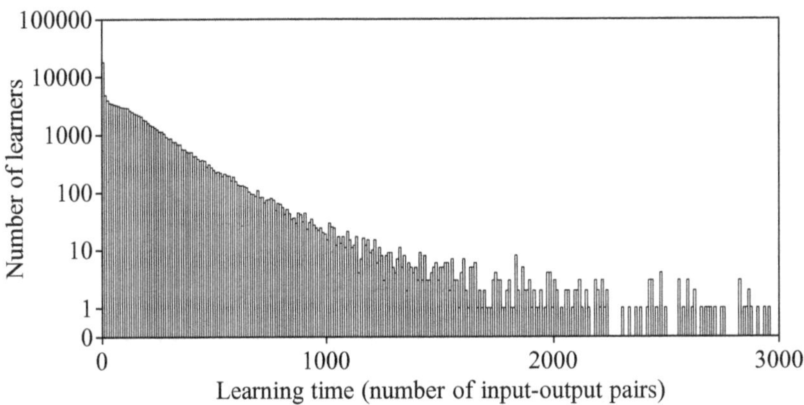

Figure 13.5: Histogram of convergence times of 100,000 Noisy HG learners

All 100,000 Noisy HG learners converge, in times slightly longer (on average) than those of the non-noisy HG learners of Section 13.3.4. The slowest learner needed 82,440 input-output pairs.

To get an idea of what a typical learning path looks like, we follow in detail one learner who has $M = 20$ possible inputs. Initially, this learner has all constraints weighted at 10.0 (like all others); for some of the 20 possible inputs, the learner initially already computes a correct output almost 100% of the time, if her evaluation noise is 2.0; for some other inputs, she scores close to 0% correct. For i_{19}, the learner initially scores 12.6% correct (this we measure by feeding the learner's grammar

1,000,000 tokens of i_{19}, and dividing the number of correct outputs by 1,000,000). Thus, 12.6% is the learner's initial proficiency for i_{19}. We then feed the learner 3,000 learning data randomly drawn from the set of 20 'correct' input-output pairs. After each learning datum, we compute the learner's proficiency in the way just described, i.e. with 1,000,000 tokens of i_{19} and an evaluation noise with a standard deviation of 2.0. Figure 13.6 shows how the proficiency develops in time. This figure, then, shows this learner's *learning curve* for i_{19}.

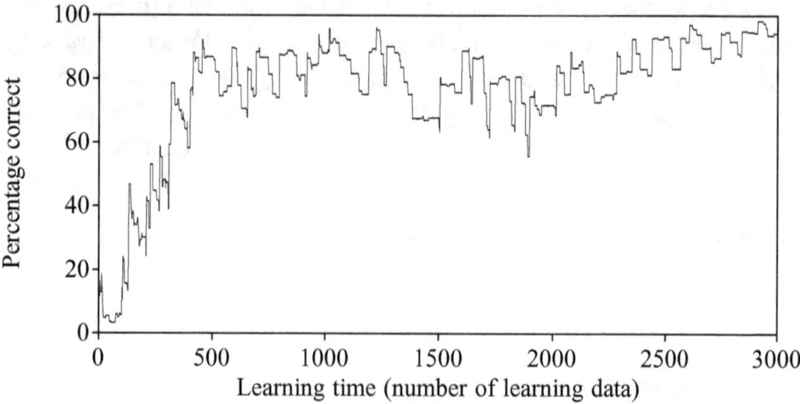

Figure 13.6: The learning curve for one learner, for one input

The learning is jumpy: there are large jumps up when the learner makes an error on i_{19}, and typically small steps up or (more often) down when the learner makes an error on any other input. Some U-shaped learning can also be discerned. In the end, however, the learner moves towards 100% correctness on i_{19}, as well as on all the other inputs.

We now examine *Noisy Exponential HG* learners. The harmony function for Noisy Exponential HG combines (30) with (32) and is given in (36).

(36)
$$H_{mn} \equiv \sum_{k=1}^{K} s_{mnk} e^{w_k + \mathbf{N}_k}$$

In other words, noise is added to the 'weights' w_k at evaluation time. Even if this sum results in a negative value, the subsequent exponentiation ensures that the violation score is multiplied by positive number. In order to have a moderate variation in the contribution of a constraint to the total harmony of a form, we set the standard deviation of the evaluation noise at 0.2; the plasticity is accordingly set at 0.01. When testing 100,000 Noisy Exponential HG learners on the same OT-generated data as in Section 13.3.5, it turns out that all learners are successful. The convergence times are longer on average than for the HG learners of Figure 13.5, although the slowest learner requires only 26,730 input-output pairs.

Finally, we examine Stochastic OT learners with OT-GLA. OT-GLA differs from HG-GLA in only two ways. First, the grammar model differs in that each time the numerical constraint values are used to evaluate a candidate set, they are converted to a corresponding ranking. Second, the learning procedure differs in the formulation of the update rule. When OT-GLA changes the ranking value of a constraint, it simply moves up or down by an amount ε (if the correct form does worse on the constraint than the incorrect optimum, the constraint moves down; if it does better, the constraint moves up). That is, the HG update rule in (6) is replaced by the OT update rule in (37).

(37) $$w_k' = w_k + \varepsilon \cdot \operatorname{sgn}\left(s_{pqk} - s_{prk}\right)$$

In the HG update rule the amount of change is proportional to the size of the difference between the satisfaction of the constraint by the correct form and by the incorrect optimum. This difference in the update rules is parallel to the difference in how the grammar models work. Under OT's strict domination, the number of violations of a lower ranked constraint is irrelevant if a candidate is preferred by a higher ranked one, whereas in HG the number of violations of a constraint with lower weight can matter, as illustrated in (1).

We thus test Stochastic OT learners with OT-GLA on 100,000 datasets generated by a random total OT ranking: again up to 20 inputs, up to 20 candidates per tableau, up to 20 constraints (all initially ranked at the same height), evaluation noise during learning 2.0, plasticity 0.1. It turns out that only 99.6% of the learners converge. This incomplete convergence of OT-GLA with Stochastic OT is no surprise: it was expected on the basis of a case of nonconvergence discovered by Pater (2008).

One might wonder if the higher convergence rates of Noisy HG with HG-GLA as compared to Stochastic OT with OT-GLA can be attributed solely to a difference in either the grammar models or the learning algorithms. To test this, one can switch around the learning algorithms between grammar models. It turns out that Stochastic OT has more misconvergence with the HG-GLA update rule than with the OT-GLA update rule, and that with the OT-GLA update rule, Noisy HG learners start to show some misconvergences. Apparently, HG-GLA is the 'best' on-line learning algorithm for Noisy HG, and OT-GLA is the 'best' learning algorithm for Stochastic OT (until a fully convergent alternative is found).[5]

The fact that HG-GLA continues to converge when noise is added to the learner's grammar model allows it to combine the strengths of OT-GLA (representation of variable languages, learning under noisy data, gradual learning) with the strength of EDCD (guaranteed convergence on non-varying languages). Further research will be required to better understand the exact nature of the differences between the varying languages produced by Stochastic OT, by Noisy HG, and by the Maximum Entropy approach to stochastic HG. In addition, further work is required to extend our convergence proof to the case of learning varying languages (as Fischer 2005 did for Maximum Entropy learning). However, in preliminary tests of probability matching that we have performed, we found no noticeable differences between HG-GLA, OT-GLA, and Jäger's (2007) Maximum Entropy learners (see also Coetzee and Pater 2011).

13.5 Learning with Imperfect Knowledge of Right and Wrong

In this section, we test the ability of HG-GLA to find a correct grammar when some of the linguistic structure is 'hidden', using techniques and a benchmark learning problem developed by Tesar and Smolensky (2000). We then go on to discuss the differences in success amongst variants of HG-GLA and similar OT algorithms. To illustrate the general shape of the problem, we start with a much-discussed small portion of it: foot structure ambiguity for a medially stressed trisyllable.

In general, surface representations have more phonological structure than our examples in section 13.2. For instance, if a learner hears the

5 Magri (2012) fixes the GLA in such a way that it is guaranteed to converge for non-varying data. Our preliminary tests with it, however, indicate that Magri's algorithm, unlike OT-GLA and HG-GLA, does not succeed with probabilistic data (see the next paragraph), so the search is still on.

overt phonetic form [batáma], she will not just assign the syllable structure [.ba.tá.ma.] to it (as in section 13.2), but also assign foot structure to it, and in this case the foot structure is ambiguous: [.ba.tá.ma.] can represent either the fully structured surface form [(ba.tá)ma.], with a right-headed foot (iamb) at the left edge of the word, or the fully structured surface form [.ba(tá.ma)], with a left-headed foot (trochee) at the right edge of the word.

Tableau 38 shows the relevant constraint violations for these two forms and several others. ALIGN-R and ALIGN-L are the constraints demanding that feet align with the right and left edge of the word respectively (McCarthy and Prince 1993), and TROCHEE and IAMB are the constraints demanding each kind of foot. The full structure in (38a) is consistent with a language in which a bisyllabic word is stressed on the final syllable (38e), because both require the same ranking or weighting of IAMB above TROCHEE. Similarly, the full structure in (38b) is consistent with a language in which a bisyllabic word is stressed initially (38f), since both satisfy TROCHEE.

(38) The bata-batama universe

Overt form	Full structure	ALIGN-R	ALIGN-L	TROCHEE	IAMB
[batáma]	a. [(ba.tá)ma.]	−1		−1	
	b. [.ba(tá.ma)]		−1		−1
[bátama]	c. [(bá.ta)ma.]	−1			−1
[batamá]	d. [.ba(ta.má)]		−1	−1	
[batá]	e. [(ba.tá)]			−1	
[báta]	f. [(bá.ta)]				−1

Given that foot structure is not directly marked in the phonetic signal, and that foot structure is needed for the learner to be able to assess constraint violations, it becomes relevant to ask the question of how a learner determines which full structure, (38a) or (38b), she has to assign to the incoming overt form [batáma]. Tesar (1997) and Tesar and Smolensky (2000) propose a general solution for such hidden structure problems called *robust interpretive parsing* (RIP), which fully stays within the spirit of OT and HG: of the two relevant candidate structures (38a) and (38b), choose the one that minimally violates the constraints as ranked (or weighted) in the learner's current grammar hypothesis. Tesar and Smolensky show that with this procedure, convergence is not guaranteed. Below we explain how RIP works, and how it sometimes fails, using an example from within the small universe of the two candidate sets in (38).

The first step in the learning procedure, robust interpretive parsing, uses the current grammar to assign the optimal full structure to an overt form. In our case of an incoming overt [batáma], the learner will choose between (38a) and (38b) depending on which is preferred by the current state of the grammar. Let us assume that in the initial state all four constraints are ranked equally, as in (39a). Since both of the candidates violate two of the constraints, they are equivalently optimal.[6] We further assume that in a case of tie between optima, one is chosen randomly. Alongside each grammar in the learning sequence in (39) we include the overt learning data and the structures constructed by the learner that lead to a change in the state of the grammar. In (39a) we have overt [batáma], and interpretive parsing provides the fully structured surface form ('interpretation') [.ba(tá.ma)] – in (39) and elsewhere, the asterisk indicates that the choice was made from a set of tied optima. The learner also recovers the underlying form /batama/, which is the surface form minus prosodic structure. Learning then proceeds in the general error-driven fashion described in earlier sections. In 'production-oriented parsing', the learner finds the optimum from the entire candidate set for the underlying form /batama/, including now those that diverge from the overt form, which are (38c) and (38d). Since in our initial state all four candidates (38a–d) contain the same number of violations of equally ranked constraints, the choice is again made randomly, which could yield [(bá.ta)ma.], which is included in the production column of (39a). The full structure constructed in interpretive parsing ([.ba(tá.ma)]) does not match the result of production-oriented parsing ([(bá.ta)ma.]), so the learner updates the constraint ranks or weights. Error-Driven Constraint Demotion (EDCD; Tesar and Smolensky 1998,

6 For compatibility with Tesar and Smolensky's (1998, 2000) original presentation, we here adopt their version of EDCD, which posits stratified hierarchies, in which tied constraints pool their violations. See Boersma (2009) for a revised EDCD that maintains a total order on the constraints, as in Prince and Smolensky (1993/2004). In that version, the learner could also fall into this trap through chance, selecting an unlucky random initial ranking of the constraints. In the simulations below, we use Boersma's (2009) EDCD, which unlike the original version, is guaranteed to converge on a correct total order in the absence of hidden structure. For results with the original version, see Tesar and Smolensky (2000) and Boersma (2003), but note that if the final strata of the 'successful' learners in those papers were turned into totally ranked hierarchies, they would be incorrect half of the time, achieving lower success rates than we report here.

2000) places any constraint that prefers the learner's incorrect optimum [(bá.ta)ma.] (here only ALIGN-L, as can be seen by comparing 38b with 38c) beneath the highest ranked constraint that prefers the allegedly correct form [.ba(tá.ma)] (here ALIGN-R). It will thus form the grammar in (39b), with ALIGN-L in a new lower stratum beneath ALIGN-R.

(39) A path into a hidden structure trap for RIP/EDCD

Grammar	Overt	Interpret.	Production
a. {ALIGN-L, ALIGN-R, TROCH, IAMB}	[batáma]	[.ba(tá.ma)]*	[(bá.ta)ma.]*
b. {ALIGN-R, TROCH, IAMB} ≫ ALIGN-L	[batá]	[(ba.tá)]	[(bá.ta)]*
c. {ALIGN-R, IAMB} ≫ {ALIGN-L, TROCH}	[batáma]	[.ba(tá.ma)]*	[.ba(ta.má)]
d. ALIGN-R ≫ {ALIGN-L, TROCH} ≫ IAMB	[batá]	[(ba.tá)]	[(bá.ta)]
e. ALIGN-R ≫ ALIGN-L ≫ IAMB ≫ TROCHEE	[batáma]	[.ba(tá.ma)]	[.ba(ta.má)]
f. ALIGN-R ≫ ALIGN-L ≫ TROCHEE ≫ IAMB	[batá]	[(ba.tá)]	[(bá.ta)]

Along with [batáma], our toy language includes bisyllabic [batá], which is presented to the learner next. The grammar in (39b) incorrectly judges [(ba.tá)] and [(bá.ta)] as equivalently optimal, so the learner may choose [(bá.ta)] in production and (after comparing 38e with 38f) demote TROCHEE beneath IAMB, as she has done in (39c). The new ranking now prefers iambs, which can lead to an error on [batáma], optionally interpretively parsed as [.ba(tá.ma)] under the tie between ALIGN-R and IAMB. Since production-oriented parsing picks iambic [.ba(ta.má)], IAMB will be demoted beneath TROCHEE. This ranking is now incorrect for [batá], and when the learner is presented with it, it will again demote TROCHEE beneath IAMB. The learner has now entered an endless cycle between the rankings in (39e) and (39f). The dominance of ALIGN-R over ALIGN-L and IAMB has removed the correct interpretive parse [(ba.tá)ma.] from consideration, leaving the learner with access only to [.ba(tá.ma)], which is inconsistent with [(ba.tá)].

With a different set of random choices, EDCD would have succeeded on the above case. To assess how well RIP/EDCD performs on more realistic cases, Tesar and Smolensky (2000) tested it on similar metrical learning problems that involved 12 constraints and 62 overt forms for each of 124 languages. Here we extend Boersma's (2003) comparison of OT learners using EDCD and OT-GLA on these problems to also include HG and Exponential HG learners using HG-GLA. For each combination of grammar and learning algorithm, we also vary whether the grammar incorporates noise on the constraint values (except for EDCD, which is incompatible with noise). Here we only report the results; for further details of the simulations, see Tesar and Smolensky (2000) and

Boersma (2003). Tesar and Smolensky ran their simulations with various assumptions about the initial ranking; our simulations always have the constraints start out at an equal ranking or weighting (a value of 10.0). The non-noisy learners have an evaluation noise of zero and a plasticity of 1.0; the noisy learners have an evaluation noise of 2.0 and a plasticity of 0.1 (for the Exponential HG learners, all these values are divided by 10, as in Section 13.4.2).

We test each combination of grammar type and learning algorithm ten times on each of the 124 languages, providing each of the 1240 learners with a maximum of 1 million overt forms randomly drawn from the 62 possible overt forms of the language. As in Section 13.4.2, we judge a learner successful once its grammar, *if the evaluation noise is set to zero*, renders uniquely optimal all of the correct forms of the target language. The results are shown in (40).

(40) Results of learning simulations with hidden metrical structure

Learner grammar	Learning Algorithm	Performance of non-noisy learners	Performance of noisy learners
OT	EDCD	46.94%	–
OT	OT-GLA	55.40%	58.95%
HG	HG-GLA	82.02%	88.63%
Exponential HG	HG-GLA	70.89%	88.95%

The relative success rates of EDCD and OT-GLA replicate the findings of Tesar and Smolensky (2000) and Boersma (2003). HG-GLA learners turn out to have higher success rates than OT learners.[7]

For every combination of grammar type and learning algorithm, we see that noisy learners have a higher probability of convergence than non-noisy learners. Apparently, evaluation noise not only allows the grammar to model variable languages, but it can also help to solve difficult learning problems (see below on the connection to local optima and simulated annealing). The noisy learners also have more consistent outcomes than the non-noisy ones: for 108 languages, all 10 Noisy HG

7 Other learning algorithms can be considered. Jarosz (2013a) reports that an OT learner that randomly chooses a new ranking upon each error performs 100% correct on this test. However, the convergence time for this random algorithm scales exponentially with the number of independent constraints, so we do not consider this further here. See also Jarosz (2013b), who considers alternatives to robust interpretive parsing while maintaining the same general approach to hidden structure learning we pursue here.

learners converge, i.e. these languages can be considered learnable; for 11 languages all 10 learners fail, i.e. these languages are unlearnable; and for only five languages, the number of converging learners is between one and nine, i.e. these languages are only sometimes learned (the real-life situation that this partial learnability corresponds to is perhaps a language stage that is historically unstable). For non-noisy HG learners, the number of languages falling into each category is 68, 23, and 33: most of the non-fully learnable languages are still learned by some. The relative consistency of noisy learning allows a better diagnostic of the difficulty of particular languages, which may be useful in amending the grammatical system or the learning algorithm, or in using this difficulty to explain typological gaps (see Boersma 2003).

We emphasize that we have provided only an initial investigation of the hidden structure problem in HG. Even for just this instance, further research is required to understand the sources of the relative success rates of the learners in (40), and the nature of the problems posed by these metrical systems. Nonetheless, for the case of error-driven learning of hidden structure problems, it seems that the 'symmetric' OT-GLA and HG-GLA algorithms, which both promote and demote constraints, have an advantage over demotion-only EDCD (see also Apoussidou 2007: chs. 4, 5), and that a learner with an HG grammar has an advantage over one using OT-ranked constraints.

To show how symmetric update, noise, and weighted constraints can aid in hidden structure learning, we return to the *batá-batáma* example of (39), where a dominant ALIGN-R constraint caused a trap. Recall that when ALIGN-R determines the outcome of RIP, the resulting trochaic [.ba(tá.ma)] is inconsistent with iambic [(ba.tá)], which for EDCD leads to an inescapable cycle of demoting TROCHEE and IAMB beneath one another. A GLA learner, however, is not necessarily trapped by these circumstances. Take for example the learning sequence in (41), which results from the same data sequence and initial random choices that led EDCD into the trap in (39), but with the GLA update rule. Constraints have an initial value of 10, and the learning rate is 1; the results would be identical with either HG or OT as the model of grammar and either HG-GLA or OT-GLA as the learning algorithm. The first step yields the potentially dangerous dominance of ALIGN-R (41b). ALIGN-R has been promoted above the other constraints because the GLA update rule is symmetric: the values of the constraints preferring the form taken as correct are raised, alongside the lowering of the values of constraints preferring the learner's incorrect optimum.

(41) A GLA learner with the same data sequence and initial random choices as (39)

	Align-L	Align-R	Trochee	Iamb	Overt	Interpret.	Product.
a.	10	10	10	10	[batáma]	[.ba(tá.ma)]*	[(bá.ta)ma.]*
b.	9	11	10	10	[batá]	[(ba.tá)]	[(bá.ta)]*
c.	9	11	9	11	[batáma]	[.ba(tá.ma)]*	[.ba(ta.má)]
d.	9	11	10	10	[batá]	[(ba.tá)]	[(bá.ta)]*
e.	9	11	9	11	[batáma]	[(ba.tá)ma.]*	[.ba(ta.má)]
f.	10	10	9	11	[batáma]	[(ba.tá)ma.]	[.ba(ta.má)]
g.	11	9	9	11			

The next step brings Iamb up to the same level as Align-R, since it prefers [(ba.tá)] over [(bá.ta)]. With a tie between Iamb and Align-R, the GLA learner at (41c) is in the same position as the EDCD learner in (39c) of still allowing either [.ba(tá.ma)] or [(ba.tá)ma.] in interpretive parsing. As in (39c), we maliciously choose the problematic [.ba(tá.ma)]. The crucial difference is in what follows now. The EDCD learner in (39) went on to demote Iamb (and Trochee) to the bottom of the hierarchy, thus rendering the desired [(ba.tá)ma.] subsequently inaccessible. The symmetric update rule, on the other hand, leaves Trochee and Iamb with the same average value as Align-R and Align-L, thus allowing Iamb to rise again to the same level as Align-R. Eventually, the learner will randomly choose [(ba.tá)ma.] rather than [.ba(tá.ma)] in interpretive parsing, as in (41e). Once this happens, the learner is on the path toward the correct grammar, as shown in (41f) and (41g). In simulations, we find that a GLA with equal initial constraint values always converges on a correct grammar for *batá-batáma*, as opposed to the occasional failure of EDCD. We suspect that the difference between symmetric update and demotion-only that we have illustrated here is the cause of the better performance of the GLA learners found in the larger simulations.

It is possible to trap a GLA learner by placing Align-R sufficiently high above the other constraints. An OT-GLA learner without evaluation noise will fail to converge if the distance between Align-R and the other constraints is greater than the size of the learning step encoded in the learning rate. For example, with a learning rate of 1, the grammar in (42) is such a trap.

(42) A hidden structure trap for a no-noise OT-GLA learner

ALIGN-L	ALIGN-R	TROCHEE	IAMB
10	12	10	10

Starting with these constraint values, the noise-free OT-GLA learner will consistently pick [.ba(tá.ma)] in interpretive parsing, and never find a correct grammar. Adding noise to the grammar model changes this outcome because it gives some probability to [(ba.tá)ma.]. Once [(ba.tá)ma.] is chosen in interpretive parsing (and some other form is chosen in production), the constraint values are updated in its favor, and the learner is on its way out of the trap. In simulations, we find that the noise-free OT-GLA learner with the initial constraint values in (42) and a learning rate of 1 never converges on a correct grammar, while an OT-GLA learner with noise (standard deviation = 2) always does. More generally, adding a stochastic component to a model of learning or parsing is a commonly employed means of avoiding local minima (cf. on *simulated annealing* in neural networks and linguistics: Ackley, Hinton and Sejnowski 1985; Hinton and Sejnowski 1986; Smolensky 1986; Legendre, Sorace and Smolensky 2006; Biró 2006). The grammar in (42) is a local minimum for the noise-free OT-GLA learner because it yields the minimum error within the space of constraint values that the learner will explore.

The grammar in (42) is not a local minimum for an HG-GLA learner, even without noise. Since that grammar has a tie between IAMB and TROCHEE, the learner could make an error on [batá], leading to the grammar in (43). For the HG-GLA learner, this change has brought [(ba.tá)ma.] back into contention in interpretive parsing, since its harmony is now tied with [.ba(tá.ma)] – the weighted sum of violations of ALIGN-R and TROCHEE is the same as that of ALIGN-L and IAMB. For the OT-GLA learner, ALIGN-R still determines the outcome.

(43) A step out of the trap in (42) for a no-noise HG-GLA learner

ALIGN-L	ALIGN-R	TROCHEE	IAMB
10	12	9	11

In simulations, we find that in contrast with OT-GLA, a noise-free HG-GLA learner always succeeds when started with the weights in (42). The advantage of constraint weighting illustrated here is that there is a wider range of constraint values that lead to the result of [(ba.tá)ma.] being chosen as the optimal interpretation for overt [batáma].

We suspect that as in this example, HG learners generally more readily escape the potential local minima introduced by hidden structure because constraint weighting provides closer 'stepping stones' out of these traps than ranking does. It is sometimes taken as a truism in generative linguistics that a learner is aided by minimizing the number of grammars that generate a given outcome. For hidden structure learning, this is at least sometimes false.

In sum, this section has shown that the performance of Tesar and Smolensky's (2000) procedure for learning with hidden structure, RIP, can be optimized by replacing demotion-only EDCD with a symmetric GLA, adding noisy evaluation, and replacing constraint ranking with weighting. We present these results as a useful baseline for further research on hidden structure learning. Even within the present approach of on-line learning with Harmonic Grammar, there are a number of avenues for further exploration, including changing the degree of noise over a training schedule (cf. simulated annealing), testing different models of stochastic HG, and different update rules. In research subsequent to Tesar and Smolensky (2000), Tesar (2000) proposes the Inconsistency Detection Learner, an algorithm for the learning of hidden structure that is guaranteed to converge, but at the cost of having to maintain multiple simultaneous grammar hypotheses; also, as Tesar notes (p. 37), it utterly fails when trying to learn variable languages. The overall success of the noisy HG learners on the test cases examined here provides reason to be optimistic that this approach to learning can deal with both hidden structure and variable languages, using a simple learning algorithm that maintains a single grammar hypothesis at a time.

13.6 Conclusion

This paper provides a set of basic learning results for Harmonic Grammar. We have shown that a simple on-line gradual learning algorithm, HG-GLA, is guaranteed to converge on a correct set of weights for any non-varying set of HG-generated language data, if it is supplied with full knowledge of the structure of the language data. We have shown that this result holds for both non-noisy and noisy learners. Like Stochastic OT, Noisy HG therefore provides a grammar model that can represent variation, is expected to be robust against mistakes in the learning data (Boersma and Escudero 2008), and exhibits gradual learning curves; unlike in Stochastic OT, these attributes do not come at the cost of convergence failures. In addition to these core results, we have discussed an initial investigation into the hidden structure problem in HG learning, which provide some indications that there may be advantages for HG over OT in this domain. These results provide a foundation for further research on the formal properties of this approach to learning, as well as research on the modelling of human language acquisition. The further development of HG learning theory has a large body of extant research to draw upon, since HG's linear model is widely used in neural-modeling and statistical approaches to learning. Our results draw on that strength, especially in the adaptation of the perceptron convergence proof.

These results add to a growing body of evidence that HG should be taken seriously as a model of generative grammar – see Jesney this volume, Pater this volume, and the references cited therein, as well as Boersma and Escudero (2008) on HG's advantages with positive constraint satisfactions.

References

Ackley, David H., Geoffrey E. Hinton and Terrence J. Sejnowski (1985) A learning algorithm for Boltzmann machines. *Cognitive Science* 9: 147–169.

Anderson, James and Edward Rosenfeld (eds) (1989) *Neurocomputing: Foundations of Research*. Cambridge, MA: MIT Press.

Anttila, Arto (1997) Deriving variation from grammar. In Frans Hinskens, Roeland van Hout, and Leo Wetzels (eds) *Variation, Change and Phonological Theory*, 35–68. Amsterdam: John Benjamins.

Anttila, Arto (2007) Variation and optionality. In Paul de Lacy (ed.) *The Cambridge Handbook of Phonology*, 519–536. Cambridge: Cambridge University Press.

Apoussidou, Diana (2007) *The learnability of metrical phonology*. Doctoral dissertation, University of Amsterdam.

Becker, Michael and Joe Pater (2007) OT-Help user manual. In Michael Becker (ed.) *University of Massachusetts Occasional Papers in Linguistics 36: Papers in Theoretical and Computational Phonology*, 1–12. Amherst, MA: GLSA Publications. Available on Rutgers Optimality Archive, ROA-928.

Biró, Tamás (2006) *Finding the right words: implementing Optimality Theory with simulated annealing*. Doctoral dissertation, Groningen University.

Bishop, Christopher (1995) *Neural Networks for Pattern Recognition*. Oxford: Oxford University Press.

Block, H. D. (1962) The perceptron: a model for brain functioning. *Reviews of Modern Physics* 34: 123–135. [Reprinted in Anderson and Rosenfeld 1989]

Boersma, Paul (1997) How we learn variation, optionality, and probability. *Proceedings of the Institute of Phonetic Sciences* 21: 43–58. University of Amsterdam.

Boersma, Paul (1998) *Functional phonology: Formalizing the interactions between articulatory and perceptual drives.* Doctoral dissertation, University of Amsterdam.

Boersma, Paul (2003) Review of Tesar and Smolensky (2000): *Learnability in Optimality Theory. Phonology* 20: 436–446.

Boersma, Paul (2009) Some correct error-driven versions of the Constraint Demotion algorithm. *Linguistic Inquiry* 40: 667–686.

Boersma, Paul, Joost Dekkers and Jeroen van de Weijer (2000) An introduction to Optimality Theory: Phonology, syntax, and acquisition. In Joost Dekkers, Frank van der Leeuw, and Jeroen van de Weijer (eds) *Optimality Theory: Phonology, Syntax, and Acquisition*, 1–44. Oxford: Oxford University Press.

Boersma, Paul and Paola Escudero (2008) Learning to perceive a smaller L2 vowel inventory: An Optimality Theory account. In Peter Avery, Elan Dresher and Keren Rice (eds) *Contrast in Phonology: Theory, Perception, Acquisition*, 271–301. Berlin: Mouton de Gruyter.

Boersma, Paul and Bruce Hayes (2001) Empirical tests of the Gradual Learning Algorithm. *Linguistic Inquiry* 32: 45–86.

Boersma, Paul and Clara Levelt (2000) Gradual constraint-ranking learning algorithm predicts acquisition order. In Eve V. Clark (ed.), *Proceedings of Child Language Research Forum 30*, 229–237. Stanford, CA: CSLI Publications.

Boersma, Paul and Clara Levelt (2003) Optimality Theory and phonological acquisition. *Annual Review of Language Acquisition* 3: 1–50.

Boersma, Paul and Joe Pater (2007) Constructing constraints from language data: The case of Canadian English raising. Paper presented at North East Linguistic Society (NELS) 38, Ottawa. [Available at http://people.umass.edu/pater/boersma-pater-nels.pdf]

Boersma, Paul and David Weenink (1992–2013) Praat, a system for doing phonetics by computer. http://www.praat.org

Coetzee, Andries and Joe Pater (2011) The place of variation in phonological theory. In John Goldsmith, Jason Riggle, and Alan Yu (eds) *The Handbook of Phonological Theory* (2nd ed.), Malden, MA: Blackwell. Available on Rutgers Optimality Archive, ROA-946.

Collins, Michael (2002) Discriminative training methods for Hidden Markov Models: Theory and experiments with perceptron algorithms. *2002 Conference on Empirical Methods in Natural Language Processing.* [Available at http://people.csail.mit.edu/mcollins/papers/tagperc.pdf]

Curtin, Suzanne and Kie Zuraw (2002) Explaining constraint demotion in a developing system. In Anna H.-J. Do, Laura Domínguez and Aimee Johansen (eds) *BUCLD 26: Proceedings of the 26th annual Boston University Conference on Language Development.* Somerville, MA: Cascadilla Press.

Davidson, Lisa, Paul Smolensky and Peter Jusczyk (2004) The initial and final states: Theoretical implications and experimental explorations of Richness of the Base. In René Kager, Joe Pater and Wim Zonneveld (eds) *Fixing Priorities: Constraints in Phonological Acquisition*, 321–368. Cambridge: Cambridge University Press.

Demuth, Katherine (1995) Markedness and the development of prosodic structure. In Jill Beckman (ed.) *Proceedings of the North-East Linguistics Society* 25, 13–25. Amherst, MA: GLSA Publications. Available on Rutgers Optimality Archive, ROA-50.

Farris-Trimble, Ashley (2008) *Cumulative faithfulness effects in phonology.* Doctoral dissertation, Indiana University, Bloomington IN. Available on Rutgers Optimality Archive, ROA-991.

Fischer, Markus (2005) A Robbins-Monro type learning algorithm for an entropy maximizing version of stochastic Optimality Theory. Master's thesis, Humboldt University, Berlin. Available on Rutgers Optimality Archive, ROA-767.

Fodor, Janet Dean (1998) Unambiguous triggers. *Linguistic Inquiry* 29: 1–36.

Gibson, Edward and Kenneth Wexler (1994) Triggers. *Linguistic Inquiry* 25: 407–454.

Gold, Mark (1967) Language identification in the limit. *Information and Control* 10: 447–474.

Goldrick, Matt and Robert Daland (2009) Linking grammatical principles with experimental speech production data: insights from Harmonic Grammar networks. *Phonology* 26: 147–185.

Goldwater, Sharon and Mark Johnson (2003) Learning OT constraint rankings using a maximum entropy model. In Jennifer Spenader, Anders Eriksson and Östen Dahl (eds) *Proceedings of the Workshop on Variation within Optimality Theory,* 111–120. Stockholm University.

Hayes, Bruce (2004) Phonological acquisition in OT: The early stages. In René Kager, Joe Pater and Wim Zonneveld (eds) *Fixing Priorities: Constraints in Phonological Acquisition,* 158–203. Cambridge: Cambridge University Press. Available on Rutgers Optimality Archive, ROA-327.

Hayes, Bruce, Bruce Tesar and Kie Zuraw (2003) OTSoft 2.1. Software package, http://www.linguistics.ucla.edu/people/hayes/otsoft/

Hinton, Geoffrey E. and Terrence J. Sejnowski (1986) Learning and relearning in Boltzmann machines. In David E. Rumelhart, James L. McClelland and the PDP Research Group (eds) *Parallel Distributed Processing: Explorations in the Microstructure of Cognition,* Vol. 1, 282–317. Cambridge, MA: MIT Press.

Jäger, Gerhard (2007) Maximum entropy models and Stochastic Optimality Theory, In Annie Zaenen, Jane Simpson, Tracy Holloway King, Jane Grimshaw, Joan Maling and Chris Manning (eds) *Architectures, Rules, and Preferences: Variations on Themes by Joan W. Bresnan,* 467–479. Stanford, CA: CSLI.

Jäger, Gerhard and Anette Rosenbach (2006) The winner takes it all – almost: Cumulativity in grammatical variation. *Linguistics* 44: 937–971.

Jarosz, Gaja (2013a) Naive parameter learning for Optimality Theory – The hidden structure problem. In Seda Kan, Claire Moore-Cantwell and Robert Staubs (eds) *Proceedings of the North East Linguistic Society* 40, Vol. 2, 1–14. Amherst, MA: GLSA Publications.

Jarosz, Gaja (2013b) Learning with hidden structure in optimality theory and harmonic grammar: Beyond robust interpretive parsing, *Phonology* 30, 27–71.

Jesney, Karen (2007) The locus of variation in weighted constraint grammars. Poster presented at the *Workshop on Variation, Gradience and Frequency in Phonology.* Stanford University, July 2007. [Available at http://www.stanford.edu/dept/linguistics/linginst/nsf-workshop/Jesney–Poster.pdf]

Jesney, Karen (2011) *Cumulative constraint interaction in phonological acquisition and typology.* Doctoral dissertation, University of Massachusetts, Amherst.

Jesney, Karen (This volume) Positional constraints in Optimality and Harmonic Grammar.

Jesney, Karen and Anne-Michele Tessier (2011) Biases in Harmonic Grammar: The road to restrictive learning. *Natural Language and Linguistic Theory* 29: 251–290.

Johnson, Mark (2002) Optimality-theoretic Lexical Functional Grammar. In Suzanne Stevenson and Paola Merlo (eds) *The Lexical Basis of Syntactic Processing: Formal, Computational and Experimental Issues*, 59–73. Amsterdam: John Benjamins.

Kager, René (1999) *Optimality Theory*. Cambridge: Cambridge University Press.

Keller, Frank (2000) *Gradience in grammar: experimental and computational aspects of degrees of grammaticality*. Doctoral dissertation, University of Edinburgh. Available on Rutgers Optimality Archive, ROA-677.

Keller, Frank and Ash Asudeh (2002) Probabilistic learning algorithms and Optimality Theory. *Linguistic Inquiry* 33: 225–244.

Keller, Frank (2006) Linear Optimality Theory as a model of gradience in grammar. In Gisbert Fanselow, Caroline Féry, Ralph Vogel and Matthias Schlesewsky (eds) *Gradience in Grammar: Generative Perspectives*, 270–287. Oxford: Oxford University Press. Available on Rutgers Optimality Archive, ROA-679.

Kiparsky, Paul (1993) An OT perspective on phonological variation. Handout from *Rutgers Optimality Workshop 1993*, also presented at New Ways of Analyzing Variation (NWAV) 1994, Stanford University. [Available at http://www.stanford.edu/~kiparsky/Papers/nwave94.pdf].

Legendre, Géraldine, Yoshiro Miyata and Paul Smolensky (1990) Can connectionism contribute to syntax? Harmonic Grammar, with an application. In M. Ziolkowski, M. Noske and K. Deaton (eds) *Proceedings of the 26th Regional Meeting of the Chicago Linguistic Society*, 237–252. Chicago, IL: Chicago Linguistic Society.

Legendre, Géraldine, Antonella Sorace and Paul Smolensky (2006) The Optimality Theory – Harmonic Grammar connection. In Smolensky and Legendre (2006), 903–966.

Levelt, Clara (1995) Unfaithful kids: Place of articulation patterns in early vocabularies. Colloquium presented at the University of Maryland, College Park.

Magri, Giorgio (2007) The Multiplicative GLA. Paper presented to the Northeast Computational Phonology circle, University of Massachusetts Amherst, November 10, 2007.

Magri, Giorgio (2012) Convergence of error-driven ranking algorithms. *Phonology* 29: 213–269.

McCarthy, John J. (2002) *A Thematic Guide to Optimality Theory*. Cambridge: Cambridge University Press.

McCarthy, John J. (2008) *Doing Optimality Theory*. Malden, MA and Oxford: Blackwell.

Minsky, Marvin L. and Seymour A. Papert (1969) *Perceptrons*. Cambridge, MA: MIT Press.

Nagy, Naomi and Bill Reynolds (1997) Optimality Theory and word-final deletion in Faetar. *Language Variation and Change* 9: 37–55.

Novikoff, Albert B.J. (1962) On convergence proofs on perceptrons. In *Proceedings of the Symposium on the Mathematical Theory of Automata* 12, 615–622. Polytechnic Institute of Brooklyn.

Ohala, Diane K. (1996) *Cluster reduction and constraints in acquisition*. Doctoral dissertation, University of Arizona.

Pater, Joe (2008) Gradual learning and convergence. *Linguistic Inquiry* 39: 334–345.

Pater, Joe (2009a) Weighted constraints in generative linguistics. *Cognitive Science* 33: 999–1035 [ROA-982]

Pater, Joe (2009b) Review of Smolensky and Legendre (2006). *Phonology* 26: 217–226. Available on Rutgers Optimality Archive, ROA-983.

Pater, Joe (This volume) Universal Grammar with weighted constraints.

Potts, Christopher, Joe Pater, Karen Jesney, Rajesh Bhatt and Michael Becker (2010) Harmonic Grammar with linear programming: From linear systems to linguistic typology. *Phonology* 27: 77–117.

Prince, Alan (2002) Anything goes. In Takeru Honma, Masao Okazaki, Toshiyuki Tabata and Shin-ichi Tanaka (eds.) *New Century of Phonology and Phonological Theory*, 66–90. Tokyo: Kaitakusha. Available on Rutgers Optimality Archive, ROA-536.

Prince, Alan (2007a) The pursuit of theory. In Paul de Lacy (ed.) *The Cambridge Handbook of Phonology*, Cambridge: Cambridge University Press.

Prince, Alan (2007b) Let the decimal system do it for you: A very simple utility function for OT. Ms, Rutgers University. Available on Rutgers Optimality Archive, ROA-943.

Prince, Alan and Paul Smolensky (1993/2004). *Optimality Theory: Constraint Interaction in Generative Grammar*. Technical Report, Rutgers University and University of Colorado at Boulder, 1993. Revised version published by Blackwell, 2004. Available on Rutgers Optimality Archive, ROA-537.

Prince, Alan and Paul Smolensky (1997) Optimality: From neural networks to universal grammar. *Science* 275: 1604–1610.

Prince, Alan and Bruce Tesar (2004) Learning phonotactic distributions. In René Kager, Joe Pater and Wim Zonneveld (eds), *Fixing Priorities: Constraints in Phonological Acquisition*, 245–291. Cambridge: Cambridge University Press.

Reynolds, William (1994) *Variation and phonological theory*. Doctoral dissertation, University of Pennsylvania, Philadelphia.

Riggle, Jason (2004) *Generation, recognition and learning in finite state Optimality Theory*. Doctoral dissertation, UCLA.

Rosenblatt, Frank (1958) The perceptron: A probabilistic model for information storage and organization in the brain. *Psychological Review* 65: 386–408. [Reprinted in Anderson and Rosenfeld 1989]

Rosenblatt, Frank (1962) *Principles of Neurodynamics: Perceptrons and the Theory of Brain Mechanisms*. Washington, DC: Spartan.

Samek-Lodovici, Vieri and Alan Prince (1999) *Optima*. Ms, Rutgers University. [ROA-363]

Samek-Lodovici, Vieri and Alan Prince (2005) Fundamental properties of harmonic bounding. Ms, Rutgers University. Available on Rutgers Optimality Archive, ROA-785.

Smolensky, Paul (1986) Information processing in dynamical systems: Foundations of Harmony Theory. In David E. Rumelhart, James L. McClelland and the PDP Research Group (eds) *Parallel Distributed Processing: Explorations in the Microstructure of Cognition*, Vol. 1, 194–281. Cambridge, MA: MIT Press.

Smolensky, Paul (1996) The initial state and 'Richness of the Base' in Optimality Theory. Ms, Johns Hopkins University. Available on Rutgers Optimality Archive, ROA-154.

Smolensky, Paul (2006) Optimality in phonology II: Harmonic completeness, local constraint conjunction, and feature domain markedness. In Smolensky and Legendre (2006), 27–160.

Smolensky, Paul and Géraldine Legendre (2006) *The Harmonic Mind: From Neural Computation to Optimality-Theoretic Grammar*. Cambridge, MA: MIT Press.

Soderstrom, Melanie, Donald Mathis and Paul Smolensky (2006) Abstract genomic encoding of Universal Grammar in Optimality Theory. In Smolensky and Legendre (2006), 403–471.

Tesar, Bruce (1995) *Computational Optimality Theory*. Doctoral dissertation, University of Colorado, Boulder.

Tesar, Bruce (1997) An iterative strategy for learning metrical stress in Optimality Theory. In Elizabeth Hughes, Mary Hughes and Annabel Greenhill (eds) *Proceedings of the 21st Annual Boston University Conference on Language Development*, 615–626. Somerville, MA: Cascadilla.

Tesar, Bruce (2000) Using inconsistency detection to overcome structural ambiguity in language learning. Technical Report RuCCS-TR-58, Rutgers Center for Cognitive Science, Rutgers University, New Brunswick, NJ. Available on Rutgers Optimality Archive, ROA-426.

Tesar, Bruce (2007) A comparison of lexicographic and linear numeric optimization using violation difference ratios. Ms, Rutgers University. Available on Rutgers Optimality Archive, ROA-939.

Tesar, Bruce and Paul Smolensky (1998) Learnability in Optimality Theory. *Linguistic Inquiry* 29: 229–268.

Tesar, Bruce and Paul Smolensky (2000) *Learnability in Optimality Theory*. Cambridge, MA: MIT Press.

Wexler, Kenneth and Peter Culicover (1980) *Formal Principles of Language Acquisition*. Cambridge, MA: MIT Press.

Wilson, Colin (2006) Learning phonology with substantive bias: an experimental and computational study of velar palatalization. *Cognitive Science* 30: 945–982.

Yang, Charles (2002) *Knowledge and Learning in Natural Language*. New York: Oxford University Press.

Subject index

Language index